AF581640

Biomolecular Data Science—in Honor of Professor Philip E. Bourne

Biomolecular Data Science—in Honor of Professor Philip E. Bourne

Editors

Cameron Mura
Lei Xie

Basel • Beijing • Wuhan • Barcelona • Belgrade • Novi Sad • Cluj • Manchester

Editors

Cameron Mura
Data Science
University of Virginia
Charlottesville
United States

Lei Xie
Computer Science
The City University of New York
New York City
United States

Editorial Office
MDPI
St. Alban-Anlage 66
4052 Basel, Switzerland

This is a reprint of articles from the Special Issue published online in the open access journal *Biomolecules* (ISSN 2218-273X) (available at: www.mdpi.com/journal/biomolecules/special_issues/Honor_Phil).

For citation purposes, cite each article independently as indicated on the article page online and as indicated below:

Lastname, A.A.; Lastname, B.B. Article Title. *Journal Name* **Year**, *Volume Number*, Page Range.

ISBN 978-3-0365-8611-3 (Hbk)
ISBN 978-3-0365-8610-6 (PDF)
doi.org/10.3390/books978-3-0365-8610-6

Contents

About the Editors

Cameron Mura

Cameron Mura is a senior scientist and co-director, with Prof Bourne, of a Biomolecular Data Science Lab at the Univ of Virginia (UVa). After earning a BS degree at Georgia Tech, Mura was an NSF Graduate Fellow at UCLA, where he received a PhD for crystallographic, biochemical, biophysical and bioinformatic analyses of archaeal RNA–associated systems. After then training in molecular biophysics and computational biology, as a Sloan/DOE Postdoctoral Fellow at UCSD, Mura joined UVa as an assistant professor, where he received an NSF Career award for his lab's work in the structural and computational biology of RNA-associated systems. Mura's early scientific contributions were two structural 'firsts'—one static (first structure of an intact 'Sm'ring), one dynamic (first µsec–scale simulation of DNA, including a model for base-flipping). At UVa, Mura's research group also made contributions to scientific software development and pursued several pedagogy/education efforts. Over 20 undergraduates have trained with Mura thus far, and he has advised five masters degrees and five PhD dissertations. Mura's general interests lie in the realms of structural and computational biology, particularly as regards molecular evolution and the intersection of these areas with data science (e.g., explainable AI, alongside physical/mechanistic theories from natural sciences, as a way to illuminate the black-box of machine learning models). Some of Mura's recent work, with Robert Preissner and colleagues (Charité, Berlin), has explored what one can glean about diseases (and therapies) by leveraging data science approaches with electronic health records (e.g., in connection with COVID). Mura's most recent interests focus on deep learning strategies for exploratory analyses of the protein universe, particularly in light of a new 'Urfold' model of protein structure.

Lei Xie

Dr. Lei Xie is currently a professor in Computer Science at Hunter College, and Ph.D. program at Computer Science, Biochemistry, and Biology at the Graduate Center, The City University of New York. He is also an Adjunct Professor in Neuroscience at Weill Cornell Medicine, Cornell University. His research focuses on developing new methods in machine learning, systems biology, and biophysics for multi-scale modeling of drug actions and causal genotype-phenotype associations, and applying them to drug discovery and precision medicine. From 2001 to 2011, he was a principle scientist at San Diego Supercomputer Center (SDSC), research scientist in pharmaceutical company Hoffmann-La Roche and biotechnology start-up Eidogen. He was trained in Computational Biology and Biophysics as a postdoctoral fellow at Columbia University and Howard Hughes Medical Institute from 2000 to 2001. He obtained his Ph.D. in Medicinal Chemistry and M.S. in Computer Science from Rutgers University, and B.S. in Polymer Physics from University of Science and Technology of China.

MDPI

Editorial

A Tribute to Phil Bourne—Scientist and Human

Cameron Mura [1,*], Emma Candelier [1] and Lei Xie [2]

1 School of Data Science, University of Virginia, Charlottesville, VA 22903, USA
2 Department of Computer Science, Hunter College, The City University of New York, New York, NY 10065, USA
* Correspondence: cmura@virginia.edu; Tel.: +1-434-249-3035

This Special Issue of *Biomolecules*, commissioned in honor of Dr. Philip E. Bourne, focuses on a new field of biomolecular data science. In this brief retrospective paper, we consider the arc of Phil's 40-year scientific and professional career, particularly as it relates to the origins of this new field.

Phil, as he is known to all—from students to university presidents and beyond—is the founding Dean of the School of Data Science (SDS) at the University of Virginia (UVA). He previously served as the first Associate Director for Data Science at the U.S. National Institutes of Health (NIH), where he led a novel *Big Data to Knowledge* initiative [1]. Prior to the NIH, Phil had a highly productive and impactful 20-year career at the University of California, San Diego (UCSD), with close ties to the San Diego Supercomputer Center and the Protein Data Bank (which he co-directed). At UCSD, Phil was also a Professor of Pharmacology, and ultimately an Associate Vice Chancellor.

This tribute, which accompanies an interview in this Special Issue, does not seek to delineate Phil's curriculum vitae or detail his many honors and achievements—e.g., serving as an early President of the International Society for Computational Biology and as the first Editor-in-Chief of *PLoS Computational Biology*—but rather to highlight the several ways in which Phil's contributions and leadership in multiple, disparate fields have coalesced as part of a new field of biomolecular data science. For details, note that a brief autobiographical account of Phil is available [2], as are his Wikipedia profile [3], his Ph.D. dissertation [4], and a list of the many scientists [5] whom Phil has trained, mentored and advised over the past four decades (this information is also available as a taxonomic tree [6], fittingly enough). Also, we would be remiss were we not to mention that one can learn what Phil, Monty Python, X-ray crystallography, and the county of Yorkshire, England all have in common by visiting ref [7]. Here, we intentionally intertwine the personal and the professional—as one can gather from even just brief interactions with him, Phil-the-human and Phil-the-scientist are refreshingly one and the same (Figure 1).

Currently a Professor of Biomedical Engineering and the Stephenson Dean of the School of Data Science at UVA, Phil spent much of his career exploring and helping *define* the intersection of biomolecules and computation—as a practicing scientist and as a leader [8] in academia, in open-access academic publishing [9], in the broader open-science movement [10,11], and in conjunction with government and industry (Phil's role as an associate vice chancellor at UCSD concerned "innovation and industrial alliances"). Over the span of Phil's four-decade career, our knowledge of biomolecular structures, dynamics, functions and evolution (in both health and disease) has rapidly advanced, often exponentially. *What enabled this?* The staggering advances were enabled, in no small part, by Phil's highly collaborative and foundational work, where three pervasive themes have been: (i) a **structural approach** to biological systems, including knowing when to be reductionist and when not to be; (ii) the development and application of core **computational methodologies**; and (iii) **multidisciplinarity**, to an extreme.

Citation: Mura, C.; Candelier, E.; Xie, L. A Tribute to Phil Bourne—Scientist and Human. *Biomolecules* **2023**, *13*, 181. https://doi.org/10.3390/biom13010181

Received: 7 December 2022
Accepted: 5 January 2023
Published: 16 January 2023

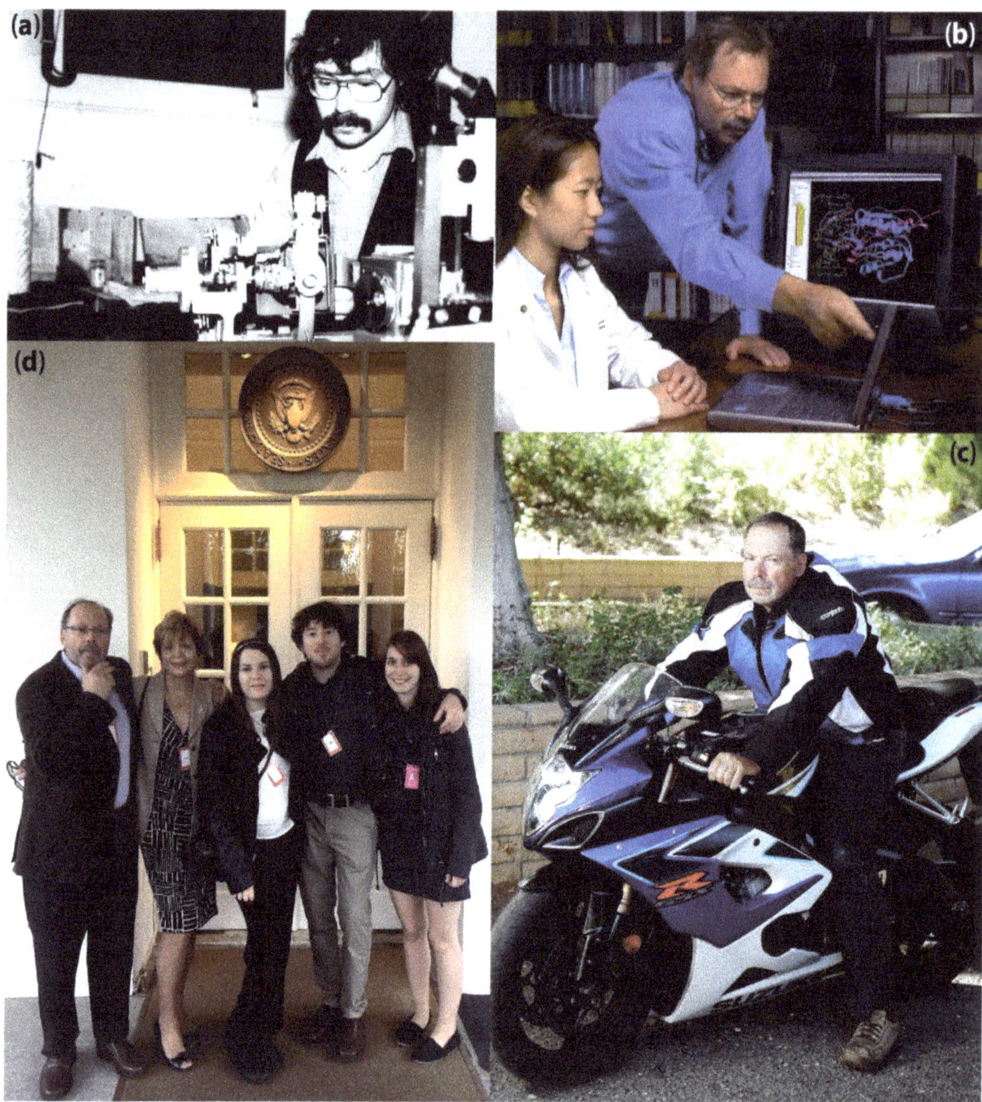

Figure 1. Phil's life in science started (**a**) very hands on, progressed to (**b**) mentoring, and then he finally (**c**) took off for the (**d**) White House with his family. While he's been a scientist for over 40 years, Phil's been an avid cyclist for even longer; at UVA, he's a founding member of the Hells Administrators (https://www.youtube.com/watch?v=ZgtNp1ditzE (accessed on 5 January 2023)).

To elaborate these three points—structure, computation, multidisciplinarity—we note that from the start of his career, first in small-molecule crystallography [12] and then in (very) large-molecule structural biology [13], Phil embraced the *key role of three-dimensional structure* [2] as an information-rich bridge between a biomolecule's sequence and its function. (Phil's *Structural Bioinformatics* text is a mainstay on many researchers' bookshelves [14].) As regards point (ii), a hallmark of Phil's research programs over the years has been the development and application of *computational methodologies & resources*, including state-of-the-art databases (most notably the Protein Data Bank [15]) and associated data standardization, dictionary and exchange approaches, such as the macromolecular crystallographic information file (mmCIF) [16]. Along the way, Phil and his teams created data standards and interoperable tools that were freely disseminated, before this was appreciated and accepted as scientific best practice, and they developed algorithms and software, such as the widely used combinatorial extension (CE) method for 3D structure alignment [17] and a novel approach to using "sequence order-independent profile–profile alignment" to examine protein functional sites across vast evolutionary distances [18]. Finally, as regards point (iii), computational biology and related areas are well-understood to be *highly* interdisciplinary [19], and here we simply reiterate that Phil was a pioneer in these fields from their inception (before they were 'a thing'). As an extreme example that is specific to Phil, not many scientists have both published research on "ancient shifts in trace metal geochemistry" [20] and written a book on Unix [21]!

In addition to foundational 'basic research' advances, Phil's work and its applications have had significant impact across a vast array of biological and biomedical domains, including early-stage drug discovery [22], molecular evolution [23], immunology [24], and more—resulting in over 350 papers, several books, and nearly 75,000 citations of his work [25]. In recent years, Phil's attention has turned to considering what is possible at the junction of data science and structural biology [26,27]; notably, Phil's receipt of Microsoft's *Jim Gray Award for eScience* (2010) foretold his move into this area, as this award cited his "*groundbreaking accomplishments in data–intensive science*". All throughout these career milestones, Phil has been unwavering in his support of public service in government and academia, in open scholarship, in research best practices [28], and in the professional development of all who have crossed his path, from students to peers to colleagues. Indeed, as regards professional development, many readers are likely familiar with the *Ten Simple Rules* (TSR) series that Phil conceived of and initiated 20 years ago. There are now well over 1000 rules [29], covering everything from strategically forging one's career path in academia, government and industry [30,31], to winning a Nobel Prize [32], to focused guides on leveraging Git/GitHub [33], to avoiding and resolving conflicts with your colleagues [34]. The full collection of TSRs, which is freely available at ref. [35] and organized by topical areas/categories (*Career development*, *Education & mentoring*, etc.), is a testament to how Phil empowers scientists to more effectively navigate the world of very-human scientific activities (papers, talks, careers) that begin where the data-collection and number-crunching end.

Those who have worked with Phil have likely noticed that a pronounced trait in his approach to biosciences, and now data science, is that it is expansive and forward-looking, with a healthy dose of irreverence and provocation [36]—in a word, *visionary*. Phil's interests in recent years have converged upon "biomedical data sciences", which can be viewed as a natural evolution (and synthesis) of bioinformatics, computational biology, structural biology, biophysics, systems biology, and other allied fields [36]. In a real sense, the intense multidisciplinarity of Phil's career foreshadowed a field such as biomedical data science. This Special Issue honors Phil by trying to capture his vision as it relates to biomolecules—how this vision arose and what it can encompass, as expressed in a collection of original research papers, perspectives and reviews. We hope that the breadth and depth of the contributions in this Special Issue convey the spirit of Phil's vision.

Finally, as we honor Phil in this Special Issue, recognizing his role today as the Dean of the UVA School of Data Science—the first of its kind in the nation—we close by noting that Phil's vision of biomedical data science can be mapped to four core elements of data science: *Systems*, *Analysis*, *Design* and *Value*. For example, *Systems*, in our context of biomolecular data science, relates to the underlying infrastructure, such as data structures, ontologies, software libraries and tools, that enables discovery. With respect to biomolecules, *Analysis* has been largely dominated by machine learning approaches such as deep learning, for which robust systems and frameworks to access and efficiently utilize training data are critical (e.g., [37]). *Design*, which can refer to human–computer interaction, visualization and so on, has played a vital role throughout the history of structural and computational biology, and now biomolecular data science. Finally, the *Value* element seeks to optimize the benefit of research for those it serves, from society at large to local communities; here, clear links exist between drug and therapeutic development, health disparities research, and other realms at the heart of biomolecular and biomedical data sciences.

The papers in this Special Issue exemplify what a field of biomolecular data sciences can represent, as a fitting tribute to someone who has moved the field forward via his own work and by his steadfast support of many research communities, biomolecular and beyond. In keeping with Phil's mantra, *'Onwards!'* ...

Author Contributions: All authors contributed to the writing and editing of the text. All authors have read and agreed to the published version of the manuscript.

Acknowledgments: We thank Phil Bourne for providing the photographs used in Figure 1.

Conflicts of Interest: The authors declare no conflict of interest. The funders had no role in the design of the study; in the collection, analyses, or interpretation of data; in the writing of the manuscript; or in the decision to publish the results.

References

1. Margolis, R.; Derr, L.; Dunn, M.; Huerta, M.; Larkin, J.; Sheehan, J.; Guyer, M.; Green, E.D. The National Institutes of Health's Big Data to Knowledge (BD2K) initiative: Capitalizing on biomedical big data. *J. Am. Med. Inform. Assoc.* **2014**, *21*, 957–958. [CrossRef] [PubMed]
2. Bourne, P.E. Life Is Three-dimensional, and It Begins with Molecules. *PLoS Biol.* **2017**, *15*, e2002041. [CrossRef] [PubMed]
3. Philip Bourne–Wikipedia. Available online: https://en.wikipedia.org/wiki/Philip_Bourne (accessed on 5 January 2023).
4. Bourne, P.E. Crystal Structure Analyses: Metal Complexes of Biological Interest and the Stereochemistry of Substituted Phenylbicyclo[2,2,2]octanes. Ph.D. Thesis, Flinders University of South Australia, Adelaid, Australia, 1979. Available online: https://www.worldcat.org/title/708090572 (accessed on 5 January 2023).
5. Bourne, P.E. Bourne Lab Alumni. Available online: https://tinyurl.com/BourneLabAlumni (accessed on 16 November 2022).
6. Computational Biology Family Tree—Philip, E. Bouine. Available online: https://academictree.org/compbio/tree.php?pid=157978 (accessed on 16 November 2022).
7. Nicholls, A.; Bourne, P.E.; Banquet, H. *Macromolecular Crystallography Computing School*; IUCr: Bellingham, Washington, USA, 1996; Available online: https://tinyurl.com/IUCrBanquetHumour (accessed on 5 January 2023).
8. Bourne, P.E. Ten Simple Rules for Good Leadership. *PLoS Comput. Biol.* **2022**, *18*, e1010133. [CrossRef] [PubMed]
9. McKiernan, E.C.; Bourne, P.E.; Brown, C.T.; Buck, S.; Kenall, A.; Lin, J.; McDougall, D.; Nosek, B.A.; Ram, K.; Soderberg, C.K.; et al. How Open Science Helps Researchers Succeed. *Elife* **2016**, *5*. [CrossRef] [PubMed]
10. Bourne, P.E.; Bonazzi, V.; Brand, A.; Carroll, B.; Foster, I.; Guha, R.V.; Hanisch, R.; Keller, S.A.; Kennedy, M.L.; Kirkpatrick, C.; et al. Playing Catch-up in Building an Open Research Commons. *Science* **2022**, *377*, 256–258. [CrossRef]
11. Wilkinson, M.D.; Dumontier, M.; Aalbersberg, I.J.; Appleton, G.; Axton, M.; Baak, A.; Blomberg, N.; Boiten, J.W.; da Silva Santos, L.B.; Bourne, P.E.; et al. The FAIR Guiding Principles for Scientific Data Management and Stewardship. *Sci. Data* **2016**, *3*, 160018. [CrossRef]
12. Bourne, P.E.; Taylor, M.R. The Structure of Aqua[3-ethoxy-2-oxobutyraldehyde bis(thiosemicarbazonato)]zinc(II). *Acta Cryst. Sect. B* **1980**, *36*, 2143–2145. [CrossRef]
13. Clegg, G.A.; Stansfield, R.F.; Bourne, P.E.; Harrison, P.M. Helix Packing and Subunit Conformation in Horse Spleen Apoferritin. *Nature* **1980**, *288*, 298–300. [CrossRef]
14. Bourne, P.E.; Gu, J. *Structural Bioinformatics*, 2nd ed.; Wiley-Blackwell: Hoboken, NJ, USA, 2009; p. 1096.
15. Burley, S.K.; Berman, H.M.; Duarte, J.M.; Feng, Z.; Flatt, J.W.; Hudson, B.P.; Lowe, R.; Peisach, E.; Piehl, D.W.; Rose, Y.; et al. Protein Data Bank: A Comprehensive Review of 3D Structure Holdings and Worldwide Utilization by Researchers, Educators, and Students. *Biomolecules* **2022**, *12*, 1425. [CrossRef]
16. Bourne, P.E.; Berman, H.M.; McMahon, B.; Watenpaugh, K.D.; Westbrook, J.D.; Fitzgerald, P. Macromolecular Crystallographic Information File. In *Methods in Enzymology*; Academic Press: Cambridge, MA, USA, 1997; Volume 277, pp. 571–590.
17. Shindyalov, I.N.; Bourne, P.E. Protein Structure Alignment by Incremental Combinatorial Extension (CE) of the Optimal Path. *Protein Eng.* **1998**, *11*, 739–747. [CrossRef]
18. Xie, L.; Bourne, P.E. Detecting Evolutionary Relationships Across Existing Fold Space, Using Sequence Order-independent Profile-profile Alignments. *Proc. Natl. Acad. Sci. USA* **2008**, *105*, 5441–5446. [CrossRef]
19. Eddy, S.R. "Antedisciplinary" Science. *PLoS Comput. Biol.* **2005**, *1*, e6. [CrossRef]
20. Dupont, C.L.; Yang, S.; Palenik, B.; Bourne, P.E. Modern Proteomes Contain Putative Imprints of Ancient Shifts in Trace Metal Geochemistry. *Proc. Natl. Acad. Sci. USA* **2006**, *103*, 17822–17827. [CrossRef]
21. Bourne, P.E.; Holstein, R.; McMullen, J. *UNIX for OpenVMS Users*, 3rd ed.; Digital Press: Oxford, UK, 2003; p. 562.

22. Kinnings, S.L.; Liu, N.; Buchmeier, N.; Tonge, P.J.; Xie, L.; Bourne, P.E. Drug Discovery Using Chemical Systems Biology: Repositioning the Safe Medicine Comtan to Treat Multi-drug and Extensively Drug Resistant Tuberculosis. *PLoS Comput. Biol.* **2009**, *5*, e1000423. [CrossRef]
23. Scheeff, E.D.; Bourne, P.E. Structural Evolution of the Protein Kinase-like Superfamily. *PLoS Comput. Biol.* **2005**, *1*, e49. [CrossRef]
24. Ponomarenko, J.V.; Bourne, P.E. Antibody-protein Interactions: Benchmark Datasets and Prediction Tools Evaluation. *BMC Struct. Biol.* **2007**, *7*, 64. [CrossRef]
25. Philip, E. Bourne—Google Scholar. Available online: https://tinyurl.com/BourneGoogleScholar (accessed on 16 November 2022).
26. Mura, C.; Draizen, E.J.; Bourne, P.E. Structural Biology Meets Data Science: Does Anything Change? *Curr. Opin. Struct. Biol.* **2018**, *52*, 95–102. [CrossRef]
27. Bourne, P.E.; Draizen, E.J.; Mura, C. The Curse of the Protein Ribbon Diagram. *PLoS Biol.* **2022**, *20*, e3001901. [CrossRef]
28. Borgman, C.L.; Bourne, P.E. Why It Takes a Village to Manage and Share Data. *Harv. Data Sci. Rev.* **2022**, *4*. [CrossRef]
29. Bourne, P.E.; Lewitter, F.; Markel, S.; Papin, J.A. One Thousand Simple Rules. *PLoS Comput. Biol.* **2018**, *14*, e1006670. [CrossRef]
30. Searls, D.B. Ten Simple Rules for Choosing Between Industry and Academia. *PLoS Comput. Biol.* **2009**, *5*, e1000388. [CrossRef] [PubMed]
31. Bourne, P.E. Ten Simple Rules in Considering a Career in Academia Versus Government. *PLoS Comput. Biol.* **2017**, *13*, e1005729. [CrossRef] [PubMed]
32. Roberts, R.J. Ten Simple Rules to Win a Nobel Prize. *PLoS Comput. Biol.* **2015**, *11*, e1004084. [CrossRef] [PubMed]
33. Perez-Riverol, Y.; Gatto, L.; Wang, R.; Sachsenberg, T.; Uszkoreit, J.; Leprevost Fda, V.; Fufezan, C.; Ternent, T.; Eglen, S.J.; Katz, D.S.; et al. Ten Simple Rules for Taking Advantage of Git and GitHub. *PLoS Comput. Biol.* **2016**, *12*, e1004947. [CrossRef]
34. Lewitter, F.; Bourne, P.E.; Attwood, T.K. Ten Simple Rules for Avoiding and Resolving Conflicts with Your Colleagues. *PLoS Comput. Biol.* **2019**, *15*, e1006708. [CrossRef]
35. Ten Simple Rules Collection. Available online: https://collections.plos.org/collection/ten-simple-rules/ (accessed on 5 January 2023).
36. Bourne, P.E. Is "Bioinformatics" Dead? *PLoS Biol.* **2021**, *19*, e3001165. [CrossRef]
37. Draizen, E.J.; Murillo, L.F.R.; Readey, J.; Mura, C.; Bourne, P.E. PROP3D: A Flexible, Python-based Platform for Protein Structural Properties and Biophysical Data in Machine Learning. *bioRxiv* **2023**, Submitted. [CrossRef]

 *biomolecules*

Article

Using GPT-3 to Build a Lexicon of Drugs of Abuse Synonyms for Social Media Pharmacovigilance

Kristy A. Carpenter [1] **and Russ B. Altman** [1,2,*]

[1] Department of Biomedical Data Science, Stanford University, Stanford, CA 94305, USA
[2] Departments of Bioengineering, Genetics, and Medicine, Stanford University, Stanford, CA 94305, USA
* Correspondence: russ.altman@stanford.edu

Abstract: Drug abuse is a serious problem in the United States, with over 90,000 drug overdose deaths nationally in 2020. A key step in combating drug abuse is detecting, monitoring, and characterizing its trends over time and location, also known as pharmacovigilance. While federal reporting systems accomplish this to a degree, they often have high latency and incomplete coverage. Social-media-based pharmacovigilance has zero latency, is easily accessible and unfiltered, and benefits from drug users being willing to share their experiences online pseudo-anonymously. However, unlike highly structured official data sources, social media text is rife with misspellings and slang, making automated analysis difficult. Generative Pretrained Transformer 3 (GPT-3) is a large autoregressive language model specialized for few-shot learning that was trained on text from the entire internet. We demonstrate that GPT-3 can be used to generate slang and common misspellings of terms for drugs of abuse. We repeatedly queried GPT-3 for synonyms of drugs of abuse and filtered the generated terms using automated Google searches and cross-references to known drug names. When generated terms for alprazolam were manually labeled, we found that our method produced 269 synonyms for alprazolam, 221 of which were new discoveries not included in an existing drug lexicon for social media. We repeated this process for 98 drugs of abuse, of which 22 are widely-discussed drugs of abuse, building a lexicon of colloquial drug synonyms that can be used for pharmacovigilance on social media.

Keywords: large language models; pharmacovigilance; social media; drugs of abuse

Citation: Carpenter, K.A.; Altman, R.B. Using GPT-3 to Build a Lexicon of Drugs of Abuse Synonyms for Social Media Pharmacovigilance. *Biomolecules* **2023**, *13*, 387. https://doi.org/10.3390/biom13020387

Academic Editors: Cameron Mura and Lei Xie

Received: 11 January 2023
Revised: 9 February 2023
Accepted: 16 February 2023
Published: 18 February 2023

1. Introduction

The opioid epidemic is a growing crisis, driving a drastic rise in deaths attributed to drug overdose over the past several years in the United States [1,2]. Of the nearly 92,000 overdose deaths in 2020, over 56,000 involved synthetic opioids such as fentanyl [3]. It is imperative for researchers to understand the past, present, and future of drug abuse in order to combat this national emergency.

Pharmacovigilance is the detection, assessment, and analysis of the usage and effects of drugs [4]. Monitoring trends in the opioid epidemic and the abuse of other drugs is a critical first step in reducing the number of deaths from drug overdoses [5]. Several international and national agencies, such as the World Health Organization (WHO), the European Medicines Agency (EMA), the U.S. Food and Drug Administration (FDA), the U.S. Centers for Disease Control and Prevention (CDC), the U.S. National Institutes of Health (NIH), the U.S. Drug Enforcement Administration (DEA), and the U.S. Department of Health and Human Services (HHS), survey and monitor drug use and effects. Notable pharmacovigilance systems from these agencies include VigiBase, EudraVigilance, the FDA Adverse Event Reporting System (FAERS), the National Health and Nutrition Examination Survey (NHANES), the National Drug Early Warning System (NDEWS), the National Forensic Laboratory Information System (NFLIS), and the National Survey on Drug Use and Health (NSDUH). Concern about growing opioid abuse has driven numerous analyses on opiates' pharmacovigilance data from these systems [6–13]. In addition to analysis of statistics and metrics from reporting systems, pharmacovigilance studies have also been conducted using text mining and natural language processing (NLP) on free text notes in electronic health records (EHRs) [14–19].

There has been an increased interest over the past decade in using social media for NLP-based pharmacovigilance. While official pharmacovigilance surveys may have a latency of months to years to make results available, social media data can be queried nearly instantaneously. Drug users are also willing to freely post their experiences with drugs online, sharing information that they may not want to make accessible to federal agencies, making social media a valuable resource for surveillance of illicit drug use [20]. The pseudo-anonymity offered by various social media platforms facilitates this openness. Social media pharmacovigilance is believed to have first appeared in 2010 [21] and has gained traction since, with most studies using the social media platforms Twitter [22–30], Facebook [28–30], and Reddit [31–33] for tasks such as adverse drug reaction extraction and off-label drug usage analysis.

While social media holds promise for improving pharmacovigilance efforts, it also brings unique challenges. Social media data are fundamentally different from that of the FDA, WHO, CDC, NIH, or even clinical notes in an EHR in that it is by nature casual and unstandardized, and therefore rife with misspellings and slang. There is abundant missingness, as not all posts will contain geographic or demographic information. Misinformation is rampant on social media, with the ease of posting and incentivization of viral content leading to the easy spread of rumors, conspiracy theories, and misleading interpretations of scientific results, both intentionally and unintentionally [34,35]. In many cases, it is impossible to verify the validity of any information posted. The contents of social media are also heavily influenced by politics, current events, and pop culture. As such, analyses of social media could be considered "unscientific", or at least in violation of the traditional standards of epidemiological studies. There is much improvement to be desired from social media pharmacovigilance efforts [36], which are still very much in their infancy. However, despite these limitations, multiple meta-analyses and qualitative reviews have found that social media pharmacovigilance efforts are able to extract some meaningful signal pertaining to drug use and effects [37–39].

One method that addresses some of the outstanding problems of social media pharmacovigilance is RedMed [40], a word embedding model based on continuous bag-of-words modeling [41] and trained on archived comments on health- and drug-related Reddit forums. After training the model to cluster similar terms, RedMed can discover candidate terms with significant cosine similarity to an index term from DrugBank [42] and subsequently verify those terms with filters related to edit distance, phonetics, pill impressions, and Google search results. RedMed produced a lexicon of drug synonyms that included misspellings and slang terms, enabling better retrieval of pharmacovigilance-relevant text from social media sources; it was subsequently used for quantification of adverse drug reaction severity [32].

We propose to extend RedMed without training a new embedding model by using pre-trained large language models—specifically, Generative Pretrained Transformer 3 (GPT-3) [43]. GPT-3 is the third installation of a generative pre-trained transformer from the company OpenAI that has been trained on the entire internet. It is an autoregressive language model of unprecedented size, with 175 billion parameters. Generally, language models are probability distributions over sequences (typically of words) that can identify if a given sequence is likely or generates likely sequences. Transformers are a machine learning architecture that is built around the attention mechanism [44], and have sparked great advances in language modeling. GPT-3 garnered much discussion upon its release in 2020 due to its performance in few-shot learning; given only a few examples, it is able to produce desired text output that closely resembles real human writing. Typical examples of GPT-3 tasks are question-answering, story completion, translation, and summarization. Researchers have also explored using GPT-3 in a medical context, on tasks such as EHR summarization or supporting a medical chatbot, but no such models have been deemed ready for deployment in the clinic [45–47]. The reception of GPT-3 has not all been positive, and due to its training on the entire internet, it is prone to generating text that perpetuates harmful stereotypes or promotes dangerous activity [43,48,49]. OpenAI has made GPT-3 available as an application programming interface (API), allowing researchers to leverage its capabilities without needing to train a massive language model themselves.

We argue that GPT-3 is valuable for social media pharmacovigilance as it is able to generate text that closely resembles common writing patterns used on the internet at large. In this work, we make the following contributions:

- We introduce a novel method to repeatedly query GPT-3 for drug synonyms and filter the generated terms to create a lexicon enriched for likely synonyms, all in an automated fashion. We make the code for the method publicly available to build similar lexicons, facilitating interpretable pharmacovigilance on messy, casually-written social media data that does not require training a new large machine learning model;
- We present a lexicon of GPT-3 synonyms for 98 drugs of abuse, including 22 widely-discussed drugs of abuse, which can be used to easily flag text likely to be related to drug abuse from a large corpus of informal language in an interpretable manner;
- Finally, we also demonstrate of the capabilities of GPT-3, and similar models, for practical contributions to pharmacovigilance.

Code and data are available on GitHub (https://github.com/kristycarp/gpt3-lexicon).

2. Materials and Methods

2.1. Datasets

2.1.1. RedMed

We use index terms from the RedMed lexicon to provide seed terms to GPT-3. The RedMed lexicon is comprised of index terms from DrugBank, their respective associated known drug terms (AKDTs) (e.g., brand names), and their respective synonymous terms generated by a word embedding model and subsequently filtered. The terms in the RedMed lexicon data frame are organized into columns to indicate how the term was validated: because it is an AKDT (`known`), within close edit distance (`edOne`, `edTwo`), within close phonetic edit distance (`misspellingPhon`), a pill impression (`pillMark`), validated by Google search (`google_ms`, `google_title`, `google_snippet`), or present in a slang-specific database (`ud_slang`). To improve quality of our results, we only sample from the RedMed synonyms that are a single-word AKDT, within close edit distance, within close phonetic edit distance, and a pill impression. The choice to limit inclusion of AKDTs to only those comprised of a single word followed from the observation that many multi-word AKDTs were simply short phrases containing the seed term (e.g., for alprazolam, commonly known as xanax, the multi-word AKDTs include "started taking alprazolam", "xanax works great", and "quit taking xanax"). When these phrases are presented to GPT-3, more such phrases are generated, which are not useful for our task as they already contain a known drug synonym and therefore add no information. We also excluded RedMed synonyms from the `google_ms`, `google_title`, `google_snippet`, and `ud_slang` columns, as these tended to include a higher rate of false positives, and presenting GPT-3 with irrelevant examples leads to generation of more irrelevant terms.

2.1.2. Drugs of Abuse

The DEA maintains a list of controlled substances, which are defined to be drugs with high potential for abuse. As of July 2022, there are 543 DEA-controlled substances [50]. We took the intersection of the 543 controlled substances and the 2997 index terms in RedMed, resulting in 131 controlled index terms. Of these, 33 contained fewer than three terms in our selected columns of RedMed; we eliminated these index terms as they did not have sufficient RedMed synonyms for GPT-3 prompt generation. This resulted in a final set of 98 controlled index terms to input into the GPT-3 query pipeline.

2.1.3. Widely-Discussed Drugs of Abuse

Some of the 98 selected index terms are more widely-discussed online than others, and therefore would likely have more synonyms than less widely-discussed index terms. In order to better evaluate how our pipeline performs on these drugs, we took approximately the top 25% of index terms with respect to discussion on Reddit and created a subset of widely-discussed drugs of abuse. For each of the 98 index terms, we used Google to search for exact matches to the index term on Reddit. We choose to limit to Reddit to reduce noise and because Reddit is a popular platform for discussing drug use [40,51,52]. The cutoff for the top 25% of index terms was approximately 10,000 Reddit hits (Figure A1), so for simplicity we used 10,000 hits on Reddit webpages as the cutoff to determine if an index term should be included in the "widely-discussed" subset. Of the 98 selected index terms, 22 are "widely-discussed." We note that this subset is only intended to demonstrate pipeline performance on drugs which are discussed more frequently on relevant discussion forums, as they are more likely downstream applications of our lexicon and pipeline than less prominent drugs.

2.2. External Models

2.2.1. GPT-3

We accessed the GPT-3 model [43] through the OpenAI API. We used the `text-davinci-002` engine for all queries.

2.2.2. Google Search API

We used the Custom Search JSON API from Google's Programmable Search Engine to automate Google searches of generated terms.

2.3. Terminology

Terminology coined in this manuscript (or in [40] and key to this study) is defined in Table 1.

Table 1. Definitions of new terminology used throughout this manuscript.

Term	Abbreviation	Definition
Associated known drug term	AKDT	As defined in [40], known terms used synonymously for a given drug, most often brand names.
Controlled index term	-	A *controlled substance* that is an *index term.*
Controlled substance	-	A substance (i.e., drug) that is deemed to have a high potential for abuse by the DEA and is therefore controlled.
Generated term	-	See *GPT-3 generated term.*
GPT-3 generated term	-	A term generated from a GPT-3 query as a candidate synonym for the corresponding index term used in the prompt.
GPT-3 synonym	-	A *GPT-3 generated term* that has been automatically labeled as a synonym following a filtering scheme.

Table 1. *Cont.*

Term	Abbreviation	Definition
Index term	-	The identifying term of a drug as indexed in RedMed; also the generic name of a drug as indicated in DrugBank.
Novel GPT-3 synonym	-	A *GPT-3 synonym* that is not already present in RedMed as a *RedMed synonym*.
Non-synonym	-	A *generated term* that has been manually labeled as not synonymous for the corresponding queried *index term*.
RedMed synonym	-	A term listed in RedMed as synonymous for a given *index term*.
Synonym	-	A *generated term* that has been manually labeled as synonymous for the corresponding queried *index term*.
Unique novel GPT-3 synonym	UNGS	Equivalent to a *novel GPT-3 synonym* but specifying that each unique *novel GPT-3 synonym* is only counted once no matter how many times it has been generated.
Widely-discussed	-	Specifying that a drug appears relatively more frequently on Reddit, suggesting higher rates of online discussion, more synonymous terms, and potentially greater interest for pharmacovigilance.

2.4. Methods

2.4.1. Overview of Query Pipeline

An iteration of the query pipeline begins by uniformly sampling three RedMed synonyms for the queried index drug term. We insert the index term and the sampled RedMed synonyms into a prompt template (further described below), which we provide to GPT-3 as a Completion query. Because we use an enumerative list in our prompt templates, and because GPT-3 is easily able to pick up on enumerative formatting, nearly all results returned by GPT-3 will also be formatted in an enumerative list. We automatically parse the listed results to extract the GPT-3 generated terms. We repeat this process to build a set of GPT-3 generated terms for the queried index term. We also pass the generated terms through filters described below. A schematic of this GPT-3 querying pipeline is depicted in Figure 1.

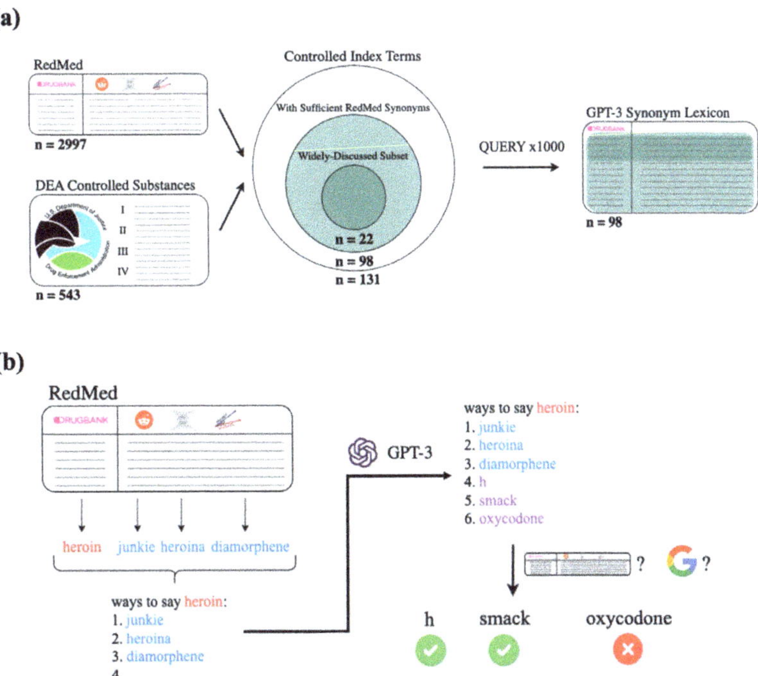

Figure 1. Overview of the selection steps and overall pipeline. (**a**) preprocessing. The intersection of RedMed and DEA controlled substances are taken as index terms to feed through the pipeline. Each index term goes through the pipeline for 1000 iterations, resulting in a lexicon of GPT-3 synonyms. Approximately one-quarter of the index terms put through the pipeline are designated as "widely-discussed" and are used to examine performance on terms with many synonyms and of high relevance to pharmacovigilance; (**b**) an example of a single iteration through the GPT-3 querying pipeline. For the desired index term (red), we uniformly sample three RedMed synonyms (blue) to insert into the prompt template. We present the prompt to the GPT-3 Completions API and parse the returned result for generated terms (purple). We use a Google search filter and a drug name filter to determine whether to classify generated terms as GPT-3 synonyms (green checkmark) or not (red x).

2.4.2. GPT-3 Prompt Templates

We experimented with a variety of prompt templates at a small scale in a sandbox environment when constructing the format of GPT-3 queries. Examples included asking for synonymous terms with and without examples, asking for synonymous terms in a colloquial manner (using slang and misspellings in the prompt), and writing the prompt as a conversation between two drug users discussing slang terms. We observed that prompts formulated as an enumerative list most often led to GPT-3 completions that continued the list, facilitating automated parsing of generated terms. Other types of prompts (such as asking in a colloquial manner or framing the prompt as a conversation) led to responses that were too varied to easily extract sets of generated terms at scale. In addition, because GPT-3 is specialized for few-shot learning [43], we know that it works very well for a desired task when given a few examples of desired output, and saw this reflected in our small-scale prompt experiments. These observations resulted in the choice of the following prompt template:

"ways to say [index term]:
1. [RedMed synonym 1]
2. [RedMed synonym 2]
3. [RedMed synonym 3]
4."

The hanging "4." indicates to the model that it should continue filling in the list.

We chose to include three example synonyms in the prompt template because we observed in our small scale experiments that GPT-3 tended to complete the enumerated list until there were three or ten items in the list. Therefore, using three example synonyms often led to seven additional terms being generated, maximizing the number of generated terms when this pattern was followed. Because we did not observe a drastic change in the number of generated terms beyond this pattern, we chose to not further investigate varying the number of synonyms presented in the prompt template, though this could become an area of future work. We note that GPT-3 queries limit the number of tokens in the prompt and response combined, meaning that listing a large number of example synonyms could impact the number of terms able to be generated.

We also observed that, with this formulation, GPT-3 tends to generate the names of drugs that are different from the index term but have the same indications (e.g., generating the names of other anti-anxiety medications when prompted for alprazolam terms). We hypothesized that providing counterexamples in the prompt might reduce this phenomenon. Our prompt template with counterexamples is as follows:

"these are not synonyms for [index term]:
1. [counterexample 1]
2. [counterexample 2]
3. [counterexample 3]
4. [counterexample 4]
but these are synonyms for [index term]:
1. [RedMed synonym 1]
2. [RedMed synonym 2]
3."

In our parameter search experiments, we used hand-picked counterexamples; for the index term of alprazolam, our counterexamples were ativan, zoloft, lexapro, and klonopin.

2.4.3. GPT-3 Parameter Search

The GPT-3 query API allows for the specification of model parameters, which include temperature, frequency penalty, and presence penalty. Temperature indicates how much the model should prioritize high-likelihood answers over providing diverse answers and ranges from 0 to 1; a low temperature leads to the model prioritizing high-likelihood answers, and a high temperature leads to the model prioritizing diverse answers. The frequency penalty controls how likely the model is to generate the same tokens verbatim and ranges from -2 to 2, with more positive numbers increasing the penalty of this verbatim repetition. In this context, a token is a sequence of characters (often full words, though a word can also be comprised of multiple tokens) commonly found in the training corpus of GPT-3. GPT-3 functions by learning the statistical relationships between tokens [43]. The presence penalty controls how likely the model is to generate text about new topics and ranges from -2 to 2, with more positive numbers increasing the penalty of topic repetition. We sought to identify the model parameters, as well as the prompt template that would maximize the number of unique novel GPT-3 synonyms (UNGSes), which we define as generated terms that pass the post-query filters and are not already present in RedMed. We only want to count each unique generated term once, as generating the same term multiple times does not add new information to the lexicon. We do not want to count terms that are already RedMed synonyms because these were already known and available. We ran 1000 iterations of the query pipeline on one index term for each possible combination of the following parameter settings: temperatures of 0.0, 0.3, 0.6, and 1.0; frequency penalties of 0.0, 0.5, and 1.0; presence penalties of 0.0, 0.5, and 1.0; and the prompt templates with and without counterexamples. We chose to only investigate these settings, rather than conduct a full automated parameter sweep, due to budget constraints (both the OpenAI API and the Google Search API incur costs per query) and the rationale that a grid search of values spanning the ranges of each parameter would be sufficient to identify settings useful for downstream application. We selected alprazolam as the index term for the initial parameter sweep experiment because it is a common drug of abuse that is discussed widely online and therefore a representative example of the type of term for which we would like good performance. We confirmed the observed trends from alprazolam by additionally running 1000 iterations of the query pipeline on two more drugs on a smaller set of parameter setting combinations: temperatures of 0.0, 0.5, and 1.0; frequency penalties of 0.0 and 1.0; presence penalties of 0.0 and 1.0; and only the prompt template without counterexamples. We selected heroin and benzphetamine as our index terms for these follow-up experiments because these are both drugs of abuse that the DEA classifies as having higher and lower potential for abuse, respectively, than alprazolam. Additionally, we would expect heroin to be discussed at a rate similar to or higher than that of alprazolam, whereas we would expect much less discussion of benzphetamine.

2.4.4. Google Filter

We used Google searches to automate an approximate validation of whether generated terms were synonymous with the index term. Upon extraction of each generated term from the GPT-3 response, we made a series of Google searches: the generated term alone, the generated term with "pill" appended, the generated term with "drug" appended, and the generated term with "slang" appended. The rationale behind the searches with appended keywords is that some drug slang terms have multiple meanings and a search of only the term itself may yield non-drug-related results; appending "drug", "pill", or "slang" makes it more likely to yield results with the drug-related context. We processed the search results through a specified maximum depth (e.g., a specified maximum depth of 10 would entail processing the top 10 search results), recording whether there is a search result within the maximum depth that has an instance of the index term appearing in its title or content snippet, and if so, the depth of the first result for which it does. Because the Google API limits the rate and daily number of API queries, we terminated the Google searching process for a term once one search contained a result with the index term. We also made the searching process more efficient with memoization.

2.4.5. Drug Name Filter

We filtered out generated terms if they appeared in the set of RedMed index terms and were not the same as the queried index term. This choice was informed by the observation that GPT-3 tends to generate the name of different drugs with the same indications as the queried index term.

2.4.6. Final Pipeline Parameters

After our parameter search experiments, we ran the final version of the pipeline on the set of 98 controlled index terms. We used a temperature of 1.0, a frequency penalty of 0.0, and a presence penalty of 0.0 for all GPT-3 queries. We used the Google filter with depth 10 and the drug name filter. We conducted 1000 iterations of the pipeline for each index term.

2.4.7. Manual Labeling

To be able to evaluate pipeline performance, we manually labeled the terms generated by our pipeline for alprazolam and fentanyl. We chose to manually label these two index terms because these drugs are very widely abused and discussed and therefore of high interest for pharmacovigilance efforts; in addition to informing this study, generating gold standard labels for alprazolam and fentanyl may be useful for later pharmacovigilance research. For each unique generated term, a human labeler performs internet searches, searches directly on substance-related Reddit forums, and cross-references with compiled lists of known drug slang terms to determine if, by their best judgment, the generated term was a valid slang term, misspelling, brand name, or other synonym. All labelers had previous experience in drug-related informatics and were very familiar with the domain. We instructed labelers to mark a generated term as a synonym if they found at least one instance online of a person using that term in a context where it was apparent that they were referring to the index term or if it was the brand name of the index term in any country. This includes terms that have both drug meanings and non-drug meanings. For example, "bars" could refer to alprazolam, a long rod, an establishment serving alcoholic drinks, or the action of prohibiting something. Even though in many contexts, "bars" does not refer to alprazolam, it would be labeled as a synonym because there are contexts in which "bars" indisputably does refer to alprazolam. Terms in other languages were also accepted. For example, "alprazolan" is Spanish for alprazolam and is therefore a synonym of alprazolam. We acknowledge that it is possible that some manual labels may be incorrect, but given the expertise of the reviewers, we believe that such errors are scarce enough to not majorly impact the conclusions we draw from our results.

2.4.8. Evaluation Criteria

We quantify the performance of our pipeline on the two index terms that we manually labeled by calculating precision (Equation (1)) and recall (Equation (2)). The F score is a metric to quantify the trade-off between precision and recall; the F1 score weights the two equally (Equation (3)), whereas the F2 score favors high recall over high precision (Equation (4)). We denote number of true positives by TP, number of false positives by FP, and number of false negatives by FN:

$$\text{precision} = \frac{\text{TP}}{\text{TP} + \text{FP}} \tag{1}$$

$$\text{recall} = \frac{\text{TP}}{\text{TP} + \text{FN}} \tag{2}$$

$$\text{F1 score} = 2 * \frac{\text{precision} * \text{recall}}{\text{precision} + \text{recall}} \tag{3}$$

$$\text{F2 score} = 5 * \frac{\text{precision} * \text{recall}}{2 * \text{precision} + \text{recall}} \tag{4}$$

In the context of this method, we prefer high recall to high precision when evaluating different filtering schemes. We do this because the primary use case for the lexicons produced by our pipeline is to scan social media posts for drug-related terms in order to identify which posts are likely about the drug of interest. In this context, it is better to flag irrelevant posts as relevant than to miss relevant posts because it is possible to use manual inspection or other automated models to further filter the posts, whereas the size of social media corpora makes it intractable to identify false negative posts. Therefore, we consider both the F1 and F2 scores in our evaluation.

We note that some applications of this lexicon or pipeline may require higher specificity or precision than provided by our current criteria, in which case subsequent filters will be needed to remove false positives. However, we maintain that favoring recall in an initial evaluation is important because, while it is possible to filter out likely false positives from an existing lexicon, it is much more difficult to introduce likely false negatives into an existing lexicon.

3. Results

3.1. Parameter Search

We first examined parameter trends from the results of the pipeline parameter sweep for the index term of alprazolam. We saw a clear relationship between increased temperature and increased number of UNGSes (Figure 2a). We also saw a large difference between the two prompt templates; the prompt template without counterexamples had dramatically more UNGSes than that with counterexamples (Figure 2b). There were not obvious relationships between frequency penalty or presence penalty and number of UNGSes, though for both penalties, we saw that the the maximum of the range of UNGSes per iteration tended to decrease as the penalties increased (Figure 2c,d). We found that the temperature trend was consistent when examining the results of the heroin and benzphetamine iterations (Figure 3a); we did not repeat the prompt variation as the alprazolam results were so stark. These two sets of iterations also showed a slight average decrease in number of UNGSes when either the frequency penalty or presence penalty was increased (Figure 3b,c). We therefore decided that a temperature of 1.0, a frequency penalty of 0.0, a presence penalty of 0.0, and the prompt template without counterexamples were the best parameter settings to use going forward. Further solidifying the decision to use these parameter settings, we observed that, for each drug, the set of 1000 iterations that generated the most UNGSes was one with all, or almost all, parameter settings matching our choices (Figure 4).

3.2. Google Search Depth Analysis

We used the manually labeled alprazolam and fentanyl data to both determine an appropriate cutoff for the maximum search depth and to examine if there is a relationship between search depth and manual label. We used a maximum search depth of 30 for the 1000 alprazolam iterations and saw a sharp decrease in the proportion of generated terms that are synonyms to generated terms that are not synonyms around a search depth of 10, in addition to an overall decrease in the number of unique terms generated (Figure 5a). This informed a lower maximum search depth of 10 for the 1000 fentanyl iterations, which displayed a similar power-law-like decrease in number of synonyms, non-synonyms, and unique terms generated overall as the depth increased (Figure 5b).

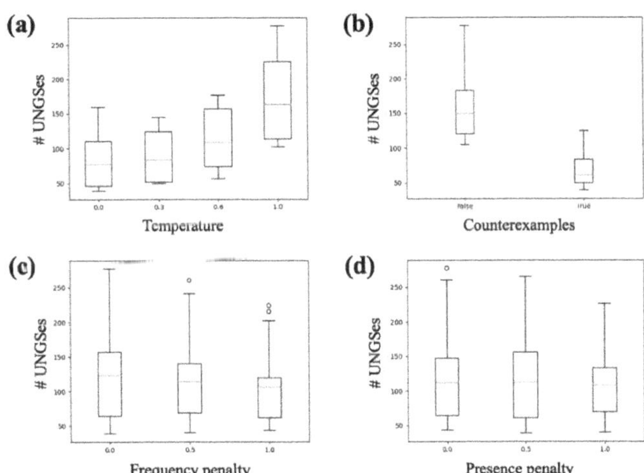

Figure 2. Parameter search using alprazolam as index term. Each combination of temperature, prompt template, frequency penalty, and presence penalty was used to conduct 1000 iterations of the query pipeline. The number of unique novel GPT-3 synonyms (UNGSes) generated by the 1000 iterations was recorded for each parameter set. Each subfigure shows the distribution of UNGSes for each value of (**a**) temperature; (**b**) prompt template; (**c**) frequency penalty; and (**d**) presence penalty.

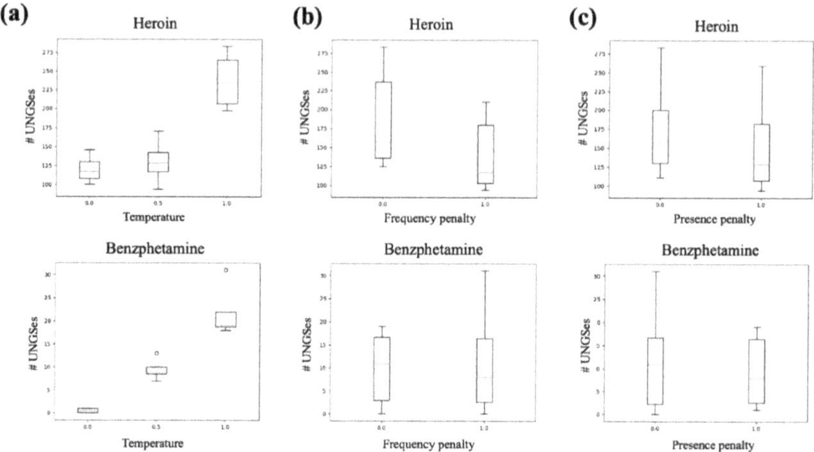

Figure 3. Smaller parameter search using heroin and benzphetamine as index terms. Each combination of temperature, frequency penalty, and presence penalty was used to conduct 1000 iterations of the query pipeline. The number of UNGSes generated by the 1000 iterations was recorded for each index term and parameter set. Each subfigure shows the distribution of UNGSes for both heroin and benzphetamine for each value of (**a**) temperature; (**b**) frequency penalty; and (**c**) presence penalty.

3.3. Generation Frequency Analysis

Across the 1000 iterations of the query pipeline, GPT-3 tended to generate many terms more than once. Notably, some very common colloquial names for the index terms used in these initial experiments appeared at a very high rate. We sought to investigate if a generated term's frequency of generation could be used to estimate how likely it is to be a true synonymous term. For both the set of alprazolam iterations and the set of fentanyl iterations, we observed that terms generated only once or twice were overwhelmingly manually labeled as non-synonyms, and that most, but not all, terms generated more than 15 times were manually labeled as synonyms (Figure 6a,b). While it is possible that not all manual labels are correct, the trend still holds even if there are some erroneous labels. We examined the five most frequently generated terms for both alprazolam and fentanyl. The five most frequent alprazolam terms were "xanax", "ativan", "zoloft", "alprazolan", and "xanor". Both "xanax" and "xanor" are common brand names of alprazolam, and "alprazolan" is both a common misspelling of alprazolam and the Spanish word for alprazolam. "Ativan" and "zoloft" are brand names of lorazepam and sertraline, respectively, which are distinct from alprazolam but share its anxiolytic effects. These two terms, which are not synonyms of alprazolam, were not caught by the drug name filter as they are brand names. The five most frequent fentanyl terms were "sublimaze", "duragesic", "fentanil", "fentanylum", and "fentora", all of which are either brand names or common misspellings of fentanyl, and are therefore fentanyl synonyms.

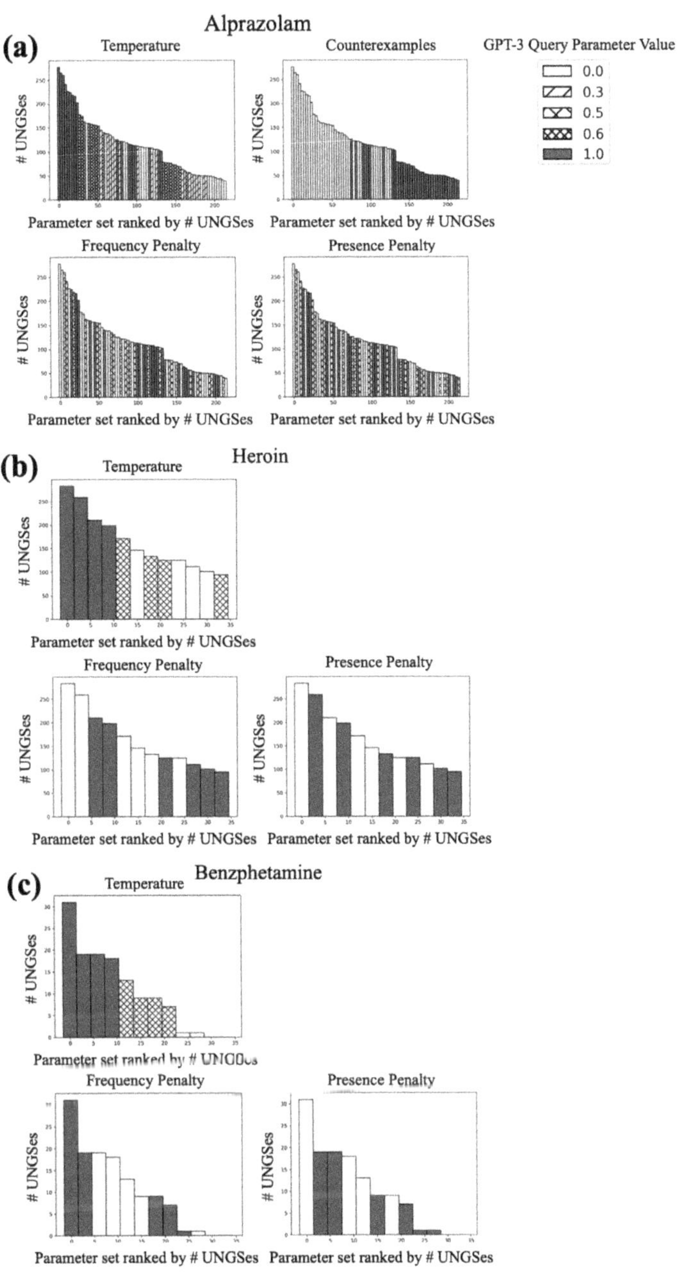

Figure 4. Number of UNGSes per each combination of temperature, prompt template, frequency penalty, and presence penalty for (**a**) alprazolam; (**b**) heroin; and (**c**) benzphetamine. The bar shading in each subplot represents the value of the parameter indicated in the title of that subplot. Each bar represents a different parameter set used for 1000 iterations of the pipeline.

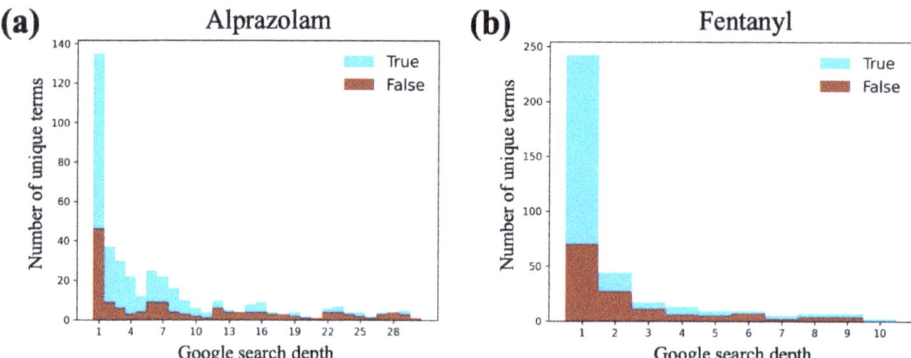

Figure 5. Histogram of the number of unique terms generated at each depth in the Google search for both (**a**) alprazolam and (**b**) fentanyl. At each search depth, the count of synonyms (true examples) is shown as blue bars, and the count of non-synonyms (false examples) is shown as red bars. The blue bars are stacked on top of the red bars (i.e., they do not continue behind the red bars). The alprazolam queries allowed a maximum search depth of 30, whereas the fentanyl queries were limited to a maximum search depth of 10 as utility drops after the tenth result.

After applying the Google search filter to the generated terms, we observed a reduction in the number of non-synonyms, most notably at the low end of the frequency range. Without the Google search filter, there were 137 alprazolam synonyms, 571 alprazolam non-synonyms, 168 fentanyl synonyms, and 907 fentanyl non-synonyms generated once (Figure 6a,b). With the Google search filter, there were 128 alprazolam synonyms, 115 alprazolam non-synonyms, 125 fentanyl synonyms, and 152 fentanyl non-synonyms generated once (Figure 6c,d). While, without the Google search filter, we may have discarded the terms only generated once or twice due to their high proportion of non-synonyms, we see that the proportion evens out after applying the Google search filter. We also note that, on the high-frequency end of the spectrum, the two non-synonyms present in the top five most frequently generated alprazolam terms do not pass the Google search filter, while the three synonyms do. Because of this effect, as well as the fact that there are synonyms at all frequency levels, we choose to not include a frequency-based filter into our query pipeline.

3.4. *Pipeline Performance*

We evaluated the performance of the lexicon generation pipeline using the manual labels for both alprazolam and fentanyl generated terms as a proxy for ground truth. In doing so, we sought to characterize the ability of each filter setup to automatically identify manually labeled synonyms and to determine which filters to run the pipeline with on a larger set of index terms.

As a baseline, we analyzed the performance when predicting that all terms generated by GPT-3 are synonyms (Figure 7a). This demonstrated how many synonyms were generated by GPT-3 for both alprazolam and fentanyl (269 and 314, respectively), but also showed how many non-synonyms are generated (750 and 1114, respectively). Despite the perfect recall in both cases (due to never assigning negative predicted labels), the low precision (0.264 and 0.220, respectively) supports our decision to filter the GPT-3 outputs.

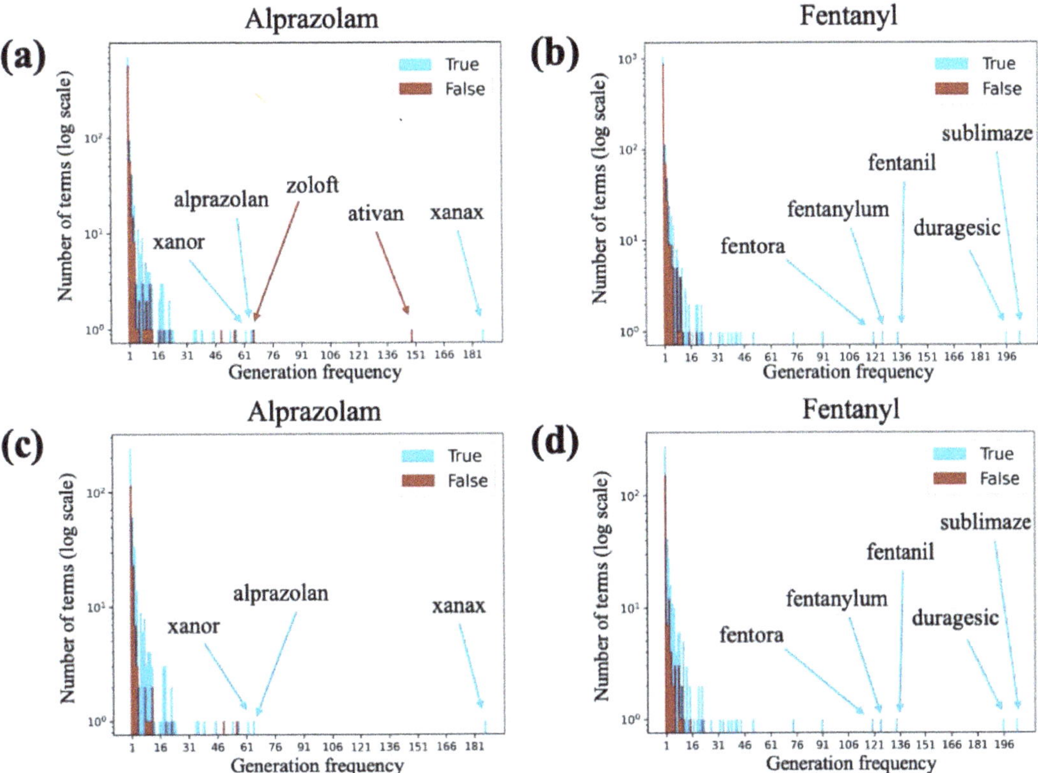

Figure 6. Histogram of the number of terms generated at different frequencies for (**a**) alprazolam without the Google search filter; (**b**) fentanyl without the Google search filter; (**c**) alprazolam with the Google search filter; and (**d**) fentanyl with the Google search filter. Plots use a logarithmic scale. At each search depth, the count of synonyms (true examples) is shown as blue bars, and the count of non-synonyms (false examples) is shown as red bars. The blue bars are stacked on top of the red bars (i.e., they do not continue behind the red bars). The top five most generated terms for each drug are labeled. All plots omit generated terms that do not pass the drug name filter.

We also analyzed the performance when only predicting a generated term as a synonym if it was already present in RedMed (Figure 7b). This led to perfect precision for both alprazolam and fentanyl (i.e., no false positives) but a low recall (0.178 and 0.115, respectively). The low recall is an indication of how many new terms that GPT-3 is generating that were not previously included in RedMed.

We analyzed multiple combinations of filters for the prediction of synonyms. The three filters assessed were the drug name filter, the frequency filter, and the Google search filter. The drug name filter removes generated terms that match any index term besides the queried index term. The frequency filter removes generated terms that are generated only once. We found that increasing the frequency threshold beyond one increased precision, but decreased recall; as previously stated, we prefer to maximize recall. The Google search filter removes generated terms if the corresponding index term does not appear in the first 10 Google search results for the term alone or with "pill", "drug", or "slang" appended. The precision, recall, and F1 and F2 scores for all filter combinations tested are shown in Table 2.

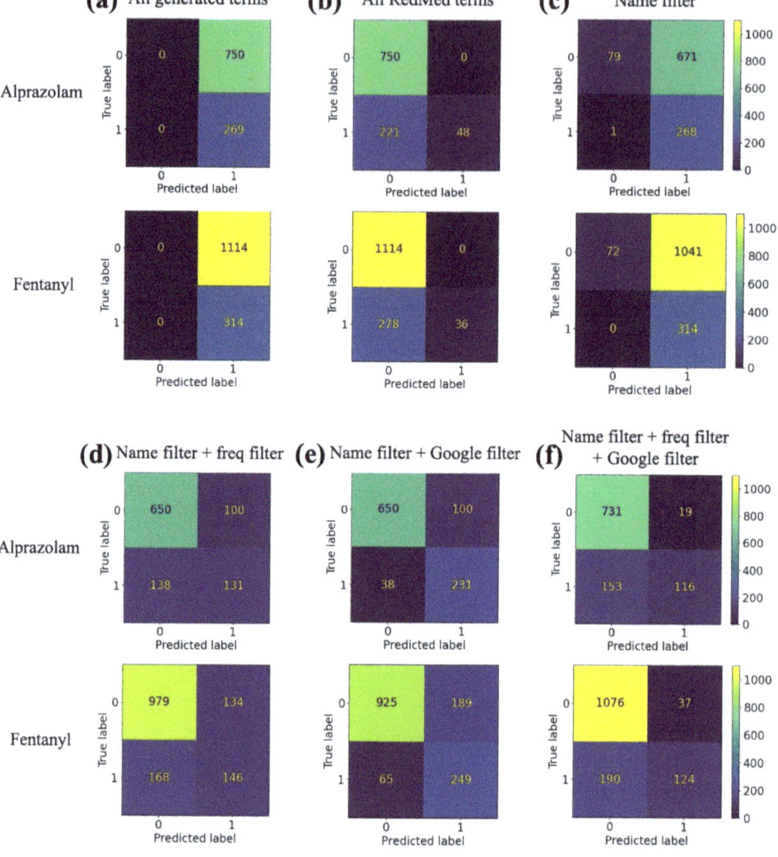

Figure 7. Confusion matrices for both alprazolam and fentanyl queries. True labels are determined by manual labeling. Predicted labels are determined by (**a**) classifying all generated terms as true; (**b**) classifying all generated terms that appear in RedMed as true; (**c**) classifying all generated terms that pass the drug name filter as true; (**d**) classifying all generated terms that pass the drug name filter and the generation frequency filter as true; (**e**) classifying all generated terms that pass the drug name filter and the Google search filter as true; and (**f**) classifying all generated terms that pass the drug name filter, the generation frequency filter, and the Google search filter as true. In each confusion matrix, a 0 denotes a negative classification, which is a non-synonym, and a 1 denotes a positive classification, which is a synonym.

Because it generated both the highest F1 and F2 scores on the manual labels for both alprazolam and fentanyl, we used the classification scheme of the drug name filter and the Google search filter (but not using the frequency filter) to build the final lexicon for all drugs of abuse. We made this decision under the assumption that high F1 and F2 scores on the manual labels would correlate with high F1 and F2 scores on the (unknown) ground truth.

Table 2. Evaluation metrics when using different classification schemes for GPT-3 synonyms and using manual labels as a proxy for ground truth.

Index Term	GPT-3 Synonym Criteria	Precision	Recall	F1 Score	F2 Score
Alprazolam	All generated terms	0.264	1.000	0.418	0.642
Fentanyl	All generated terms	0.220	1.000	0.361	0.585
Alprazolam	All RedMed terms	1.000	0.178	0.302	0.213
Fentanyl	All RedMed terms	1.000	0.115	0.206	0.140
Alprazolam	Drug name filter	0.285	0.996	0.443	0.664
Fentanyl	Drug name filter	0.232	1.000	0.377	0.602
Alprazolam	Drug name & frequency filters	0.567	0.487	0.524	0.501
Fentanyl	Drug name & frequency filters	0.521	0.465	0.491	0.475
Alprazolam	Drug name & Google filters	0.698	0.859	0.770	0.821
Fentanyl	Drug name & Google filters	0.568	0.793	0.662	0.735
Alprazolam	Drug name, frequency, & Google filters	0.859	0.431	0.574	0.479
Fentanyl	Drug name, frequency, & Google filters	0.770	0.395	0.522	0.438

3.5. Drugs of Abuse Lexicon

We conducted 1000 iterations of the query pipeline on each the 98 index terms. On average, each index term had 3880 total generated terms and 1426 unique generated terms over the 1000 iterations, though this varies widely per drug (Figure 8a,b). All generated terms that passed the drug name filter and the Google search filter were compiled into a lexicon of GPT-3 synonyms for drugs of abuse. Each index term had an average of 141 unique GPT-3 synonyms in the lexicon (Figure 8c) and an average of 132 UNGSes (Figure 8d).

When only considering widely-discussed drugs, the observed distributions of the aforementioned counts shift. Widely-discussed drugs yielded more total generated terms on average (4063 per index term; Figure 8e) but fewer unique generated terms on average (1259 per index term; Figure 8f). They also yielded more unique GPT-3 synonyms on average (293 per index term; Figure 8g) and more UNGSes on average (268 per index term; Figure 8h).

We include Google search, drug name matching, and frequency information in the full lexicon to enable the addition or removal of filters in future applications.

4. Discussion

In this study, we demonstrate that GPT-3, a large language model trained on the entire internet and used extensively for few-shot text generation, is able to generate drug synonyms to facilitate pharmacovigilance based on social media. With automated API queries and simple automated filters, we create a lexicon of slang terms, misspellings, brand names, and other synonyms of drugs identified by the DEA as drugs of abuse with minimal manual intervention. We offer both the lexicon and the code used to create the lexicon for use in identifying drug-related social media posts and characterizing large-scale trends in drug abuse and overdoses.

Our lexicon allows researchers conducting pharmacovigilance on social media (or other text source that uses colloquial language without a controlled vocabulary) to easily scan a large amount of text data and flag posts that contain terms synonymous with a drug of interest. Not only is this approach very accessible, as it does not require the machine learning expertise or computational resources needed for advanced language models, but it also provides interpretability as it is clear which term is responsible for flagging each post. This interpretability can aid the removal of false positive examples. We hope that our lexicon enables pharmacovigilance to be more efficient and have lower latency, due to the ability to utilize social media data and the lack of a need to develop complicated machine learning models. Additionally, our pipeline can be used for easy synonym generation tasks in areas beyond pharmacovigilance.

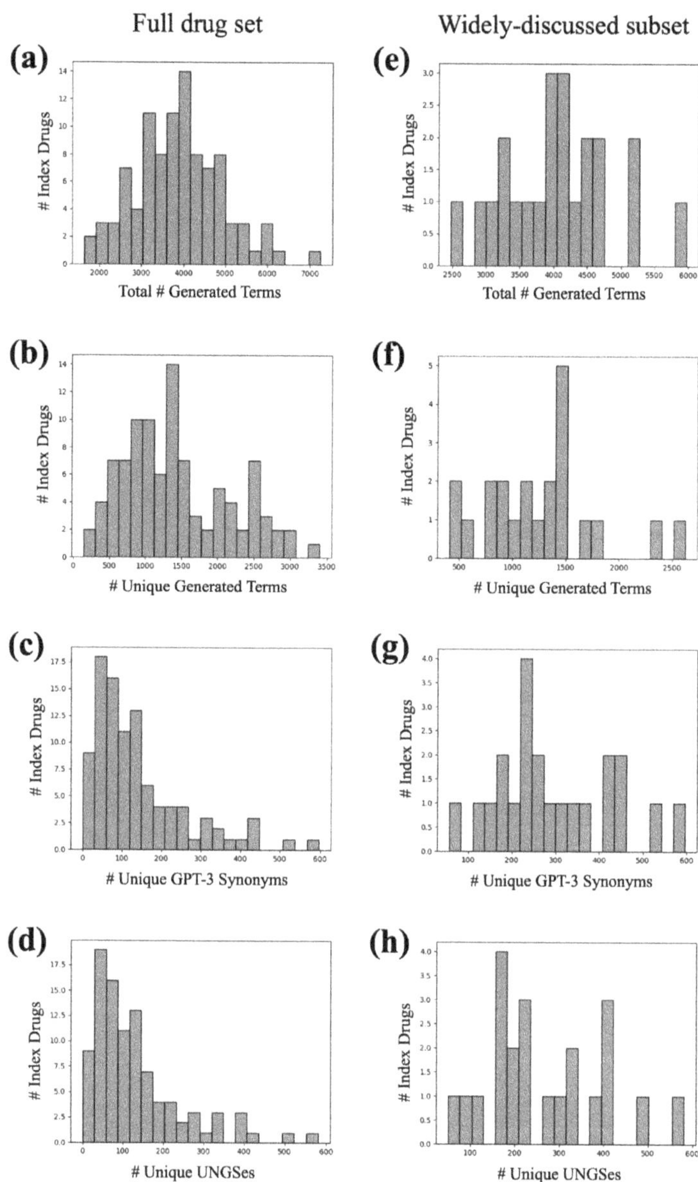

Figure 8. Histograms showing various distributions of quantities for each index term in the full lexicon of 98 drugs of abuse (**a–d**) and the subset of 22 widely-discussed drugs of abuse (**e–g**). Quantities depicted are total generated terms (**a**,**e**), unique generated terms (**b**,**f**), unique GPT-3 synonyms (generated terms passing filters) (**c**,**g**), and UNGSes (generated terms passing filters and not present in RedMed) (**d**,**h**).

While we found that GPT-3 generated hundreds of terms identified to be synonyms by manual labeling, the raw outputs also contained a large number of false positives, demonstrating the need for post-processing. We have shown that the drug name filter in combination with the Google filter yields the highest recall of all the filtering schemes. On average, our lexicon contains 141 GPT-3 synonyms per index term, and on average 132 of these are novel discoveries not found in RedMed. Importantly, these numbers increase for widely-discussed drugs that are more likely to be the focus of pharmacovigilance research. If we assume that the precision of the pipeline when generating fentanyl synonyms (the less precise of the two manually-labeled examples) holds for all index terms, then our lexicon contains 80 synonyms on average per index term, and 166 synonyms on average per widely-discussed index term. Notably, because GPT-3 is available as a pre-trained model, the process of querying GPT-3 and filtering the results to obtain these tens to hundreds of real synonyms requires relatively little effort, in direct contrast to RedMed's word embedding model, which required its own training and tuning.

Our pipeline has some limitations. For example, our choice to prioritize high recall over high precision means that the resulting lexicon is likely to contain many false positives. The number of false positives may be additionally increased by our broad definition of positive examples in the manual labeling process (e.g., labeling "bars" as a positive example/synonym for alprazolam, when in many contexts it would be not be a synonym for alprazolam). If the application for which the lexicon is being used requires higher precision, then additional filters will need to be applied to remove false positives. Alternatively, one could generate a new lexicon using different pipeline parameters than those specified above.

GPT-3 is unlikely to be able to predict new drug slang. The version of GPT-3 that we use in our experiments completed training in late 2019. It therefore has no information about any event from 2020 and onward. While GPT-3 may produce plausible-sounding predictions of the future, it is important to remember that it is not an oracle and, unlike Google, does not have up-to-date access to the happenings of the world. Drug slang terms can shift with new media and trends in pop culture, and these shifts will not be represented in the outputs of GPT-3. Therefore, as time goes on, our pipeline may generate slang terms that become less relevant to the current state of online conversations about drug use. However, because GPT-3 is optimized for few-shot learning, it is possible to present it with recent knowledge and let it generate likely tokens from that. It is also not unlikely that OpenAI will release an updated GPT model in the future that will be trained on new internet content.

Similarly, we note that the use of the Google filter in the final version of the pipeline means that the generative capabilities of GPT-3 may be suppressed, in that a plausible novel slang term that is not yet in use online would be omitted from the final lexicon. This occurs because our current mode of pipeline evaluation depends upon online presence and would therefore also miss plausible novel terms. One may use our method without the Google search filter in an attempt to recover more such original terms, but they would need a different evaluation method or else risk an influx of false positives. However, we believe that this is not a major limitation, as the primary utility of this method in a pharmacovigilance context is that it can recover terms currently in use on the internet that may be unknown to pharmacovigilance researchers; the generation of a term that will never be used is not useful for monitoring trends in drug use.

We recognize that our method requires a set of existing synonyms (e.g., RedMed synonyms) to construct the initial prompts presented to GPT-3. In the absence of a relevant RedMed entry for a drug of interest, there are alternate ways that one can generate such a set of synonyms. First, one could use resources such as existing online slang term lists, specialized slang dictionary sites such as Urban Dictionary, social media sites such as Twitter and Reddit, or a simple Google search to manually gather a few example synonyms. Second, one could modify the GPT-3 prompt template to query GPT-3 for synonyms of the index term without providing examples, and manually validate the resulting terms through the aforementioned online resources. While either of these options would require extra manual processing, we believe that the amount of work required to obtain a few synonyms to construct a prompt pales in comparison to the amount of work saved when using our pipeline to generate novel synonyms based on that prompt, as the strength of GPT-3 is in few-shot learning.

Finally, our pipeline shares a common problem with many applications that use GPT-3: GPT-3 is so good at producing plausible outputs that it is very difficult to tell if an output is truth or fiction. Beyond our discussion of using automated filters to reduce the number of false positives, we must also address the philosophical question of whether it is appropriate to use GPT-3, or a similar generative language model, for this task at all. One could argue that the task might be better approached by training a new large language model to recognize drug-relevant text, or by using simpler AI methods than large language models. However, the incredible performance of GPT-3 across a range of text generation tasks, in addition to the evidence from our experiments with alprazolam and fentanyl terms, convinces us that there are enough synonyms produced by GPT-3 for it to be a valuable resource for social media pharmacovigilance. We encourage future users of our pipeline to carefully consider if GPT-3 is appropriate for their task of interest.

This work is primarily a proof-of-concept, and there are numerous improvements to the pipeline which could be made to further enrich the resulting lexicon. One such improvement would be additional prompt engineering to further maximize the number of synonyms generated per API query. For example, it is possible that changing the format of the synonym list from a numbered list to some other form may be beneficial. One could also consider other ways of giving GPT-3 examples of desired output (whether using the index drug or some other drug) to reduce generation of false positives. We acknowledge that the manual tuning of query parameters may have led to a suboptimal choice of parameter settings, though the fact that the three numerical query parameters were at their extremes (highest possible temperature and lowest possible nonnegative presence penalty and frequency penalty) suggests that this is unlikely to be the case. Nevertheless, future work could verify this choice with a parameter sweep conducted via an automatic optimization algorithm. Additionally, the release of ChatGPT [53], which was concurrent with the preparation of this manuscript, brings a newer model with more advanced capabilities to the research community. It is possible that using ChatGPT for this task instead of GPT-3 may yield better results.

We share these methods and results in the hopes of contributing to population-scale pharmacovigilance to combat the opioid epidemic and reduce harm from drug abuse. We do not condone the use of our lexicon or pipeline for censorship or surveillance at the individual level. We also acknowledge that our pipeline could be used to evade censorship or monitoring on online platforms, or could potentially otherwise influence the emergence of new slang. However, we believe that the chance of such influence is minor, and greatly outweighed by the potential for lexicons created from this pipeline to better inform understanding of large-scale trends in drug abuse.

Author Contributions: Conceptualization, K.A.C. and R.B.A.; methodology, K.A.C.; software, K.A.C.; validation, K.A.C.; investigation, K.A.C.; data curation, K.A.C.; writing—original draft preparation, K.A.C.; writing—review and editing, R.B.A.; visualization, K.A.C.; supervision, R.B.A.; project administration, R.B.A.; funding acquisition, K.A.C. and R.B.A. All authors have read and agreed to the published version of the manuscript.

Funding: K.A.C. is supported by the National Science Foundation Graduate Research Fellowship Program under Grant No. DGE-1656518. R.B.A. is supported by NIH GM102365 and Chan Zuckerberg Biohub.

Institutional Review Board Statement: Not applicable.

Informed Consent Statement: Not applicable.

Data Availability Statement: The drugs of abuse lexicon, accompanying filter data, and pipeline code are all available at https://github.com/kristycarp/gpt3-lexicon.

Acknowledgments: We thank Daniel Sosa, Alexander Derry, Lu Yang, Gowri Nayar, Alexander Apostolov, and Rohan Koodli for their assistance in manual data labeling. Most of the computing for this project was performed on the Sherlock cluster; we would like to thank Stanford University and the Stanford Research Computing Center for providing computational resources and support.

Conflicts of Interest: The authors declare no conflict of interest.

Abbreviations

The following abbreviations are used in this manuscript:

GPT-3	Generative Pretrained Transformer 3
WHO	World Health Organization
EMA	European Medicines Agency
FDA	U.S. Food and Drug Administration
CDC	U.S. Centers for Disease Control and Prevention
NIH	U.S. National Institutes of Health
DEA	U.S. Drug Enforcement Administration
HHS	U.S. Department of Health and Human Services
FAERS	FDA Adverse Event Reporting System
NHANES	National Health and Nutrition Examination Survey
NDEWS	National Drug Early Warning System
NFLIS	National Forensic Laboratory Information System
NSDUH	National Survey on Drug Use and Health
NLP	Natural Language Processing
EHR	Electronic Health Record
AKDT	Associated Known Drug Term
API	Application Programming Interface
UNGS	Unique Novel GPT-3 Synonym
TP	True Positive
FP	False Positive
FN	False Negative

Appendix A

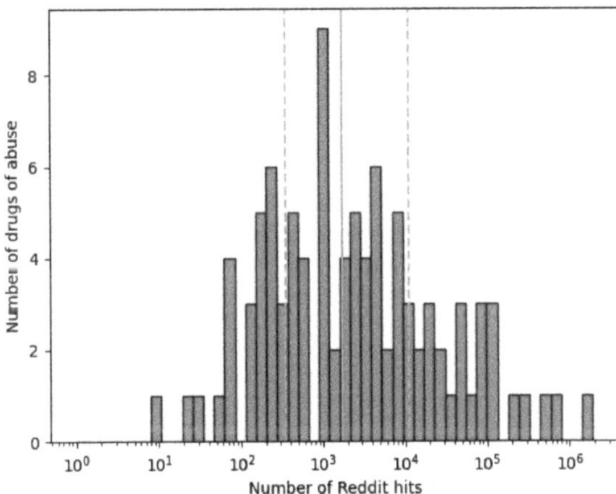

Figure A1. Histogram showing the distribution of the number of Reddit search hits for all DEA controlled substances (i.e., drugs of abuse) in RedMed. The x-axis is on the log scale. The 25th, 50th, and 75th percentiles are indicated with dashed and solid lines. All drugs in the 75th percentile are considered "widely-discussed" in this study for the purpose of examining pipeline performance for drugs with a high number of synonymous terms.

References

1. Lyden, J.; Binswanger, I.A. The United States opioid epidemic. *Semin. Perinatol.* **2019**, *43*, 123. [CrossRef] [PubMed]
2. Ciccarone, D. The Rise of Illicit Fentanyls, Stimulants and the Fourth Wave of the Opioid Overdose Crisis. *Curr. Opin. Psychiatry* **2021**, *34*, 344. [CrossRef]
3. CDC. National Center for Health Statistics. 2021. Wide-Ranging Nnline Data for Epidemiologic Research (WONDER). Available online: http://wonder.cdc.gov (accessed on 19 December 2022).
4. Beninger, P. Pharmacovigilance: An Overview. *Clin. Ther.* **2018**, *40*, 1991–2004. [CrossRef]
5. Throckmorton, D.C.; Gottlieb, S.; Woodcock, J. The FDA and the Next, Wave of Drug Abuse — Proactive Pharmacovigilance. *N. Engl. J. Med.* **2018**, *379*, 205–207. [CrossRef] [PubMed]
6. Stokes, A.; Berry, K.M.; Hempstead, K.; Lundberg, D.J.; Neogi, T. Trends in Prescription Analgesic Use Among Adults With Musculoskeletal Conditions in the United States, 1999–2016. *JAMA Netw. Open* **2019**, *2*, e1917228. . JAMANETWORKOPEN.2019.17228. [CrossRef] [PubMed]
7. Tringale, K.R.; Huynh-Le, M.P.; Salans, M.; Marshall, D.C.; Shi, Y.; Hattangadi-Gluth, J.A. The role of cancer in marijuana and prescription opioid use in the United States: A population-based analysis from 2005 to 2014. *Cancer* **2019**, *125*, 2242–2251. [CrossRef] [PubMed]
8. Veronin, M.A.; Schumaker, R.P.; Dixit, R.R.; Elath, H. Opioids and frequency counts in the US Food and Drug Administration Adverse Event Reporting System (FAERS) database: A quantitative view of the epidemic. *Drug, Healthc. Patient Saf.* **2019**, *11*, 65. [CrossRef]
9. Elmore, A.L.; Omofuma, O.O.; Sevoyan, M.; Richard, C.; Liu, J. Prescription opioid use among women of reproductive age in the United States: NHANES, 2003–2018. *Prev. Med.* **2021**, *153*, 106846. [CrossRef]
10. Robert, M.; Jouanjus, E.; Khouri, C.; Sam-Laï, N.F.; Revol, B. The opioid epidemic: A worldwide exploratory study using the WHO pharmacovigilance database. *Addiction* **2022**. [CrossRef]
11. Chiappini, S.; Vickers-Smith, R.; Guirguis, A.; Corkery, J.M.; Martinotti, G.; Harris, D.R.; Schifano, F. Pharmacovigilance Signals of the Opioid Epidemic over 10 Years: Data Mining Methods in the Analysis of Pharmacovigilance Datasets Collecting Adverse Drug Reactions (ADRs) Reported to EudraVigilance (EV) and the FDA Adverse Event Reporting System (FAERS). *Pharmaceuticals* **2022**, *15*, 675. [CrossRef]
12. Inoue, K.; Ritz, B.; Arah, O.A. Causal Effect of Chronic Pain on Mortality Through Opioid Prescriptions: Application of the Front-Door Formula. *Epidemiology* **2022**, *33*, 572. [CrossRef] [PubMed]
13. Marwitz, K.K.; Noureldin, M. A descriptive analysis of concomitant opioid and benzodiazepine medication use and associated adverse drug events in United States adults between 2009 and 2018. *Explor. Res. Clin. Soc. Pharm.* **2022**, *5*, 100130. [CrossRef] [PubMed]
14. LePendu, P.; Iyer, S.V.; Bauer-Mehren, A.; Harpaz, R.; Ghebremariam, Y.T.; Cooke, J.P.; Shah, N.H. Pharmacovigilance using Clinical Text. *AMIA Summits Transl. Sci. Proc.* **2013**, *2013*, 109. [PubMed]
15. Harpaz, R.; Callahan, A.; Tamang, S.; Low, Y.; Odgers, D.; Finlayson, S.; Jung, K.; LePendu, P.; Shah, N.H. Text mining for adverse drug events: The promise, challenges, and state of the art. *Drug Saf.* **2014**, *37*, 777–790. [CrossRef]
16. Boland, M.R.; Tatonetti, N.P. Are All Vaccines Created Equal? Using Electronic Health Records to Discover Vaccines Associated With Clinician-Coded Adverse Events. *AMIA Summits Transl. Sci. Proc.* **2015**, *2015*, 196.
17. Luo, Y.; Thompson, W.K.; Herr, T.M.; Zeng, Z.; Berendsen, M.A.; Jonnalagadda, S.R.; Carson, M.B.; Starren, J. Natural Language Processing for EHR-Based Pharmacovigilance: A Structured Review. *Drug Saf.* **2017**, *40*, 1075–1089. [CrossRef]
18. Ward, P.J.; Rock, P.J.; Slavova, S.; Young, A.M.; Bunn, T.L.; Kavuluru, R. Enhancing timeliness of drug overdose mortality surveillance: A machine learning approach. *PLoS ONE* **2019**, *14*, e0223318. [CrossRef]
19. Ward, P.J.; Young, A.M.; Slavova, S.; Liford, M.; Daniels, L.; Lucas, R.; Kavuluru, R. Deep Neural Networks for Fine-Grained Surveillance of Overdose Mortality. *Am. J. Epidemiol.* **2022**, *192*, 257–266. [CrossRef]
20. Kazemi, D.M.; Borsari, B.; Levine, M.J.; Dooley, B. Systematic review of surveillance by social media platforms for illicit drug use. *J. Public Health* **2017**, *39*, 763–776. [CrossRef]
21. Leaman, R.; Wojtulewicz, L.; Sullivan, R.; Skariah, A.; Yang, J.; Gonzalez, G. Toward Internet-Age Pharmacovigilance: Extracting Adverse Drug Reactions from User Posts to Health-Related Social Networks. In Proceedings of the of the 2010 Workshop on Biomedical Natural Language Processing, Uppsala, Sweden, 15 July 2010; pp. 117–125.
22. O'Connor, K.; Pimpalkhute, P.; Nikfarjam, A.; Ginn, R.; Smith, K.L.; Gonzalez, G. Pharmacovigilance on Twitter? Mining Tweets for Adverse Drug Reactions. *AMIA Annu. Symp. Proc.* **2014**, *2014*, 924.
23. Eshleman, R.; Singh, R. Leveraging graph topology and semantic context for pharmacovigilance through Twitter-streams. *BMC Bioinform.* **2016**, *17*. [CrossRef]
24. Cocos, A.; Fiks, A.G.; Masino, A.J. Deep learning for pharmacovigilance: Recurrent neural network architectures for labeling adverse drug reactions in Twitter posts. *J. Am. Med Inform. Assoc. JAMIA* **2017**, *24*, 813. [CrossRef]
25. Lardon, J.; Bellet, F.; Aboukhamis, R.; Asfari, H.; Souvignet, J.; Jaulent, M.C.; Beyens, M.N.; Lillo-LeLouët, A.; Bousquet, C. Evaluating Twitter as a complementary data source for pharmacovigilance. *Expert Opin. Drug Saf.* **2018**, *17*, 763–774. [CrossRef]
26. Farooq, H.; Niaz, J.S.; Fakhar, S.; Naveed, H. Leveraging digital media data for pharmacovigilance. *AMIA Annu. Symp. Proc.* **2020**, *2020*, 442.

27. Magge, A.; Tutubalina, E.; Miftahutdinov, Z.; Alimova, I.; Dirkson, A.; Verberne, S.; Weissenbacher, D.; Gonzalez-Hernandez, G. DeepADEMiner: A deep learning pharmacovigilance pipeline for extraction and normalization of adverse drug event mentions on Twitter. *J. Am. Med Inform. Assoc.* **2021**, *28*, 2184–2192. [CrossRef]
28. Pierce, C.E.; Bouri, K.; Pamer, C.; Proestel, S.; Rodriguez, H.W.; Le, H.V.; Freifeld, C.C.; Brownstein, J.S.; Walderhaug, M.; Edwards, I.R.; et al. Evaluation of Facebook and Twitter Monitoring to Detect Safety Signals for Medical Products: An Analysis of Recent FDA Safety Alerts. *Drug Saf.* **2017**, *40*, 317–331. [CrossRef]
29. Caster, O.; Dietrich, J.; Kürzinger, M.L.; Lerch, M.; Maskell, S.; Norén, G.N.; Tcherny-Lessenot, S.; Vroman, B.; Wisniewski, A.; van Stekelenborg, J. Assessment of the Utility of Social Media for Broad-Ranging Statistical Signal Detection in Pharmacovigilance: Results from the WEB-RADR Project. *Drug Saf.* **2018**, *41*, 1355. [CrossRef]
30. Hussain, Z.; Sheikh, Z.; Tahir, A.; Dashtipour, K.; Gogate, M.; Sheikh, A.; Hussain, A. Artificial Intelligence–Enabled Social Media Analysis for Pharmacovigilance of COVID-19 Vaccinations in the United Kingdom: Observational Study. *JMIR Public Health Surveill.* **2022**, *8*, e32543. [CrossRef]
31. Natter, J.; Michel, B. Memantine misuse and social networks: A content analysis of Internet self-reports. *Pharmacoepidemiol. Drug Saf.* **2020**, *29*, 1189–1193. [CrossRef]
32. Lavertu, A.; Hamamsy, T.; Altman, R.B. Quantifying the Severity of Adverse Drug Reactions Using Social Media: Network Analysis. *J. Med. Internet Res.* **2021**, *23*, e27714. [CrossRef]
33. Preiss, A.; Baumgartner, P.; Edlund, M.J.; Bobashev, G.V. Using Named Entity Recognition to Identify Substances Used in the Self-medication of Opioid Withdrawal: Natural Language Processing Study of Reddit Data. *JMIR Form. Res.* **2022**, *6*, e33919. [CrossRef] [PubMed]
34. Tasnim, S.; Hossain, M.; Mazumder, H. Impact of Rumors and Misinformation on COVID-19 in Social Media. *J. Prev. Med. Public Health* **2020**, *53*, 171. [CrossRef]
35. Suarez-Lledo, V.; Alvarez-Galvez, J. Prevalence of Health Misinformation on Social Media: Systematic Review. *J. Med. Internet Res.* **2021**, *23*, e17187. [CrossRef] [PubMed]
36. Convertino, I.; Ferraro, S.; Blandizzi, C.; Tuccori, M. The usefulness of listening social media for pharmacovigilance purposes: A systematic review. *Expert Opin. Drug Saf.* **2018**, *17*, 1081–1093. [CrossRef] [PubMed]
37. Tricco, A.C.; Zarin, W.; Lillie, E.; Jeblee, S.; Warren, R.; Khan, P.A.; Robson, R.; Pham, B.; Hirst, G.; Straus, S.E. Utility of social media and crowd-intelligence data for pharmacovigilance: A scoping review. *BMC Med. Inform. Decis. Mak.* **2018**, *18*. [CrossRef]
38. Pappa, D.; Stergioulas, L.K. Harnessing social media data for pharmacovigilance: A review of current state of the art, challenges and future directions. *Int. J. Data Sci. Anal.* **2019**, *8*, 113–135. [CrossRef]
39. Lee, J.Y.; Lee, Y.S.; Kim, D.H.; Lee, H.S.; Yang, B.R.; Kim, M.G. The Use of Social Media in Detecting Drug Safety–Related New Black Box Warnings, Labeling Changes, or Withdrawals: Scoping Review. *JMIR Public Health Surveill.* **2021**, *7*, e30137. [CrossRef]
40. Lavertu, A.; Altman, R.B. RedMed: Extending drug lexicons for social media applications. *J. Biomed. Inform.* **2019**, *99*, 103307. [CrossRef]
41. Mikolov, T.; Chen, K.; Corrado, G.; Dean, J. Efficient Estimation of Word Representations in Vector Space. *arXiv* **2013**, arXiv:1301.378.
42. Wishart, D.S.; Feunang, Y.D.; Guo, A.C.; Lo, E.J.; Marcu, A.; Grant, J.R.; Sajed, T.; Johnson, D.; Li, C.; Sayeeda, Z.; et al. DrugBank 5.0: A major update to the DrugBank database for 2018. *NUcleic Acids Res.* **2018**, *46*, D1074–D1082. [CrossRef]
43. Brown, T.B.; Mann, B.; Ryder, N.; Subbiah, M.; Kaplan, J.; Dhariwal, P.; Neelakantan, A.; Shyam, P.; Sastry, G.; Askell, A.; et al. Language Models are Few-Shot Learners. In Proceedings of the NeurIPS 2020, Virtual, 6–12 December 2020.
44. Vaswani, A.; Shazeer, N.; Parmar, N.; Uszkoreit, J.; Jones, L.; Gomez, A.N.; Łukasz K.; Polosukhin, I. Attention Is All You Need. In Proceedings of the NeurIPS 2017, Long Beach, CA, USA, 4–9 December 2017.
45. Korngiebel, D.M.; Mooney, S.D. Considering the possibilities and pitfalls of Generative Pre-trained Transformer 3 (GPT-3) in healthcare delivery. *NPJ Digit. Med.* **2021**, *4*, 93. [CrossRef]
46. Sezgin, E.; Sirrianni, J.; Linwood, S.L. Operationalizing and Implementing Pretrained, Large Artificial Intelligence Linguistic Models in the US Health Care System: Outlook of Generative Pretrained Transformer 3 (GPT-3) as a Service Model. *JMIR Med. Inform.* **2022**, *10*, e32875. [CrossRef]
47. Nath, S.; Marie, A.; Ellershaw, S.; Korot, E.; Keane, P.A.; Pearse, D.; Keane, A. New meaning for NLP: The trials and tribulations of natural language processing with GPT-3 in ophthalmology. *Br. J. Ophthalmol.* **2022**, *106*, 889–892. [CrossRef]
48. Mcguffie, K.; Newhouse, A. The radicalization risks of GPT-3 and advanced neural language models. *arXiv* **2020**, arXiv:2009.06807.
49. Abid, A.; Farooqi, M.; Zou, J. Persistent Anti-Muslim Bias in Large Language Models. *arXiv* **2021**, arXiv:2101.05783.
50. U.S. Drug Enforcement Administration, Controlled Substances—Alphabetical Order. Available online: https://www.deadiversion.usdoj.gov/schedules/orangebook/c_cs_alpha.pdf (accessed on 25 July 2022).
51. Park, A.; Conway, M. Tracking Health Related Discussions on Reddit for Public Health Applications. *AMIA Annu. Symp. Proc.* **2017**, *2017*, 1362.

52. Adams, N.; Artigiani, E.E.; Wish, E.D. Choosing Your Platform for Social Media Drug Research and Improving Your Keyword Filter List. *J. Drug Issues* **2019**, *49*, 477–492. [CrossRef]
53. OpenAI. ChatGPT: Optimizing Language Models for Dialogue. 2022. Available online: https://openai.com/blog/chatgpt/ (accessed on 16 December 2022).

biomolecules

MDPI

Article

Generalization Performance of Quantum Metric Learning Classifiers

Jonathan Kim [1] **and Stefan Bekiranov** [2,*]

1 GSK R&D Stevenage, GlaxoSmithKline, Stevenage SG1 2NY, UK
2 Department of Biochemistry and Molecular Genetics, University of Virginia, Charlottesville, VA 22908, USA
* Correspondence: sb3de@virginia.edu

Abstract: Quantum computing holds great promise for a number of fields including biology and medicine. A major application in which quantum computers could yield advantage is machine learning, especially kernel-based approaches. A recent method termed quantum metric learning, in which a quantum embedding which maximally separates data into classes is learned, was able to perfectly separate ant and bee image training data. The separation is achieved with an intrinsically quantum objective function and the overall approach was shown to work naturally as a hybrid classical-quantum computation enabling embedding of high dimensional feature data into a small number of qubits. However, the ability of the trained classifier to predict test sample data was never assessed. We assessed the performance of quantum metric learning on test ants and bees image data as well as breast cancer clinical data. We applied the original approach as well as variants in which we performed principal component analysis (PCA) on the feature data to reduce its dimensionality for quantum embedding, thereby limiting the number of model parameters. If the degree of dimensionality reduction was limited and the number of model parameters was constrained to be far less than the number of training samples, we found that quantum metric learning was able to accurately classify test data.

Keywords: quantum machine learning; quantum metric learning; kernel method; kernel classifiers

Citation: Kim, J.; Bekiranov, S. Generalization Performance of Quantum Metric Learning Classifiers. *Biomolecules* **2022**, *12*, 1576. https://doi.org/10.3390/biom12111576

Academic Editors: Cameron Mura and Lei Xie

Received: 3 October 2022
Accepted: 23 October 2022
Published: 27 October 2022

1. Introduction

Significant progress has recently been made toward the development of fault tolerant quantum computers (FTQCs) [1]. Their development would result in the speedup of many algorithms that are approaching severe limits on classical computers. The range of applications include quantum chemistry [2], search [3], cryptography [4] and machine learning [5]. These applications are relevant to many domains of study including biology and medicine. In the field of machine learning, exponential speedups on a quantum compared to classical computer have been proven [5] for implementing quantum support vector machines [6], quantum Boltzmann machines [7,8], least squares fitting [9], and quantum principal component analysis [10]. Quadratic speedups have been demonstrated [5] for classical Boltzmann machines [11], quantum reinforcement learning [12], online perceptron [13], and Bayesian inference [14,15]. However, these speedups assume a FTQC with high connectivity and hundreds to thousands, even millions for some applications, of qubits. In addition, some of these quantum algorithms require quantum RAM (qRAM) which executes a quantum coherent mapping of a classical vector into a quantum state [16,17], for their quantum advantage over classical computers. However, qRAM hardware has not been developed. Currently, quantum computing is in its noisy intermediate-scale quantum (NISQ) era [18].

A major application in which even NISQ-era quantum computers could yield advantage is kernel-based machine learning [19–22]. Broadly, two sets of approaches have recently been explored [20,21]: (1) map a large feature space into a quantum state and calculate a kernel function on a quantum computer and make use of this kernel in a classical classifier (e.g., SVM) and (2) apply a variational quantum circuit to classify data on the quantum computer in Hilbert space. Kernel-based classifiers that interfere the test and train data and effectively calculate their Euclidean distance [19,23,24] and/or inner product [23–25] have been developed and assessed on IBM quantum computers and performed close to theoretical expectations if the number of gates were kept to a relatively small number [25]. Formally, supervised quantum models have been shown to be kernel methods [22], and it has been suggested that quantum computers could enable kernel-based machine learning in a similar way that GPU-accelerated hardware enabled deep learning [22]. As a result of these developments, a number of kernel-based quantum machine learning studies have been performed in which the trainability [26–29], expressivity [30,31], robustness [32,33] and generalizability [30,31,33–35] of quantum kernel-based models implemented on NISQ-era quantum computers have been studied as well as the extent to which quantum errors can be mitigated on a classical computer [30,31].

In this work, we focus on a quantum kernel-based machine learning approach termed quantum metric learning (QML) [26]. Here, a quantum embedding is learned by maximizing the Hilbert-Schmidt distance of data samples from two classes in such a way that two classes are separated in Hilbert space. This enables a simple linear decision boundary to be implemented in Hilbert space which represents a complex decision boundary in the original feature space. This approach has all the advantages that come with kernel-based approaches mentioned above along with a number of other attractive features for NISQ-era quantum computing including: (1) simple, quantum-based cost function based on the Hilbert-Schmidt distance, (2) seamless applicability as a hybrid quantum-classical approach that reduces the dimensionality of the input feature space for quantum embedding to a small number of qubits, (3) ability to directly visualize the extent to which samples with different class labels are separated and (4) ability to be implemented on a quantum computer as a classifier using multiple swap gates [23–26]. Despite these highly promising attributes of QML, the primary manuscript detailing the method [26] only demonstrated its ability to separate training data. The ability of QML to generalize well by assessing a trained model on test data was not shown. Consequently, we fill this gap by training and testing QML with the original ImageNet Hymenoptera Dataset containing images of ants and bees [36] as well as the University of California Irvine Machine Learning Breast Cancer Wisconsin (Diagnostic) dataset [37]. The breast cancer dataset contains 30 normalized clinical features for each breast cancer patient whose tumor was diagnosed as malignant and benign. We used precision, recall and F1-score as performance metrics for test data. We also report the resulting cost function for both train and test data. We reproduced the result that for the original ant and bee image data, we were able to achieve a high level of separation on training data. However, we found that the trained classifier did not perform well on hold out test data. We noticed that the number of model parameters exceeded the number of training samples, so we hypothesized that the model was overfitting the training data. Application of principal component analysis (PCA) to reduce the input feature dimension and number of model parameters did not significantly improve test performance on this dataset. We turned to the breast cancer data which contained far fewer input features and more samples and further applied PCA as well to reduce the input feature dimensions and number of model parameters. We found that QML was able to perform well on both training and test data in this setting. Thus, when adhering to conventional bias-variance principles, namely, constraining the number of model parameters to be notably less than the number of training samples, we find that QML-based classifiers generalize well. This is true as long as the initial number of features (i.e., the number of features prior to PCA) is not too high.

2. Materials and Methods

2.1. Quantum Metric Learning Expressed as a Kernel-Based Quantum Model

In quantum metric learning, a quantum embedding,

$$|x\rangle = \Phi(x,\theta)\,|0\ldots 0\rangle\,, \tag{1}$$

is learned where $\Phi(x,\theta)$ is a feature map which maps the input data x to a quantum state $|x\rangle$ which separates the data according to class labels in Hilbert space by maximizing the Hilbert-Schmidt distance D_{hs} or, equivalently, by minimizing a cost function C defined in terms of D_{hs}, through gradient descent of the model parameters θ. The Hilbert-Schmidt distance is

$$D_{hs}(\rho,\sigma) = \mathrm{tr}[(\rho-\sigma)^2], \tag{2}$$

where ρ and σ are density matrices representing ensembles of M_a and M_b training data points a and b from class A and B, respectively:

$$\rho = \frac{1}{M_a}\sum_{a\in A} |a\rangle\langle a| \tag{3}$$

and

$$\sigma = \frac{1}{M_b}\sum_{b\in B} |b\rangle\langle b|\,. \tag{4}$$

The cost function C, whose range is $[0,1]$, that is minimized is

$$C = 1 - \frac{1}{2}D_{hs}(\rho,\sigma). \tag{5}$$

Once C is minimized, the parameters, θ, of the feature map are determined in such a way that the training data $\{a,b\}$ is separated in Hilbert space. In order to classify a test sample, x, it must first be embedded using the feature map as shown in Equation (1). A fidelity classifier [23–26] can then be defined by the difference in squared inner product between the embedded test sample $|x\rangle$ and the respective class A and B embedded training samples $\{|a\rangle,|b\rangle\}$:

$$f(x) = \frac{1}{M_a}\sum_{a\in A} |\langle a|x\rangle|^2 - \frac{1}{M_b}\sum_{b\in B} |\langle b|x\rangle|^2 \tag{6}$$

$$= \langle x|\,\rho-\sigma\,|x\rangle\,. \tag{7}$$

Equation (7) can be viewed as an expectation of a measurement, $\mathcal{M}$, where

$$\mathcal{M} = \rho - \sigma \tag{8}$$

$$= \frac{1}{M_a}\sum_{a\in A} |a\rangle\langle a| - \frac{1}{M_b}\sum_{b\in B} |b\rangle\langle b|\,. \tag{9}$$

Thus, the fidelity classifier may be expressed as follows:

$$f(x) = \langle x|\mathcal{M}|x\rangle \tag{10}$$

$$= \mathrm{tr}[|x\rangle\langle x|\,\mathcal{M}]. \tag{11}$$

Equation (11) is the definition of a quantum model (see Equation (34) of Schuld et al. [22]) which can be expressed as a quantum kernel-based model. We implement and assess the generalization performance of quantum metric learning using the following k-nearest neighbor (KNN) kernel-based classifier:

$$\hat{y} = \text{sgn}(f(x)) \tag{12}$$
$$= \text{sgn}(\sum_{a \in A} \alpha_a \kappa(a, x) - \sum_{b \in B} \alpha_b \kappa(b, x)), \tag{13}$$

where $\hat{y}$ is the prediction for test sample x and sgn denotes the sign function. Comparison of Equations (6) and (13) yields the result that $\alpha_a = 1/M_a$, $\alpha_b = 1/M_b$,

$$\kappa(a, x) = |\langle a|x\rangle|^2, \tag{14}$$

and

$$\kappa(b, x) = |\langle b|x\rangle|^2, \tag{15}$$

where $\kappa(a, x)$ and $\kappa(b, x)$ are defined as quantum kernels (see Equation (6) of Schuld et al. [22]) which are the inner product between the embedded test data, x, and training data, a and b, respectively, in the context of a KNN classifier.

2.2. The Quantum Metric Learning Embedding Circuit

Various adaptations of Lloyd et al.'s hybrid quantum metric learning embedding [26] were used throughout this work. See Figure 1A for a full illustration of the general embedding. The quantum component of the algorithm (the trainable *quantum feature map*, a repeating circuit ansatz consisting of single-qubit R_x, R_y rotation gates and two-qubit ZZ coupling gates [26,38] resulting in 12 trainable quantum parameters) was left unchanged. The classical components leading to the intermediate x_1 and x_2 inputs to the quantum feature map were replaced and varied. We note that the quantum circuit is precisely the same as that of Lloyd et al. [26]. The example ansatz in Figure 3 of [26] is for three inputs (x_1, x_2 and x_3). However, we and Lloyd et al. [26] use two inputs (x_1 and x_2) to assess QML on real world datasets.

We now describe the effects that the $R_x(x_i)$, $R_y(\theta_j)$ and $ZZ(\theta_j)$ gates have on the two-qubit state at the kth stage of the circuit, $|x_k\rangle$, where $i = 1, 2$, $j = 1, 2, ..., 12$, $k = 1, 2, ..., 14$ and

$$|x_k\rangle = \alpha_k |00\rangle + \beta_k |01\rangle + \gamma_k |10\rangle + \delta_k |11\rangle. \tag{16}$$

For example, $|x_1\rangle = |00\rangle$ with $\alpha_1 = 1$ and $\beta_1 = \gamma_1 = \gamma_1 = 0$ is the initial two-qubit state entering the circuit on the left of Figure 1A. The state $|x_{14}\rangle = |x\rangle$ is the final state shown on the right of the circuit in Figure 1A. The operation of the first $R_x(x_1)$ and $R_x(x_2)$ gates yields $|x_2\rangle$, where

$$\alpha_2 = \frac{1}{2}\cos\left(\frac{x_1 + x_2}{2}\right) + \frac{1}{2}\cos\left(\frac{x_1 - x_2}{2}\right), \tag{17}$$
$$\beta_2 = -\frac{i}{2}\sin\left(\frac{x_1 + x_2}{2}\right) + \frac{i}{2}\sin\left(\frac{x_1 - x_2}{2}\right), \tag{18}$$
$$\gamma_2 = -\frac{i}{2}\sin\left(\frac{x_1 + x_2}{2}\right) - \frac{i}{2}\sin\left(\frac{x_1 - x_2}{2}\right) \tag{19}$$

and

$$\delta_2 = -\frac{1}{2}\cos\left(\frac{x_1 - x_2}{2}\right) + \frac{1}{2}\cos\left(\frac{x_1 + x_2}{2}\right). \tag{20}$$

We see that the two-qubit state becomes angularly embedded by a combination real and complex coefficients containing sine and cosine functions. The operation of the first $ZZ(\theta_1)$ entangler gate yields $|x_3\rangle$, where

$$\alpha_3 = e^{\frac{-i\theta_1}{2}} \alpha_2, \tag{21}$$

$$\beta_3 = e^{\frac{i\theta_1}{2}} \beta_2, \tag{22}$$

$$\gamma_3 = e^{\frac{i\theta_1}{2}} \gamma_2 \tag{23}$$

and

$$\delta_3 = e^{\frac{-i\theta_1}{2}} \delta_2. \tag{24}$$

The operation of the $R_y(\theta_2)$ and $R_y(\theta_3)$ gates then yields $|x_4\rangle$, where

$$\begin{aligned} \alpha_4 = & \frac{\alpha_3 - \delta_3}{2} \cos\left(\frac{\theta_2 + \theta_3}{2}\right) + \frac{\alpha_3 + \delta_3}{2} \cos\left(\frac{\theta_2 - \theta_3}{2}\right) \\ & - \frac{\beta_3 + \gamma_3}{2} \sin\left(\frac{\theta_2 + \theta_3}{2}\right) + \frac{\beta_3 - \gamma_3}{2} \sin\left(\frac{\theta_2 - \theta_3}{2}\right), \end{aligned} \tag{25}$$

$$\begin{aligned} \beta_4 = & \frac{\beta_3 + \gamma_3}{2} \cos\left(\frac{\theta_2 + \theta_3}{2}\right) + \frac{\beta_3 - \gamma_3}{2} \cos\left(\frac{\theta_2 - \theta_3}{2}\right) \\ & + \frac{\alpha_3 - \delta_3}{2} \sin\left(\frac{\theta_2 + \theta_3}{2}\right) - \frac{\alpha_3 + \delta_3}{2} \sin\left(\frac{\theta_2 - \theta_3}{2}\right), \end{aligned} \tag{26}$$

$$\begin{aligned} \gamma_4 = & \frac{\beta_3 + \gamma_3}{2} \cos\left(\frac{\theta_2 + \theta_3}{2}\right) - \frac{\beta_3 - \gamma_3}{2} \cos\left(\frac{\theta_2 - \theta_3}{2}\right) \\ & + \frac{\alpha_3 - \delta_3}{2} \sin\left(\frac{\theta_2 + \theta_3}{2}\right) + \frac{\alpha_3 + \delta_3}{2} \sin\left(\frac{\theta_2 - \theta_3}{2}\right) \end{aligned} \tag{27}$$

and

$$\begin{aligned} \delta_4 = & -\frac{\alpha_3 - \delta_3}{2} \cos\left(\frac{\theta_2 + \theta_3}{2}\right) + \frac{\alpha_3 + \delta_3}{2} \cos\left(\frac{\theta_2 - \theta_3}{2}\right) \\ & + \frac{\beta_3 + \gamma_3}{2} \sin\left(\frac{\theta_2 + \theta_3}{2}\right) + \frac{\beta_3 - \gamma_3}{2} \sin\left(\frac{\theta_2 - \theta_3}{2}\right). \end{aligned} \tag{28}$$

In this way, we see that we get growing products of sine and cosine components (in terms of both the linear trainable parameters, x_i, and the 'quantum' trainable parameters, θ_j) in each element of the resulting vector as we progress through the circuit. As the circuit ansatz is repeated further, this results in an increase in both the 'sharpness' and the number of peaks and troughs representing the angular embedded data, allowing for the high levels of expressivity needed for effective embedding.

When working with the Hymenoptera ants and bees image dataset, the replaceable classical part of the embedding consisted of images of ants and bees that had been standardized and normalized. We explored passing them through a pre-trained ResNet-18 network (without the final layer) as well as working with them directly. The first approach resulted in 512 classical input features [26,39], while the second approach yielded 150528 classical input features. In the second approach, the features were then always dimensionally reduced via PCA to prevent there being an exceptionally high number of trainable parameters. When working with the breast cancer dataset, the replaceable classical part of the embedding corresponded to 30 normalized input clinical features. This resulted in 30 classical input features.

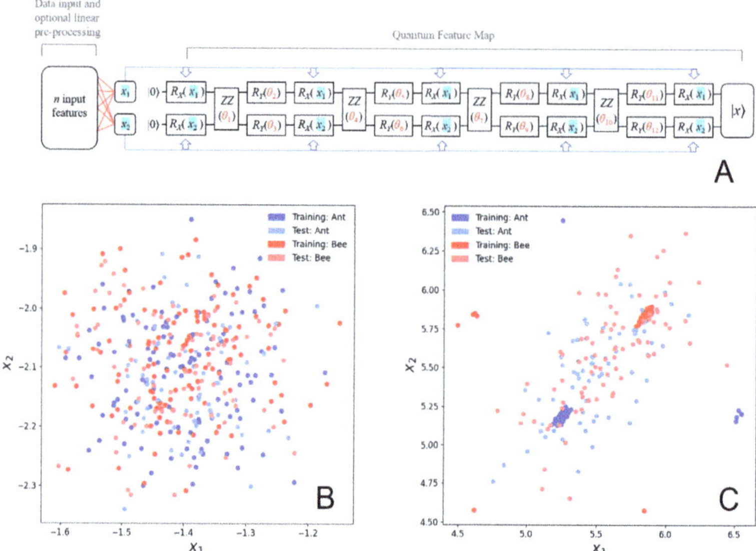

Figure 1. Diagrams illustrating the training process and results of the quantum feature map: (**A**) Diagram of the general quantum metric embedding used. The model takes n classical input features and reduces them to two intermediary values (x_1 and x_2) through matrix multiplication with a $2 \times n$ parameter matrix, whose elements behave as trainable linear parameters. Thus, n input features yield $2n$ trainable linear parameters. The resulting intermediate (x_1, x_2) values are then used as input alongside 12 trainable 'quantum' parameters (θ_1–θ_{12}) to progress through the quantum feature map. Each sample ultimately ends up in the embedded $|x\rangle$ state in which the Hilbert-Schmidt distance between different classes is maximized through iterative training of the linear and quantum parameters. The illustrated approach represents a generalized adaption of the hybrid quantum metric learning embedding used by Lloyd et al. [26] (**B**) Scatter plot of the (x_1, x_2) values of the Hymenoptera dataset with 512 ResNet features (corresponding to 1024 trainable linear parameters) after 0 steps of training. Datapoints from both the training set and the test set are depicted. We note that we used precisely the same train and test samples as in the original study [26] for the Hymenoptera data which corresponded to 61% train and 39% test. (**C**) Scatter plot of the (x_1, x_2) values of the Hymenoptera dataset with 512 ResNet features after 1500 steps of training using the PennyLane software package [40]. Datapoints from both the training set and the test set are depicted.

These n classical input features were then multiplied by a $2 \times n$ parameter matrix, resulting in $2n$ trainable linear parameters and the two inputs (x_1, x_2) to the quantum feature map. In many cases in both datasets, the initial input features also underwent dimensional reduction through principal component analysis (PCA) to yield lower values of n, so as to help minimize overfitting by the subsequent models. However, when working with this dimensional reduction approach, it was also important not to reduce the number of linear parameters too much so as to retain the expressivity of the models.

2.3. Training the Quantum Metric Learning Models

The quantum feature map itself provided $4 \times 3 = 12$ trainable quantum parameters (i.e., 4 repeated circuit ansatzes containing 3 parameters per ansatz) and as such, every model consisted of $2n + 12$ total trainable parameters. Each model was randomly initialized and trained for 1500 steps with a batch size of 10, using the root mean squared propagation (RMSProp) optimizer with a step size of 0.01. With successful training, each new (x_1, x_2) input to the model becomes embedded into a state $|x\rangle$ in Hilbert space such that the Hilbert-Schmidt distances between the embedded states of opposing classes, shown in Equation (2), are maximized or equivalently, the Hilbert-Schmidt cost function, Equation (5), is minimized. The hybrid parameter optimization steps were performed using the PennyLane software package [40] and the embedded data were subsequently classified by a k-nearest neighbor (KNN) classifier.

2.4. *ImageNet Hymenoptera Dataset*

The first dataset used to explore Lloyd et al.'s quantum metric learning embedding [26] was the ImageNet Hymenoptera image dataset [36]. This dataset consists of 397 colored images of ants and bees in various environments. Each sample can thus be assigned a class of either *ant* or *bee*. By default, the dataset is split into a training set and a test set in the approximate ratio of 3:2. This train-test split was manually changed at times, as dictated by a random seed. Each image was standardized into a resolution of 224 × 224 then normalized using the PyTorch Normalize function [41] to yield ImageNet's preferred mean pixel values of (0.485, 0.456, 0.406) and standard deviation pixel values of (0.229, 0.224, 0.225) [36,41]. Notably, this ants/bees dataset is the same dataset as the one used by Lloyd et al. in their paper [26], as well as by Mari et al. in their 2019 paper on *quantum transfer learning* [39].

2.4.1. Training QML Models with Feature Extraction Using ResNet-18

The first step in assessing the hybrid embedding was to investigate the resulting training cost, test cost, test set precision, test set recall and test set F1-score using the same embedding setup as presented in the demo code associated with Lloyd et al.'s paper [26]. This setup includes the pre-trained ResNet-18 component which converts each normalized ant or bee image into 512 input features. The $2 \times 512 = 1024$ resulting linear parameters and 12 quantum parameters of the quantum feature map were optimized as detailed above.

2.4.2. Training QML Models with Feature Extraction Using ResNet-18 Followed by PCA

To help address the potential issue of overfitting due to the high number of parameters used when training the linear half of the model, principal component analysis (PCA) was performed on the 512 output features of the ResNet. First, for each sample, the ResNet output features were normalized using the scikit-learn StandardScaler function [42], resulting in a mean of 0 and a standard deviation of 1 for each feature. The 512 normalized features were then reduced to 256, 64, 16, 4 and 2 principal components, leading also to a reduction in the number of linear model parameters. For instance, whenever the features of each sample were reduced to 256 principal components, the model would be trained with 512 linear parameters (as opposed to the original 1024 linear parameters). With a reduction to 4 principal components, the model would be trained using just 8 linear parameters. In general, n principal components were multiplied by a $2 \times n$ matrix to yield the two-dimensional (x_1, x_2) values used as input to the quantum feature map. The elements of the $2 \times n$ matrix change between each training iteration, acting as $2n$ trainable linear parameters. Other than this change to the number of trainable parameters, the training setup of the optimizer was kept the same as in the non-PCA case.

2.4.3. Training QML Models with Feature Extraction Using PCA

A more direct form of PCA was also used on the Hymenoptera image dataset. Instead of passing the images through a ResNet first, the $224 \times 224 \times 3 = 150528$ normalized pixel datapoints per image were reduced directly down to 256, 64, 16, 4 and 2 principal components. As before, these principal components were then multiplied by a $2 \times n$ parameter matrix to yield the (x_1, x_2) values and $2n$ trainable linear parameters, where n is the number of principal components. These (x_1, x_2) values were again used as input to the quantum feature map, then optimized using the optimization approach detailed above.

2.5. UCI ML Breast Cancer Wisconsin (Diagnostic) Dataset

The second dataset that we used was the University of California Irvine Machine Learning Breast Cancer Wisconsin (Diagnostic) Dataset [37]. This dataset consists of 569 breast cancer samples, each associated with 30 quantitative values such as cell radius, symmetry and smoothness. Each sample in the dataset can be classified as either *benign* or *malignant*. At different points in this work, the dataset was manually divided into different train-test splits (as determined by set random seeds), each in the ratio of 3:2.

2.5.1. Training QML Models Using All Input Features

As with the Hymenoptera dataset, the goal was to establish how well the hybrid embedding generalizes. To begin, the 30 quantitative attributes of the breast cancer dataset were normalized using the scikit-learn StandardScaler function [42], such that the mean and standard deviation of each attribute became 0 and 1, respectively. The normalized attributes were then matrix-multiplied with a 2×30 parameter matrix, resulting in a set of x_1 and x_2 values associated with each sample, as well as a set of 60 trainable linear parameters corresponding to the elements in the matrix. Mirroring the steps that were performed on the Hymenoptera dataset, the 60 linear parameters and 12 quantum parameters were then trained as detailed above. For this dataset, two sets of results were collected in separate tables. Each set of results came from a different pseudo-random train-test split of the data as determined by a random seed. Two sets of results were obtained to account for potential bias in the splits caused by chance.

2.5.2. Training QML Models with Feature Extraction using PCA

Taking the same approach as with the Hymenoptera dataset, PCA was also performed on the 30 normalized features of the breast cancer dataset to reduce the number of trainable parameters.

Two new sets of models were trained according to the same train-test splits as established in the non-PCA case. Each of these two sets consisted of models trained from 30, 16, 8, 4 and 2 principal components. Just as with the Hymenoptera dataset, the resulting principal components were multiplied by a $2 \times n$ parameter matrix where n is the number of principal components. This approach yields (x_1, x_2) values and $2n$ linear parameters needed for training and embedding. The same optimizer configuration was used as in all prior cases.

2.6. Assessing Quantum Metric Learning Model Performance

For all the QML models generated for both datasets, training costs, test costs, test set precision scores, test set recall scores and test set F1-scores resulting from each of the train-test splits were calculated. x_1, x_2 scatter plots and Hilbert space mutual data overlap matrices were generated to examine the level of expressivity of the models and to further review the ability of these models to separate and classify test data.

3. Results

3.1. Hymenoptera Dataset

As detailed in the Methods section, we trained and tested the Hymenoptera image and Breast Cancer Wisconsin (Diagnostic) datasets using the hybrid classical-quantum classifier shown in Figure 1A. We started with the Hymenoptera dataset using the same approach as Lloyd et al. [26]. In Figure 1B,C, we show a scatter plot of the inputs to the quantum circuit for train and test data before and after 1500 steps of training, respectively. Figure 1B illustrates that we recapitulate the ability of quantum metric learning (QML) to perfectly separate the Hymenoptera image training data when using the ResNet-18 layer with 512 input features in the same way as is seen in Lloyd et al.'s work [26]. With 1024 linear parameters and 12 quantum parameters, the training set (x_1, x_2) datapoints seem to cluster very well two-dimensionally after 1500 steps. In contrast, as shown in

Figure 1C, the test set datapoints remain very poorly separated. This contrast in separability suggests that the model is severely overfitting in this case.

Figure 2 illustrates the Hilbert space mutual data overlap gram matrices demonstrating the classifiability associated with the training and test results provided in Figure 1C. As expected from a case that shows a high level of overfitting, the training data is separated almost perfectly in Hilbert space (as seen in Figure 2B) while the test data remains barely separated at all (as seen in Figure 2D), demonstrating that the embedding generalizes poorly with the Hymenoptera dataset.

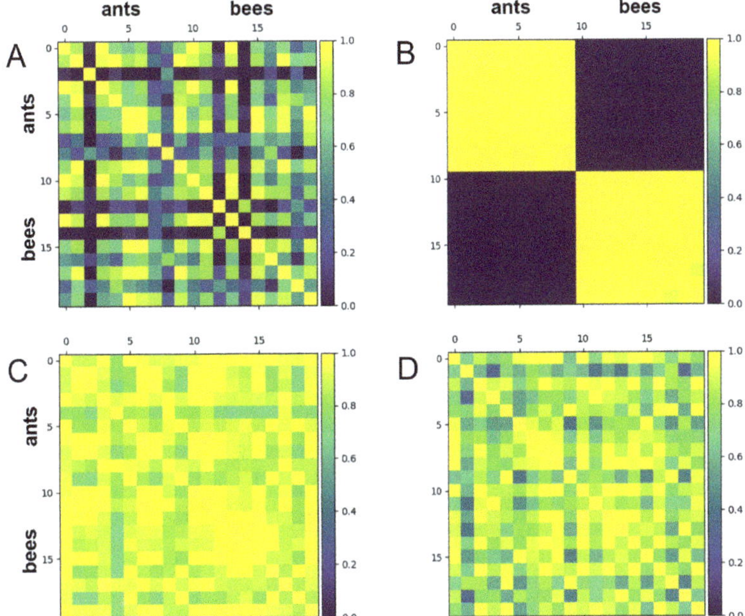

Figure 2. Gram matrices for mutual data overlap (i.e., $|\langle x|x'\rangle|^2$) in Hilbert space for 10 ant and 10 bee samples from the Hymenoptera dataset where 0 and 1 correspond to no and perfect overlap, respectively. In each case, 512 ResNet features (corresponding to 1024 trainable linear parameters) were used. The stronger the separation between the purple tiles (bees) and the yellow tiles (ants), the better the model's ability to classify. The Hymenoptera dataset's default train-test split was used for these results. The PennyLane software package was used to train the embedding [40]. (**A**) Mutual data overlap in Hilbert space for training set datapoints at optimization step 0. (**B**) Mutual data overlap in Hilbert space for training set datapoints at optimization step 1500. (**C**) Mutual data overlap in Hilbert space for test set datapoints at optimization step 0. (**D**) Mutual data overlap in Hilbert space for test set datapoints at optimization step 1500.

Summarised in Table 1 are the results of training the model on the Hymenoptera dataset in various ways. A specific random seed of '123' was used for the train-test split in every row other than the first. The first row uses the same default train-test split as was used in Lloyd et al.'s work [26]. It also corresponds to the results shown in Figures 1 and 2.

Table 1. Test set assessment outcomes for training performed on the Hymenoptera dataset's training set. Corresponding training costs are also given. In each row, training was performed for 1500 iterations using the root mean squared propagation optimizer (step size of 0.01) and a batch size of 10. All values are given to four decimal places. The features in row 1 did not undergo PCA, while the features from the rest of the rows did. A random seed of '123' used for the train-test split in every row other than the first (the first row used the default train-test split of the Hymenoptera dataset). The same random seed of '123' was used for all subsequent evaluations in all rows. The best value for each column is shown in bold.

No. of Features	ResNet (y/n)	Training Cost	Test Cost	Precision	Recall	F1-Score
512	y	**0.0141**	0.9931	**0.6184**	0.5663	**0.5912**
256	y	0.9944	0.9885	0.5326	**0.5976**	0.5632
256	n	0.9947	**0.9859**	0.4945	0.5488	0.5202
64	y	0.9756	0.9942	0.4891	0.5488	0.5172
64	n	0.9956	0.9928	0.4828	0.5122	0.4970
16	y	0.9926	0.9897	0.5000	0.5488	0.5233
16	n	0.9969	0.9892	0.4831	0.5244	0.5029
4	y	0.9909	0.9911	0.4545	0.4878	0.4706
4	n	0.9959	0.9947	0.4783	0.5366	0.5057
2	y	0.9700	0.9928	0.4545	0.4878	0.4706
2	n	0.9954	0.9965	0.4316	0.5000	0.4633

Test set F1-score and precision are maximized when using the original setup involving the full 512 output features of ResNet-18 with no further feature reduction through PCA. Training cost is minimized at 512 features, but the corresponding test cost is high, which provides further evidence of overfitting and poor generalization. This also means that the minimized training cost of 0.0141 is likely achieved only when overfitting the training data. The lowest test cost, which is achieved with 256 principal component features and no ResNet step, is hardly reduced from its maximum value of 1. The test set recall is maximized at 256 principal component features with the ResNet step.

Although the 512 feature setup and 256 principal component feature setups seemed to perform slightly better than entirely random class assignment, the resulting scores are still very poor. The highest F1-score being just 0.5912 and the lowest test cost still being as high as 0.9859. Furthermore, regardless of whether or not a ResNet step was used, subsequent feature reduction through PCA only worsened F1-score while drastically increasing training costs. Thus, after reducing the number of parameters, there seems to have been a drop in expressivity, which prevented overfitting. However, this was due to training costs becoming much worse. While it can be said than none of the models in Table 1 demonstrate good test set classification performance, the observed ability for PCA to prevent overfitting is still worth noting, despite it being achieved exclusively through increased training cost values in this case.

3.2. Breast Cancer Dataset

Figure 3A,B illustrate the effects of training the hybrid model for 1500 iterations on the breast cancer dataset. Both the training set and test set (x_1, x_2) values seem to have separated reasonably well in two dimensions, which contrasts with the Hymenoptera dataset result where only the training set separated well. However, neither set separates well enough for entirely distinct non-overlapping clusters to form (as was seen in Figure 1C). While the training set datapoints in Figure 1 separated into very tight clusters that were isolated from other surrounding clusters, the clusters in Figure 3B are much broader and less well defined. This more modest training set separation, in conjunction with the much greater similarity between the training set clusters and test set clusters indicates that the level of overfitting is much lower when using the breast cancer dataset.

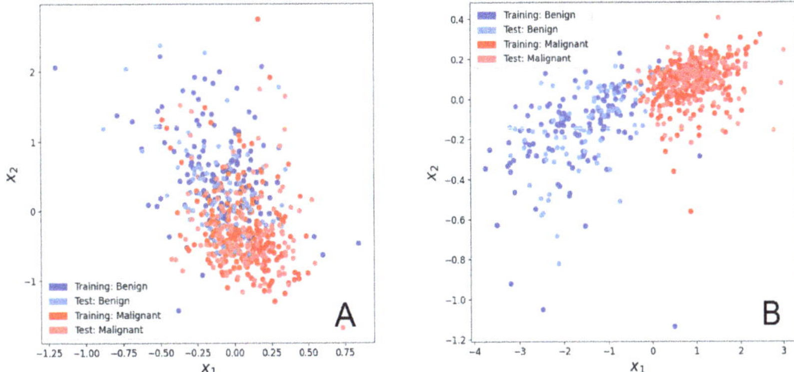

Figure 3. Scatter plots of the (x_1, x_2) values of the breast cancer dataset with 30 starting features (corresponding to 60 trainable linear parameters). Datapoints from both the training set and the test set are depicted. A random seed of '1' was used for the train-test split of this data. The PennyLane software package was used to optimize the parameters [40]. (**A**) Scatter plot of the (x_1, x_2) values after 0 training steps. (**B**) Scatter plot of the (x_1, x_2) values after 1500 training steps.

Figure 4 depicts the Hilbert space mutual data overlaps (i.e., $|\langle x|x'\rangle|^2$) associated with the training and test scatter plot results shown in Figure 3. It is clear from Figure 4B,D that both the training set embeddings and the test set embeddings separate relatively well in Hilbert space when using the trained model. The Hilbert space separation and resulting classifiability of the test set appear comparable to those of the training set, which serves as further evidence that overfitting is less of an issue with this dataset.

However, the test set is still classified observably worse than the training set, as seen by the significantly misplaced 'lines' of overlap present within Figure 4D. Thus, despite the improvements compared to the previous dataset, there is still a moderate level of overfitting occurring. Consequently, there is still room for generalization performance to be improved further.

Figure 5 demonstrates the effects of carrying out PCA on the 30 input features of the breast cancer dataset. As seen in Figure 5B,D, PCA seems to bring both training set and test set (x_1, x_2) values into tighter two-dimensional clusters compared to those seen in Figure 3B. This generally has the effect of reducing the relative surface area of the borders between neighboring clusters, which could potentially correlate with better classification after subsequent embedding.

It is worth noting that in Figure 5A (with feature reduction to 8 principal components), the (x_1, x_2) values seem to start off reasonably well separated in two-dimensions as a result of the prior PCA step. Then after 1500 steps of training, Figure 5B shows how the model is able to further separate the values such that much more distinctive, globular clusters are formed with a much lower relative surface area where the clusters meet. In contrast, Figure 5C shows that the (x_1, x_2) values resulting from 4 principal components begin in a much less well separated two-dimensional state after the initial PCA step. Despite this, the trained model is still able to separate the values into quite distinctive clusters, as shown in Figure 5D. In fact, the two-dimensional area of cluster overlap in Figure 5D still seems to be slightly smaller than the area of cluster overlap in Figure 3B. In other words, regardless of whether PCA is able to group the pre-training (x_1, x_2) values by class, the resulting post-training test set is well separated. Interestingly, the PCA-based post-training separation (Figure 5B,D) appears to be better than its non-PCA counterpart (Figure 3B). Thus, we find that feature reduction through PCA can consistently contribute to better generalization performance for this dataset.

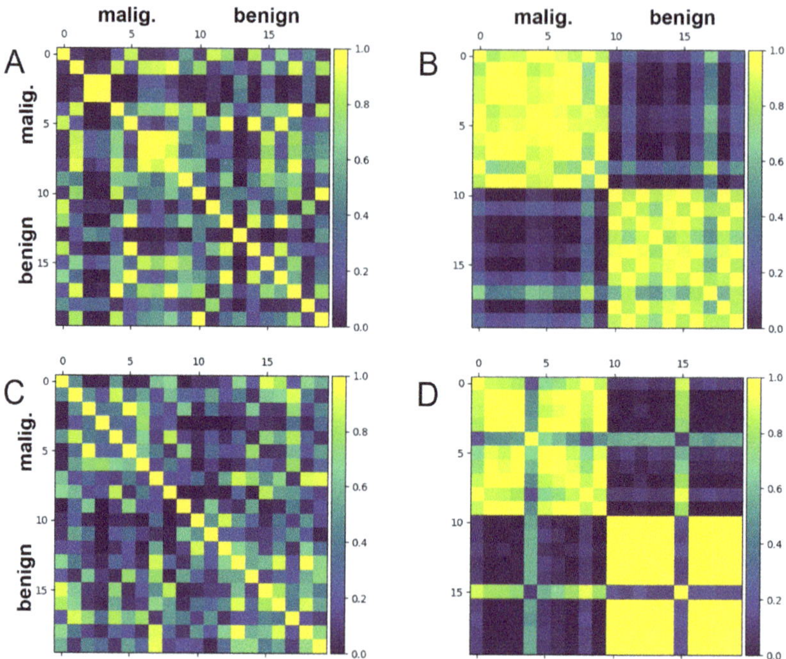

Figure 4. Gram matrices depicting mutual data overlap in Hilbert space (i.e., $|\langle x|x'\rangle|^2$) for 10 benign and malignant train and test samples from the breast cancer dataset. In each case, 30 starting features (corresponding to 60 trainable linear parameters) were used, with no subsequent PCA feature reduction. The stronger the separation between the purple tiles (benign) and the yellow tiles (malignant), the better the model's ability to classify. A random seed of '1' was used for the train-test split of this data. The PennyLane software package was used to train the embedding [40]. (**A**) Mutual data overlap in Hilbert space for training set datapoints at optimization step 0. (**B**) Mutual data overlap in Hilbert space for training set datapoints at optimization step 1500. (**C**) Mutual data overlap in Hilbert space for test set datapoints at optimization step 0. (**D**) Mutual data overlap in Hilbert space for test set datapoints at optimization step 1500.

A final observation is that the 8 principal component model (with 16 trainable linear parameters) seems to demonstrate greater expressivity than the 4 principal component model (with 8 trainable linear parameters). While the 8 principal component model moves the (x_1, x_2) values into more distinctive, globular clusters, the 4 principal component model instead moves the values into a simpler, more linear shape. It seems that having fewer trainable linear parameters can cause the model to lose expressively, leading to less well-defined clusters and perhaps worse post-embedding classification. However, as seen in Figures 1C and 3B, having too many parameters, and, thus, too much expressivity for a limited number of samples, can lead to overfitting and noisier clustering.

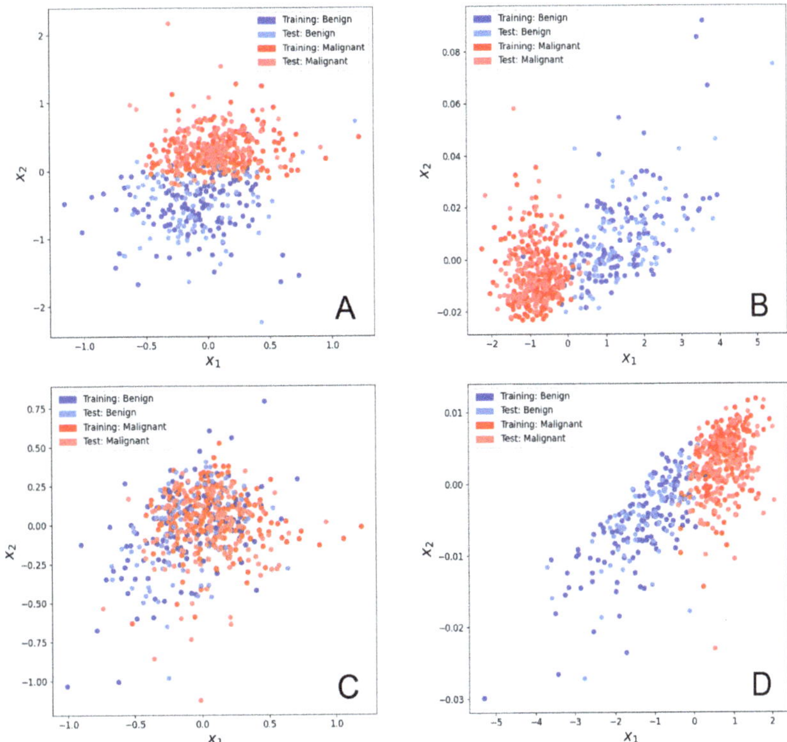

Figure 5. Scatter plots of the (x_1, x_2) values of the breast cancer dataset following feature reduction through PCA. Datapoints from both the training set and the test set are depicted. A random seed of '1' was used for the train-test split of this data. The PennyLane software package was used to optimize the parameters [40]. (**A**) Scatter plot of (x_1, x_2) values associated with 8 principal components after 0 training steps. These 8 principle components correspond to 16 trainable linear parameters. (**B**) Scatter plot of (x_1, x_2) values associated with 8 principal components after 1500 training steps. These 8 principle components correspond to 16 trainable linear parameters. (**C**) Scatter plot of (x_1, x_2) values associated with 4 principal components after 0 training steps. These 4 principle components correspond to 8 trainable linear parameters. (**D**) Scatter plot of (x_1, x_2) values associated with 4 principal components after 1500 training steps. These 4 principle components correspond to 8 trainable linear parameters.

Figure 6 illustrates the mutual test data overlaps in Hilbert space (i.e., $|\langle x|x'\rangle|^2$) that correspond to the scatter plots from Figure 5. After training, the purple and yellow tiles seem to have separated better when 8 principal components were used (Figure 6B) compared to when 4 principal components were used (Figure 6D). In particular, there are overall not as many 'lines' of misassigned overlap running across the four grouped squares in Figure 6B. This suggests that that the 8 principal component model is better at maximally separating embedded test data in Hilbert space than the 4 principal component model and is thus better at classifying new data. This aligns with the higher expressivity observed within the 8 principal component clusters of Figure 5B. Not surprisingly, there appears to be an optimal number of principal components for a given number of samples, which yields the best embedding ability, model expressivity and generalizability.

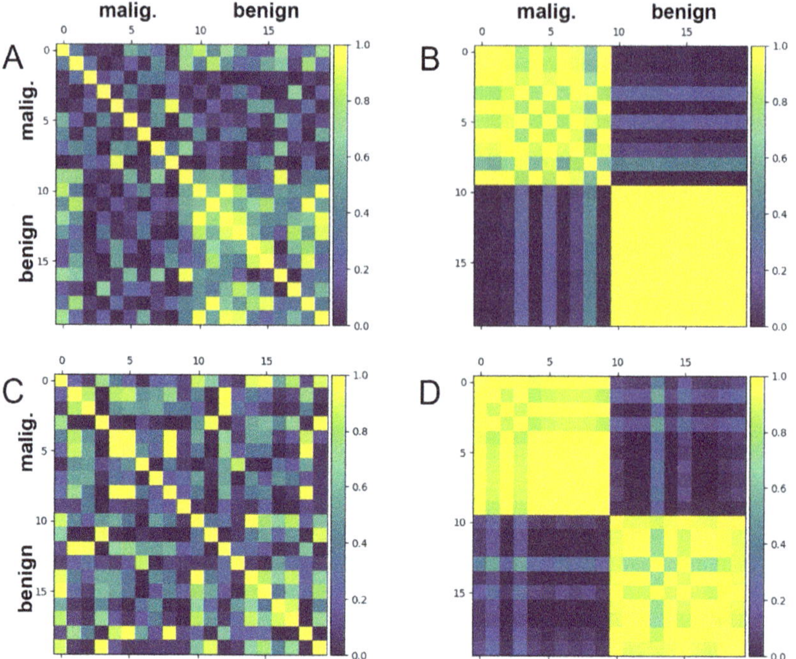

Figure 6. Gram matrices depicting mutual data overlap in Hilbert space (i.e., $|\langle x|x'\rangle|^2$) for 10 benign and 10 malignant train and test samples from the breast cancer dataset. In each case, PCA was used to reduce the number of features (and thus the number of trainable linear parameters). The stronger the separation between the purple tiles (benign) and the yellow tiles (malignant), the better the model's ability to classify. A random seed of '1' was used for the train-test split of this data. The PennyLane software package was used to train the embeddings [40]. (**A**) Mutual data overlap in Hilbert space for test set datapoints at optimization step 0, using (x_1, x_2) values generated from 8 principal components. (**B**) Mutual data overlap in Hilbert space for test set datapoints at optimization step 1500, using (x_1, x_2) values generated from 8 principal components. (**C**) Mutual data overlap in Hilbert space for test set datapoints at optimization step 0, using (x_1, x_2) values generated from 4 principal components. (**D**) Mutual data overlap in Hilbert space for test set datapoints at optimization step 1500, using (x_1, x_2) values generated from 4 principal components.

Summarised in Tables 2 and 3 are the results of training the hybrid model on the breast cancer dataset in various ways. A different random seed (for creating a pseudo-random pre-determined train-test split) was used for each table. Within each table, the random seed of choice (and thus the specific train-test split) stays consistent. Table 3 also corresponds to the results in Figures 3–6.

We emphasize that the differences between the results of Tables 2 and 3 come solely from the differences in random seeds used. In both result sets, test set F1-score is maximized and test cost is minimized when PCA is performed to produce 8 principal components. Meanwhile, training cost is minimized when all 30 principal components are used (i.e., the same as the initial number of features in the dataset). Test set precision and recall are maximized at either 8 or 16 principal components in each case and are all much higher than the Hymenoptera test set precision and recall scores from Table 1.

Based on our analysis of the breast cancer dataset, it is evident that lowering the number of input features through PCA (thus lowering the number of trainable linear parameters) reduces the level of overfitting by the trained hybrid model. This is observed in the shrinking difference between training costs and test costs. This arises from increases in training costs and is sometimes coupled with decreases in test costs, as well as improvement in test set F1-scores. However, when there are too few linear parameters, F1-scores and test costs worsen again. This is consistent with the observations made relating to Figure 5B,D, where a reduction in the number of features caused the clusters to be more linear (less globular) in shape, pertaining to a decrease in expressivity.

For this particular dataset, reducing the 30 initial features to 8 principal components (16 trainable linear parameters) seems to be the ideal compromise for good generalizability in terms of minimizing overfitting while maximizing expressivity.

Table 2. Test set assessment outcomes for training performed on the UCI ML Breast Cancer Wisconsin (Diagnostic) Dataset training set. Corresponding training costs are also given. In each row, training was performed for 1500 iterations using the root mean squared propagation optimizer (step size of 0.01) and a batch size of 10. All values are given to four decimal places. The features in row 1 did not undergo PCA, while the features from the rest of the rows did. A random seed of '123' was used in each row, for both the train-test split and for all subsequent evaluations. The best value for each column is shown in bold.

No. of Features	Training Cost	Test Cost	Precision	Recall	F1-Score
30	0.1727	0.3623	0.9032	0.9790	0.9396
30	**0.1465**	0.3751	0.9091	0.9790	0.9428
16	0.2692	0.3023	**0.9338**	0.9860	0.9592
8	0.2757	**0.2903**	0.9226	**1.0000**	**0.9597**
4	0.2569	0.3440	0.9156	0.9860	0.9495
2	0.3953	0.3817	0.8981	0.9860	0.9400

Table 3. Test set assessment outcomes for training performed on the UCI ML Breast Cancer Wisconsin (Diagnostic) Dataset training set. Corresponding training costs are also given. In each row, training was performed for 1500 iterations using the root mean squared propagation optimizer (step size of 0.01) and a batch size of 10. All values are given to four decimal places. The features in row 1 did not undergo PCA, while the features from the rest of the rows did. A random seed of '1' was used in each row, both for the train-test split and for all subsequent evaluations. The best value for each column is shown in bold.

No. of Features	Training Cost	Test Cost	Precision	Recall	F1-Score
30	0.2026	0.2791	0.9205	0.9720	0.9456
30	**0.1750**	0.2899	0.9211	0.9790	0.9492
16	0.2201	0.3101	0.9281	**0.9930**	0.9595
8	0.2497	**0.2646**	**0.9655**	0.9790	**0.9722**
4	0.2885	0.2913	0.9467	**0.9930**	0.9693
2	0.3450	0.3306	0.9517	0.9650	0.9583

4. Discussion

We implemented the hybrid classical-quantum machine learning approach termed quantum metric learning [26]. Specifically, we addressed the following gap: while the approach was shown to separate training samples perfectly on a Hymenoptera dataset containing images of ants and bees, the performance of the trained models on hold out test data was not assessed. When using the same circuit, dataset and train-test split as seen in Lloyd et al.'s paper [26], it was found that the resulting hybrid model severely overfits the training data and generalizes poorly. While almost perfect Hilbert space-embedded separation was achieved with the training data, the test data yielded very poor results with an F1-score of only 0.5912. Reducing the number of linear parameters through principal component analysis (PCA) produced even worse outcomes for both the training set and the test set. This is likely due to a decrease in model expressivity. Specifically, a drop in test set recall and F1-score was observed, along with a very steep increase in training cost. The increase in training cost was so dramatic (from 0.0141 to $\geq$0.9700) that the training cost values became comparable to those of the test. After omitting the ResNet-18 step and carrying out PCA directly on the pixel data, there were no improvements to the results. We found that no method resulted in even modest generalizability for this dataset which had a large number of features compared to the number of samples.

The breast cancer dataset consists of a significantly smaller number of features, while having a greater number of total samples. Even without carrying out PCA, the trained models seemed to generalize reasonably well for the test data, yielding high F1-scores of 0.9396 and 0.9456. However, there was still some evidence of overfitting, with training costs of 0.1727 and 0.2026 being associated with much higher test costs of 0.3623 and 0.2791, respectively. When PCA was performed on the initial features, resulting test set F1-scores were always higher than that of their non-PCA counterpart, while differences between the training costs and test costs were often much lower. Not surprisingly, we also found that test costs and F1-scores tended to worsen again if the number of principal components was too low. For the breast cancer dataset, the ideal balance of high expressivity and low overfitting needed for good generalization was found to be at 8 principal components (16 linear parameters). This yielded an F1-score as high as 0.9722 and a test cost as low as 0.2646 (with a similar training cost of 0.2497). Of course, the optimal number of principal components would vary depending on the dataset.

Quantum metric learning models appear to follow the traditional bias-variance constraints, namely, good generalization results if the number of model parameters is significantly lower than the number of training samples. The above requirements are fulfilled by the breast cancer dataset, where there are 72 initial parameters (resulting from just 30 initial features) and as many as 357 training samples. The initial 72 parameter model generalized well and parameter reduction through PCA served to improve this generalization even further, most notably after a reduction to just 28 model parameters. In contrast, the Hymenoptera dataset has as many as 1036 initial parameters (resulting from at least 512 initial features) while having only 244 training samples; the initial 1036 parameter model generalized poorly and parameter reduction through PCA offered no significant improvement.

For future explorations, it would be insightful to vary the shape of the quantum feature map (and thus the number of quantum parameters involved) and to assess the subsequent effects this has on the expressivity and overfitting observed in any resulting trained models. The quantum feature map can be varied both in its length (the number of 'horizontal' repetitions of each gate) and its width (the number of qubits used). It could be the case that varying the dimensions of the quantum feature map changes the ideal ratio between the number of initial parameters and the number of samples to achieve good generalization performance. It would also be valuable to explore methods of dimensional reduction other than PCA, such as classical or quantum auto-encoding. Comparisons in generalization performance and classification accuracy between quantum metric learning and other methods of classification (using the breast cancer dataset, as well as a broad range of other datasets) would also be insightful.

Author Contributions: Conceptualization, S.B.; methodology, J.K. and S.B.; software, J.K.; validation, J.K.; formal analysis, J.K.; investigation, J.K. and S.B.; data curation, J.K.; writing—original draft preparation, J.K. and S.B.; writing—review and editing, J.K. and S.B.; visualization, J.K.; supervision, S.B.; project administration, S.B.; funding acquisition, S.B. All authors have read and agreed to the published version of the manuscript.

Funding: S.B. was funded in part by the National Science Foundation grant number IIS-2106913. GlaxoSmithKline paid the publication costs.

Data Availability Statement: The ImageNet Hymenoptera dataset can be accessed on Kaggle: https://www.kaggle.com/datasets/melodytsekeni/hymenoptera-data, accessed on 21 October 2022. The Breast Cancer Wisconsin (Diagnostic) Data Set can be accessed through UCI: https://archive.ics.uci.edu/ml/datasets/breast+cancer+wisconsin+(diagnostic), accessed on 21 October 2022. The code used in this manuscript can be accessed on GitHub: https://github.com/Rlag1998/QML_Generalization, accessed on 21 October 2022.

Acknowledgments: We acknowledge the support of Fausto Artico and Kevin Harrigan from GlaxoSmithKline.

Conflicts of Interest: Stefan Bekiranov consults with GlaxoSmithKline on quantum computing and quantum machine learning which does not represent a conflict of interest regarding the present manuscript.

Abbreviations

The following abbreviations are used in this manuscript:

QML	quantum metric learning
PC	principal component
PCA	principal component analysis
KNN	k-nearest neighbor

References

1. Preskill, J. The Physics of Quantum Information. *arXiv* **2022**, arXiv:2208.08064.
2. Cao, Y.; Romero, J.; Olson, J.P.; Degroote, M.; Johnson, P.D.; Kieferová, M.; Kivlichan, I.D.; Menke, T.; Peropadre, B.; Sawaya, N.P.D.; et al. Quantum Chemistry in the Age of Quantum Computing. *Chem. Rev.* **2019**, *119*, 10856–10915. PMID: 31469277. [CrossRef] [PubMed]
3. Grover, L.K. A Fast Quantum Mechanical Algorithm for Database Search. In Proceedings of the Twenty-Eighth Annual ACM Symposium on Theory of Computing; Association for Computing Machinery, New York, NY, USA, 3–5 May 1996; STOC '96, pp. 212–219.
4. Shor, P. Algorithms for quantum computation: discrete logarithms and factoring. In Proceedings of the 35th Annual Symposium on Foundations of Computer Science, Santa Fe, New Mexico, 20–22 November 1994; pp. 124–134.
5. Biamonte, J.; Wittek, P.; Pancotti, N.; Rebentrost, P.; Wiebe, N.; Lloyd, S. Quantum machine learning. *Nature* **2017**, *549*, 195. [CrossRef] [PubMed]
6. Spagnolo, N.; Vitelli, C.; Sansoni, L.; Maiorino, E.; Mataloni, P.; Sciarrino, F.; Brod, D.J.; Galvão, E.F.; Crespi, A.; Ramponi, R.; et al. General Rules for Bosonic Bunching in Multimode Interferometers. *Phys. Rev. Lett.* **2013**, *111*, 130503. [CrossRef] [PubMed]
7. Amin, M.H.; Andriyash, E.; Rolfe, J.; Kulchytskyy, B.; Melko, R. Quantum Boltzmann Machine. *Phys. Rev. X* **2018**, *8*, 021050 [CrossRef]
8. Kieferová, M.; Wiebe, N. Tomography and generative training with quantum Boltzmann machines. *Phys. Rev. A* **2017**, *96*, 062327. [CrossRef]
9. Wiebe, N.; Braun, D.; Lloyd, S. Quantum Algorithm for Data Fitting. *Phys. Rev. Lett.* **2012**, *109*, 050505. [CrossRef]
10. Lloyd, S.; Mohseni, M.; Rebentrost, P. Quantum principal component analysis. *Nat. Phys.* **2014**, *10*, 631 . [CrossRef]
11. Wiebe, N.; Kapoor, A.; Svore, K.M. Quantum Deep Learning. *arXiv* **2014**, arXiv:1412.3489.
12. Dunjko, V.; Taylor, J.M.; Briegel, H.J. Quantum-Enhanced Machine Learning. *Phys. Rev. Lett.* **2016**, *117*, 130501. [CrossRef]
13. Kapoor, A.; Wiebe, N.; Svore, K. Quantum perceptron models. In Proceedings of the Advances in Neural Information Processing Systems, Barcelona, Spain, 5–10 December 2016; pp. 3999–4007.
14. Low, G.H.; Yoder, T.J.; Chuang, I.L. Quantum inference on Bayesian networks. *Phys. Rev. A* **2014**, *89*, 062315. [CrossRef]
15. Wiebe, N.; Granade, C. Can small quantum systems learn? *arXiv* **2015**, arXiv:1512.03145.
16. Giovannetti, V.; Lloyd, S.; Maccone, L. Quantum Random Access Memory. *Phys. Rev. Lett.* **2008**, *100*, 160501. [CrossRef]
17. Rebentrost, P.; Mohseni, M.; Lloyd, S. Quantum Support Vector Machine for Big Data Classification. *Phys. Rev. Lett.* **2014**, *113*, 130503. [CrossRef]
18. Preskill, J. Quantum Computing in the NISQ era and beyond. *Quantum* **2018**, *2*, 79. [CrossRef]
19. Schuld, M.; Fingerhuth, M.; Petruccione, F. Implementing a distance-based classifier with a quantum interference circuit. *EPL (Europhys. Lett.)* **2017**, *119*, 60002. [CrossRef]
20. Havlícek, V.; Córcoles, A.D.; Temme, K.; Harrow, A.W.; Kandala, A.; Chow, J.M.; Gambetta, J.M. Supervised learning with quantum-enhanced feature spaces. *Nature* **2019**, *567*, 209–212. [CrossRef]
21. Schuld, M.; Killoran, N. Quantum Machine Learning in Feature Hilbert Spaces. *Phys. Rev. Lett.* **2019**, *122*, 040504. [CrossRef]
22. Schuld, M. Supervised quantum machine learning models are kernel methods. *arXiv* **2021**, arXiv:2101.11020.
23. Blank, C.; Park, D.K.; Rhee, J.K.K.; Petruccione, F. Quantum classifier with tailored quantum kernel. *NPJ Quantum Inf.* **2020**, *6*, 41. [CrossRef]
24. Park, D.K.; Blank, C.; Petruccione, F. The theory of the quantum kernel-based binary classifier. *Phys. Lett.* **2020**, *384*, 126422. [CrossRef]
25. Kathuria, K.; Ratan, A.; McConnell, M; Bekiranov, S. Implementation of a Hamming distance–like genomic quantum classifier using inner products on ibmqx2 and ibmq_16_melbourne. *Quantum Mach. Intell.* **2020**, *2*, 7. [CrossRef]
26. Lloyd, S.; Schuld, M.; Ijaz, A.; Izaac, J.; Killoran, N. Quantum embeddings for machine learning. *arXiv* **2022**, arXiv:2001.03622.
27. Thumwanit, N.; Lortaraprasert, C.; Yano, H.; Raymond, R. Trainable Discrete Feature Embeddings for Variational Quantum Classifier. *arXiv* **2021**, arXiv:2106.09415.

28. Suzuki, Y.; Yano, H.; Gao, Q.; Uno, S.; Tanaka, T.; Akiyama, M.; Yamamoto, N. Analysis and synthesis of feature map for kernel-based quantum classifier. *Quantum Mach. Intell.* **2019**, *2*, 9. [CrossRef]
29. García, D.P.; Cruz-Benito, J.; García-Peñalvo, F.J. Systematic Literature Review: Quantum Machine Learning and its applications. *arXiv* **2022**, arXiv:2201.04093.
30. Hubregtsen, T.; Wierichs, D.; Gil-Fuster, E.; Derks, P.J.H.S.; Faehrmann, P.K.; Meyer, J.J. Training Quantum Embedding Kernels on Near-Term Quantum Computers. *arXiv* **2021**, arXiv:2105.02276.
31. Wang, X.; Du, Y.; Luo, Y.; Tao, D. Towards understanding the power of quantum kernels in the NISQ era. *Quantum* **2021**, *5*, 531. [CrossRef]
32. LaRose, R.; Coyle, B. Robust data encodings for quantum classifiers. *Phys. Rev.* **2020**, *102*, 032420. [CrossRef]
33. Easom-Mccaldin, P.; Bouridane, A.; Belatreche, A.; Jiang, R. On Depth, Robustness and Performance Using the Data Re-Uploading Single-Qubit Classifier. *IEEE Access* **2021**, *9*, 65127–65139. [CrossRef]
34. Canatar, A.; Peters, E.; Pehlevan, C.; Wild, S.M.; Shaydulin, R. Bandwidth Enables Generalization in Quantum Kernel Models. *arXiv* **2022**, arXiv:2206.06686.
35. Caro, M.C.; Huang, H.Y.; Cerezo, M.; Sharma, K.; Sornborger, A.; Cincio, L.; Coles, P.J. Generalization in quantum machine learning from few training data. *Nat. Commun.* **2022**, *13*, 4919. [CrossRef]
36. Deng, J.; Dong, W.; Socher, R.; Li, L.J.; Li, K.; Fei-Fei, L. Imagenet: A large-scale hierarchical image database. In Proceedings of the 2009 IEEE Conference on Computer Vision and Pattern Recognition, Miami, FL, USA, 20–25 June 2009; pp. 248–255.
37. Dua, D.; Graff, C. UCI Machine Learning Repository. 2019. Available online: http://archive.ics.uci.edu/ml (accessed on 1 August 2022).
38. Farhi, E.; Goldstone, J.; Gutmann, S. A Quantum Approximate Optimization Algorithm. *arXiv* **2014**, arXiv:1411.4028.
39. Mari, A.; Bromley, T.R.; Izaac, J.; Schuld, M.; Killoran, N. Transfer learning in hybrid classical-quantum neural networks. *arXiv* **2019**, arXiv:1912.08278.
40. Bergholm, V.; Izaac, J.; Schuld, M.; Gogolin, C.; Ahmed, S.; Ajith, V.; Alam, M.S.; Alonso-Linaje, G.; AkashNarayanan, B.; Asadi, A.; et al. PennyLane: Automatic differentiation of hybrid quantum-classical computations. *arXiv* **2018**, arXiv:1811.04968.
41. Paszke, A.; Gross, S.; Massa, F.; Lerer, A.; Bradbury, J.; Chanan, G.; Killeen, T.; Lin, Z.; Gimelshein, N.; Antiga, L.; et al. PyTorch: An Imperative Style, High-Performance Deep Learning Library. In *Advances in Neural Information Processing Systems 32*; Wallach, H., Larochelle, H., Beygelzimer, A., d'Alché-Buc, F., Fox, E., Garnett, R., Eds.; Curran Associates, Inc.: Red Hook, NY, USA, 2019; pp. 8024–8035.
42. Pedregosa, F.; Varoquaux, G.; Gramfort, A.; Michel, V.; Thirion, B.; Grisel, O.; Blondel, M.; Prettenhofer, P.; Weiss, R.; Dubourg, V.; et al. Scikit-learn: Machine Learning in Python. *J. Mach. Learn. Res.* **2011**, *12*, 2825–2830.

MDPI

Article

GraphSite: Ligand Binding Site Classification with Deep Graph Learning

Wentao Shi [1], Manali Singha [2], Limeng Pu [3], Gopal Srivastava [2], Jagannathan Ramanujam [1,3] and Michal Brylinski [2,3,*]

1 Division of Electrical and Computer Engineering, Louisiana State University, Baton Rouge, LA 70803, USA; wshi6@lsu.edu (W.S.); eejaga@lsu.edu (J.R.)
2 Department of Biological Sciences, Louisiana State University, Baton Rouge, LA 70803, USA; msing21@lsu.edu (M.S.); gsriva2@lsu.edu (G.S.)
3 Center for Computation and Technology, Louisiana State University, Baton Rouge, LA 70803, USA; lpu1@lsu.edu
* Correspondence: michal@brylinski.org; Tel.: +1-(225)-578-2791; Fax: +1-(225)-578-2597

Abstract: The binding of small organic molecules to protein targets is fundamental to a wide array of cellular functions. It is also routinely exploited to develop new therapeutic strategies against a variety of diseases. On that account, the ability to effectively detect and classify ligand binding sites in proteins is of paramount importance to modern structure-based drug discovery. These complex and non-trivial tasks require sophisticated algorithms from the field of artificial intelligence to achieve a high prediction accuracy. In this communication, we describe GraphSite, a deep learning-based method utilizing a graph representation of local protein structures and a state-of-the-art graph neural network to classify ligand binding sites. Using neural weighted message passing layers to effectively capture the structural, physicochemical, and evolutionary characteristics of binding pockets mitigates model overfitting and improves the classification accuracy. Indeed, comprehensive cross-validation benchmarks against a large dataset of binding pockets belonging to 14 diverse functional classes demonstrate that GraphSite yields the class-weighted F1-score of 81.7%, outperforming other approaches such as molecular docking and binding site matching. Further, it also generalizes well to unseen data with the F1-score of 70.7%, which is the expected performance in real-world applications. We also discuss new directions to improve and extend GraphSite in the future.

Keywords: structure-based drug discovery; ligand binding sites; deep learning; graph neural network

Citation: Shi, W.; Singha, M.; Pu, L.; Srivastava, G.; Ramanujam, J.; Brylinski, M. GraphSite: Ligand Binding Site Classification with Deep Graph Learning. *Biomolecules* **2022**, *12*, 1053. https://doi.org/10.3390/biom12081053

Academic Editors: Cameron Mura and Lei Xie

Received: 3 June 2022
Accepted: 20 July 2022
Published: 29 July 2022

1. Introduction

Proteins carry out numerous biological functions in the cellular environment. Interactions between proteins and other molecules, such as peptides, neurotransmitters, nucleic acids, hormones, lipids, and metabolites, are, therefore, vital to understanding the biology of the cell. In particular, interactions between proteins and small molecules, or ligands, are associated with a wide range of the functions of a living cell [1]. Ligand binding sites are typically pockets and cavities on the surface of proteins formed by spatially close amino acid residues interacting with small molecules in a specific way [2]. The ability to precisely detect and annotate these sites in protein structures is of paramount importance in modern structure-based drug discovery. It can help reveal novel targets for pharmacotherapy and support the design of biopharmaceuticals not only against the most common health issues affecting a large population worldwide [3] but also rare diseases without any treatment options currently available [4]. Numerous approaches have been developed over past years to identify and analyze ligand binding sites in proteins, including LIGSITE [5], FTSite [6], *e*FindSite [7], Fpocket [8], and SiteComp [9], to mention a few examples. A comprehensive characterization of ligand binding accounts for multiple factors of this multifaceted phenomenon, such as the conformational dynamics [10], the druggability [11], interaction hotspots [12], and the amino acid composition [13]. Despite the encouraging progress in ligand binding site detection, there is a need for a better functional characterization of the identified sites with respect to the types and properties of binding molecules.

It has been demonstrated that similar ligands can bind to evolutionary unrelated proteins [14]. Therefore, accurate methods to classify binding sites depending on the ligand information are essential to study ligand binding at a system level with a broad range of applications in polypharmacology [15], side effects prediction [16], and drug repositioning [17]. Several algorithms to predict binding sites in protein targets, given the ligand information, have been developed to date. For instance, the ProBiS-ligands web server can help figure out the types of ligands binding to the input protein structures [18]. As many ligands perform specific cellular functions important for a variety of biological processes, such as cell signaling, active transport, cell metabolism, and the regulation of the cell cycle, several algorithms focus on specific types of ligands. VitaPred employs the evolutionary information to predict residues interacting with vitamin ligands [19], SITEPred identifies nucleotide-binding residues from protein sequences [20], and HemeBIND detects heme binding residues based on the sequence and structure information [21]. Similar techniques were designed to work with other specific organic molecules, such as flavin adenine dinucleotide [22], guanosine triphosphate [23], nicotinamide adenine dinucleotide [24], and inorganic ions, such as calcium [25] and zinc [26]. Most of these methods employ traditional machine learning classifiers to predict binding residues based on the sequence, structure, and evolutionary information. However, currently available state-of-the-art deep learning approaches hold significant promise to greatly improve the accuracy of the functional annotation of ligand binding sites.

Deep learning is currently the most advanced group of machine learning techniques employing various types of multilayer artificial neural networks to learn complex patterns from the input data. Deep learning makes headway in the computer vision field, where it has successfully been applied across numerous tasks, including object detection [27], face recognition [28], and body pose estimation [29]. A key to the success of deep learning methods is the convolutional neural network (CNN), which utilizes local trainable filters to effectively learn hierarchical latent features from the Euclidean data, such as 2D and 3D images [30]. Advances in computer vision have inspired the development of deep learning tools for biology and biomedicine as well. Most approaches to predict and annotate ligand binding sites in proteins with CNNs represent pockets as either 2D or 3D images. The former group of methods includes BionoiNet, which first projects pockets onto a 2D plane encoding various physicochemical, structural, and evolutionary properties, and then employs a 2D-CNN to perform classification tasks [31]. An example of a 3D-based approach is DeepDrug3D, which deploys a 3D-CNN to accurately classify binding sites for adenosine triphosphate (ATP) and heme ligands represented as voxel-based 3D images [17]. A related method, DeeplyTough, employs a similar pocket representation as DeepDrug3D and pocket matching with a CNN to detect similar binding sites [32]. Another 3D-based predictor is DeepSite, which deploys a CNN to binding pockets represented as voxels annotated with various atomic-based pharmacophoric properties [33].

In addition to the Euclidean space, many contemporary data, such as social networks, sensor networks, biological networks, and meshed surfaces, have an underlying structure that belongs to the non-Euclidean domain. Graph neural network (GNN) is a group of deep learning models designed to work specifically with non-Euclidean graph data [34]. GNNs have been demonstrated to achieve unparalleled performance in numerous applications against non-Euclidean data, including text classification [35], traffic prediction [36], and complex physics simulations [37]. GNNs were deployed to address important problems in biology as well, for instance, to predict the quantum properties of organic molecules [38], generate molecular fingerprints [39], detect protein interfaces [40], and identify drug-target interactions [41]. These applications are based on a notion that molecular structures can conveniently be represented as graphs, in which atoms are nodes, and chemical bonds are undirected edges connecting pairs of nodes.

In this communication, we expand the repertoire of graph-based approaches in biology and biomedicine by developing GraphSite, a new method to classify ligand binding sites with a GNN. First, a large and diverse dataset of binding sites are converted into graphs preserving the physicochemical properties of local protein structures, which are then used to train a GNN classifier. In contrast to computationally more intensive methods operating in the Euclidean space, lightweight GraphSite generates the graph representations of ligand binding site on-the-fly without any pre-processing requirements. Encouragingly, it not only achieves state-of-the-art performance in multi-class classification benchmarks with respect to other approaches but also generalizes well to unseen data. A comprehensive analysis of selected predictions by GraphSite demonstrates that its high performance is a result of the ability to effectively learn the underlying patterns of various types of binding pockets. We would like to note that the current GraphSite employing a GNN model to classify ligand binding sites is distinct from another software with the same name that utilizes a graph transformer to predict DNA binding residues in protein structures [42].

2. Materials and Methods

2.1. Datasets of Ligand Binding Pockets

A non-redundant collection of 51,677 pockets were compiled in September 2019 following a protocol developed previously to construct a dataset to evaluate binding site prediction with *e*FindSite [7,43]. Binding ligands in the *e*FindSite dataset were clustered at a Tanimoto coefficient (TC) threshold of 0.7 with the SUBSET program [44]. The 30 most abundant clusters were then manually curated into 14 pocket classes, referred to as the benchmarking dataset. The benchmarking dataset was divided into training (80%) and testing (20%) subsets by randomly splitting each class at a 4:1 ratio. The unseen dataset was created by selecting ligand-bound protein structures deposited to the Protein Data Bank (PDB) [45] no earlier than October 2019. Those proteins having a sequence identity of $\geq$50% to any protein in the benchmarking dataset were excluded. Pocket classes were assigned based on the chemical similarity of binding ligands to small molecules in the benchmarking dataset at a TC threshold of 0.7. This procedure resulted in 45 unseen pockets assigned to 9 classes. Finally, as the negative dataset, we use a previously published collection of 42 surface pockets resembling binding sites but not known to bind any ligand [46].

The *e*FindSite collection of ligand binding pockets [7,43] was first clustered by ligand chemical similarity and then the 30 most abundant clusters were manually curated into a dataset of 14 pocket classes. Clusters containing ATP, adenosine diphosphate (ADP), phosphoaminophosphonic acid-adenylate ester (ANP), uridine monophosphate (UMP), thymidine monophosphate (TMP), nicotinamide adenine dinucleotide, adenosine, azamethionine-5′-deoxyadenosine, and β-D-erythrofuranosyl adenosine, were merged to form class 0 (nucleotides). Further, clusters composed of glucose, fructose, α-D galactopyranose, and manopyranose, were combined into class 2 (carbohydrates). Another merged class 5 comprises phosphocholine, bromododecanol, tetradecylpropanedioic acid, oleic acid, palmitic acid, and hexaenoic acid. Clusters containing amino acids, such as lysine, arginine, and norvaline, citric acid and its derivatives, tartaric acid, tetraglycine phosphinate, and 1,3 dihydroxyacetone phosphate were joined to class 6. Finally, class 10 includes methylbenzamide, pentanamide, hexaethylene glycol, and tetraethylene glycol. The remaining clusters were sufficiently distinct to become separate classes. The clustering procedure followed by a manual data curation resulted in the benchmarking dataset of 21,124 pockets assigned to 14 classes binding a variety of ligands listed in Table 1.

Table 1. Classes of ligand binding sites in the primary benchmarking dataset. Support is the number of pockets in the dataset.

Class	Binding Ligands	Support
0	nucleotide	7625
1	heme	1158
2	carbohydrate	3001
3	benzene ring	1054
4	chlorophyll	968
5	lipid	1890
6	essential amino/citric/tartaric acids	1663
7	S-adenosyl-L-homocysteine	602
8	coenzyme A	573
9	pyridoxal phosphate	566
10	benzoic acid	897
11	flavin mononucleotide	417
12	morpholine ring	374
13	phosphate	337

2.2. Graph Representation of Binding Sites

Ligand binding pockets are converted to graphs, which are the input for the classifier. The nodes of these graphs are atoms contacting ligands identified through the analysis of interatomic contacts with the Ligand-Protein Contacts (LPC) software [47]. Nodes are connected by undirected edges when the distance between two atoms is $\leq$4.5 Å. We employ 11 node features, 7 of which are spatial features, and the other 4 are physicochemical/evolutionary features. Spatial features defining the shape of binding pockets include atomic Cartesian coordinates (x, y, z), spherical coordinates (r, ϑ, γ), and the solvent accessible surface area (SASA). Physicochemical/evolutionary features comprising charge, hydrophobicity, binding probability, and sequence entropy have been previously used in Bionoi, a method to represent ligand binding sites as Voronoi diagrams [48]. To distinguish between various bonding and non-bonding interactions, the bond multiplicity is used as the edge attribute with the value of 1.5 for aromatic bonds and 0 for non-covalent interactions.

Figure 1 illustrates the procedure to transform pockets into graphs. Atoms of binding residues become nodes connected to neighboring nodes within a distance threshold of 4.5 Å. To distinguish between bonding and non-bonding interactions, the edge attribute is set to either the bond multiplicity if two atoms form a chemical bond or 0 for those atoms interacting non-covalently. Individual nodes are assigned two types of features, spatial features defining the shape of the binding pocket (atomic coordinates and the solvent accessible surface area) and physicochemical/evolutionary features describing various properties, such as the charge, the hydrophobicity, the binding probability, and the sequence entropy. Representing pockets as graphs captures their overall characteristics and enables the information flow between atoms during the GNN model training.

2.3. Graph Neural Network

As pockets are represented as graphs, the binding site classification task becomes a graph classification problem essentially. A general graph classification framework employing a GNN incorporates three key components, message passing, the graph readout, and the classification stage. The overall architecture of a classifier implemented in GraphSite is presented in Figure 2. The main module consists of an embedding network (Figure 2B–D) comprising message passing layers (Figure 2B), the jumping knowledge connections (Figure 2C), and a global pooling layer to perform the graph readout (Figure 2D). As illustrated in Figure 2B, the node features of the input graph are first iteratively updated by neural weighted message (NWM) passing layers h_ω taking the edge attribute $\mathbf{e}_{12}$ as input to generate $\mathbf{a}_{12}$ as the weight of a message propagating from node 2 to node 1. Subsequently, the jumping knowledge network (JK-Net) [49] connecting message passing layers is employed, allowing the model to learn the optimal number of layers for individual nodes. The generated outputs are then processed by a max pooling layer performing a feature-wise pooling. The max pooling layer is followed by a global pooling layer to reduce the node feature dimension to a fixed-size vector, which is passed to a set of fully connected layers to generate the final classification result (Figure 2E).

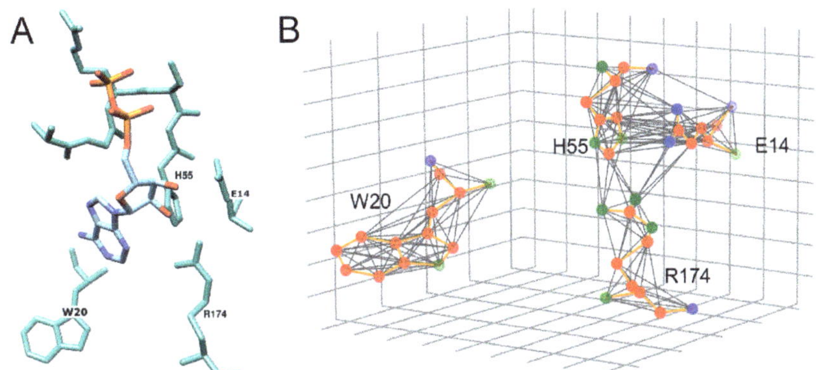

Figure 1. Example of the graph representation of a binding site. (**A**) The structure of a binding pocket for ADP in DnaA regulatory inactivator Had from *E. coli* (PDB-ID: 5x06). (**B**) The graph representation of four residues, W20, R174, E14, and R53, selected from (**A**).

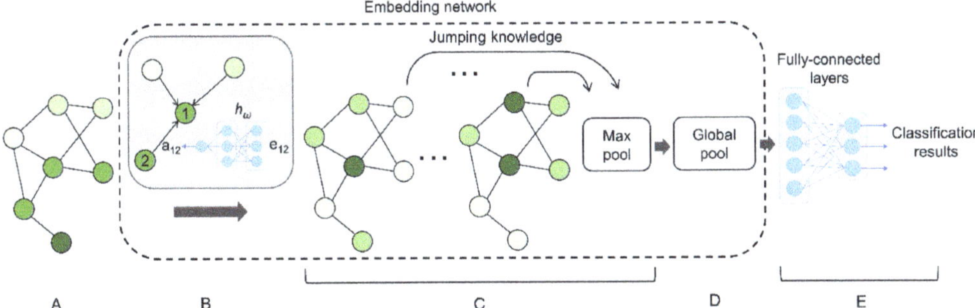

Figure 2. Architecture of the pocket classifier in GraphSite. (**A**) The input graph represents a binding site. (**B**) A neural network computing the weight for message passing from the edge attributes of the input graph. (**C**) Message passing layers of the jumping knowledge network. (**D**) A global pooling layer implementing the Set2Set model. (**E**) Fully connected layers generate the final classification results.

2.3.1. Message Passing

The role of message passing layers of the GNN is to update node features by propagating the information along edges. Node features updated with the information aggregated from neighbors contain valuable local patterns. Message passing layers in GraphSite adopt the general form of the neighborhood aggregation [50]:

$$\mathbf{x}_i^{(k)} = \lambda\left(\mathbf{x}_i^{(k-1)}, \underset{j\in\mathcal{N}(i)}{aggr}\,\phi\left(\mathbf{x}_i^{(k-1)}, \mathbf{x}_j^{(k-1)}, \mathbf{e}_{ij}\right)\right), \tag{1}$$

where ϕ is a differentiable function generating a message, $aggr$ is a permutation-invariant function aggregating all messages, and λ is the updating function. Other parameters are $\mathbf{x}_i^{(k)}$ corresponding to the output feature vector of node i in layer k, $\mathbf{x}_j^{(k)}$ representing feature vectors of the neighbors of node i, and the edge attribute $\mathbf{e}_{ij}$. To better exploit node and edge features of binding site graphs, we implemented the following single-channel NWM:

$$\mathbf{x}_i^{(k)} = h_\theta\left((1+\epsilon)\cdot\mathbf{x}_i^{(k-1)} + \sum_{j\in\mathcal{N}(i)} h_\omega\left(\mathbf{e}_{ij}\right)\cdot\mathbf{x}_j^{(k-1)}\right), \tag{2}$$

where h_ω is an MLP taking the edge attribute as the input and outputting a message weight, which is a node feature j, ϵ is a learnable scalar, and h_θ is another MLP updating the aggregated information. Edge attributes are the same for all layers and are not updated during training. The NWM message passing rule can be regarded as an extension of the graph isomorphism network (GIN) [51], an expressive message passing model that is as powerful as the Weisfeiler–Lehman test in distinguishing graph structures. Its *sum* aggregator is replaced in GraphSite by the sum of weighted messages with weights generated by a neural network h_ω. From another perspective, the NWM model belongs to the message passing neural network (MPNN) family [38]. The gated graph neural network (GGNN) is an MPNN family member whose message is formed by $\mathbf{A}_{\mathbf{e}_{ij}}\mathbf{x}_j^{(k)}$, where $\mathbf{A}_{\mathbf{e}_{ij}}$ is a square transformation matrix generated by a multilayer perceptron (MLP) from the edge attribute $\mathbf{e}_{ij}$. The GGNN can be regularized to the NWM by imposing a restriction on the matrix $\mathbf{A}_{\mathbf{e}_{ij}}$ to make it diagonal with all elements on the diagonal equal. We found empirically that the regularization of GGNN to NWM is not only computationally more efficient but also helps mitigate model overfitting.

Finally, inspired by the idea that multiple aggregators can improve the expressiveness of GNNs [52], we extended the single-channel NWM layer described by Equation (2) to a multi-channel NWM layer by concatenating the outputs of multiple aggregators:

$$\mathbf{x}_i^{(k)} = h_\theta\left(\underset{c\in Channels}{concat}\left((1+\epsilon_c)\cdot\mathbf{x}_i^{(k-1)} + \sum\nolimits_{j\in\mathcal{N}(i)} h_{\omega c}\left(\mathbf{e}_{ij}\right)\cdot\mathbf{x}_j^{(k-1)}\right)\right), \tag{3}$$

where ϵ_c and $h_{\omega c}$ represent an aggregator learned as channel c. The aggregated node features are concatenated in their last dimension so that the concatenated node features have the shape of n by $d \times |C|$, where d is the dimension of node features. The updated neural network h_θ also acts as a reduction function, decreasing the size of node features from $d \times |C|$ to d. Intuitively, the concatenation of multiple aggregators in the GNN is analogous to using multiple filters in the CNN; each aggregator corresponds to a filter, and the concatenated output is equivalent to the output feature maps in the convolution layer of the CNN.

2.3.2. Graph Readout

A graph readout function reduces the size of a graph to a single node. GraphSite employs Set2Set [53] as a global pooling function to perform graph readout. Set2Set generates fixed-sized embeddings for sets of various sizes by utilizing the attention mechanism to compute the global representation of a set. Briefly, a long short-term memory (LSTM) [54] neural network recurrently updates a global hidden state of the input set. During the recurrent process, the global hidden state is used to compute attention values associated with each element in the set, which are in turn used to update the global hidden state. After several iterations, a global graph representation is created by concatenating the global hidden state constructed by the LSTM and the weighted sum of elements in the set. The global pooling layer reduces the node feature dimension from $n \times d$ to d, where n is the number of nodes and d is dimension of the node feature vector.

2.3.3. Loss Function

The dataset of ligand binding pockets is imbalanced, meaning that some classes, such as nucleotide, have many more data points than other classes. Consequently, a training mini batch contains mostly the data from major classes, which could bias a typical loss function utilizing the cross-entropy. To mitigate this problem, GraphSite employs the focal loss (FL) function adding a damping factor $(1 - p_t)^\gamma$ to the cross-entropy loss [55]:

$$FL(p_t) = -(1 - p_t)^\gamma \log(p_t), \tag{4}$$

where p_t is the predicted probability generated by the softmax function, and $\gamma \geq 0$ is a tunable hyperparameter. With this damping factor, dominating predictions with high probabilities are suppressed, while those predictions having low probabilities are assigned higher weights. This approach has been shown to minimize the problem of imbalanced classes.

2.4. Other Methods to Classify Pockets

A docking-based approach employs a small library of 14 ligands, each representing one class of pockets listed in Table 1. These compounds are docked to a query pocket with a molecular docking program smina [56] and the class of a molecule with the best docking score is assigned to that pocket. A pocket matching-based approach scans a query pocket against a small library of 14 representative pockets for all classes in Table 1 with a local structure alignment program G-LoSA [57]. The query pocket is then assigned a class from the library pocket having the best matching score. A random classifier randomly assigns the query pocket with a class according to the frequencies of individual classes in the dataset.

3. Results

3.1. Classification Performance against the Benchmarking Dataset

The performance of GraphSite is compared to that of several other approaches, GIN, molecular docking, pocket matching, and a random classifier. The GIN is an expressive message passing model, shown to be as powerful as the Weisfeiler–Lehman algorithm in distinguishing graph structures [51]. As the GIN employs a sum aggregator ignoring edge attributes, it constitutes an appropriate baseline to demonstrate the benefit of taking advantage of edge attributes in GraphSite with the NWM model. To conduct a fair comparison, the configurations of GraphSite and GIN are identical, except for the architecture of GNN layers. In addition to GNN-based classifiers, we also include docking- and pocket matching-based approaches. The former method employs smina [56], a fork of AutoDock Vina [58] featuring improved scoring and minimization, whereas pocket matching is conducted with G-LoSA, a tool to align protein local structures in a sequence order independent way [57].

After training, Graphsite and GIN is tested on the testing split of the dataset. Training the GraphSite classifier on Nvidia V100 GPU for 200 epochs took about 5 h. The classification performance of all tested methods on the testing subset is reported in Table 2. GraphSite achieves the best overall classification accuracy with a high recall of 81.3% and F1-score of 81.7%. Both recall and F1-score for the GIN are lower, therefore, utilizing edge attributes with multi-channel NWM layers indeed improves the classification accuracy over GIN layers. The performance of docking- and pocket matching-based approaches assessed by the recall and F1-score is comparable to that of a random classifier. Despite this low sensitivity, both techniques achieve relatively high precision, corresponding to a high fraction of correctly classified instances among all pockets. We note that docking and pocket matching were executed with default parameters because it is impractical to apply these algorithms exhaustively to increase the classification accuracy further.

Figure 3 shows the confusion matrix calculated for GraphSite predictions against the benchmarking dataset, in which numbers on the diagonal are recall values for ligand classes. Although GraphSite correctly predicted most classes, it misclassified a few pockets as well. There are two main reasons for these misclassifications. First, the support for some pocket classes across the dataset is low; for instance, only 1.8% of instances belong to class 12 and 1.6% to class 13 (Table 1). As more gradients are generated for the majority of classes during training, the model learns these classes more efficiently. Although this issue can partially be mitigated by employing the focal loss [55], the performance of minority classes is still going to be somewhat lower compared to those classes having stronger support. The second reason is that ligands binding to pockets belonging to different classes can, in fact, contain similar chemical moieties. We discuss several representative examples of these misclassifications in the following section.

Table 2. Classification performance against the benchmarking dataset. GraphSite is compared to the graph isomorphism network (GIN), molecular docking with smina, pocket matching with G-LoSA, and a random classifier. Precision, recall, and F1-score are class-weighted.

Method	Recall	Precision	F1-Score
GraphSite	81.3%	82. 3%	81.7%
GIN	75.1%	74.3%	74.3%
Smina	16.7%	43.4%	16.1%
G-LoSA	14.8%	34.4%	15.9%
Random	17.8%	17.7%	17.7%

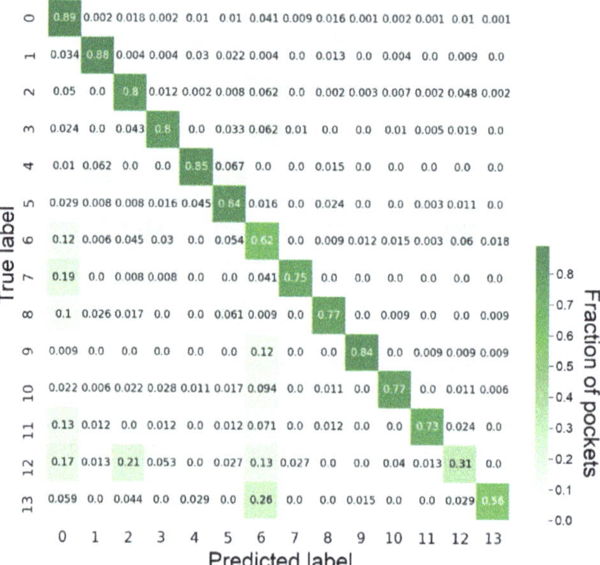

Figure 3. Confusion matrix for classification with GraphSite on the benchmarking dataset. Each row of the confusion matrix is normalized. Numbers on the diagonal correspond to the recall of each class, while other numbers indicate the fraction of misclassified pockets.

3.2. Examples of Misclassified Pockets

Class 12 comprises pockets binding ligands containing morpholine rings, 17% of which are misclassified as nucleotides (Figure 3). Examples of these molecules are commonly used organic buffering agents [59], such as piperazine-N,N′-bis(2-ethanesulfonic acid) (PIPES). GraphSite classified a binding site in centromere-associated protein E (CENP-E) complexed with PIPES (PDB-ID: 1t5c) [60] as a nucleotide-binding pocket with a confidence score of 0.96. This prediction can be validated by structurally aligning the CENP-E pocket with a known nucleotide binding site. Here, we selected the ATP binding site in phosphoribosylformylglycinamidine (FGAM) synthase II (PDB-ID: 2hs0) [61], whose sequence identity with CENP-E is only 21%. Ligand binding sites in both proteins were aligned with PocketAlign, which employs shape descriptors in the form of geometric perspectives, supplemented by chemical group classification, to compute sequence order-independent alignments [62]. Figure 4A shows the superposition of binding sites in CENP-E (purple) and FGAM synthase II (yellow). Encouragingly, the root-mean-square deviation over Cα atoms (Cα-RMSD) of 9 equivalent residues is as low as 1.6 Å. Generally, values below 3.0 Å indicate that the aligned pockets are structurally similar [62].

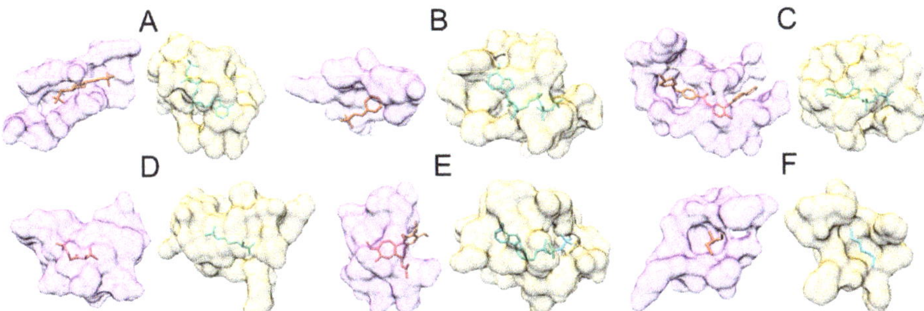

Figure 4. Structure alignments between misclassified pockets and those belonging to the predicted class. (**A**) PIPES (orange sticks) binding site in CENP-E (purple surface) and ATP (cyan sticks) binding site in FGAM synthase II (yellow surface). (**B**) MES (orange sticks) binding site in zitR (purple surface) and ATP (cyan sticks) binding site in FGAM synthase II (yellow surface). (**C**) Imatinib (orange sticks) binding site in ANC-AS (purple surface) and ATP (cyan sticks) binding site in FGAM synthase II (yellow surface). (**D**) (3R)-3-hydroxy-2,4-dioxopentyl dihydrogen phosphate (orange sticks) binding site in LsrF (purple surface) and arginine (cyan sticks) binding site in AT (yellow surface). (**E**) Colchicine (orange sticks) binding site in BRD4 (purple surface) and ATP (cyan sticks) binding site in FGAM synthase II (yellow surface). (**F**) Tromethamine (orange sticks) binding site in MAT (purple surface) and di(hydroxyethyl)ether (cyan sticks) binding site in BtR318A (yellow surface).

Another example is 2-(N-morpholino)ethanesulfonic acid (MES) containing the morpholine ring that is structurally related to the piperazine ring with one nitrogen atom replaced by oxygen [63]. GraphSite classified a binding pocket in zinc transport transcriptional regulator (zitR) complexed with MES (PDB-ID: 5yhz) [64] as a nucleotide-binding pocket with a confidence score of 0.97. Figure 4B shows that this pocket (purple) is structurally related to the ATP binding site in FGAM synthase II (yellow) with 1.5 Å Cα-RMSD over 6 equivalent residues reported by PocketAlign. Note that the global sequence identity between zitR and FGAM is only 20%. Piperazine and morpholine rings are often used to develop molecules competing with nucleotides. For instance, morpholinos, nucleotide analogs blocking mRNA splicing and translation [65], contain the morpholine ring replacing the sugar group of a nucleotide [66]. Further, morpholine-containing pyrazolopyrimidines are selective and potent ATP-competitive inhibitors of mTOR, showing anti-cancer properties in xenograft tumor models [67]. ATP-competitive inhibitors often contain piperazine rings to increase their aqueous solubility [68] and to form favorable interactions with the hinge region of protein kinases [69].

An example of the ATP-competitive inhibitor containing piperazine is imatinib, a widely used chemotherapeutic to treat certain types of cancer [70]. Piperazine and benzene rings in imatinib are required for their inhibitory activity against leukemia cell lines [10]. A binding site in Src-Abl tyrosine kinase ancestor (ANC-AS) complexed with imatinib (PDB-ID: 4csv) [71] was classified by GraphSite as a nucleotide-binding pocket with a confidence of 0.99. Despite a low sequence identity between ANC-AS and FGAM synthase II of 23%, PocketAlign aligned their binding sites with a Cα-RMSD of 1.8 Å over 17 equivalent residues (Figure 4C, ANC-AS is purple and FGAM yellow), indicating that both pockets can bind similar ligands. Indeed, ANC-AS has also been co-crystallized with ATP (PDB-ID: 4ueu); therefore, the classification by GraphSite is, in fact, correct. This is an example of a pocket capable of binding multiple, chemically dissimilar ligands, which may belong to more than one class.

GraphSite classified 26% of pockets binding alkyl phosphates belonging to class 13 as binding sites for essential amino acids (Figure 3). For instance, a binding site in a coenzyme A-dependent thiolase LsrF bound to (3R)-3-hydroxy-2,4-dioxopentyl dihydrogen phosphate (PDB-ID: 4p2v) [72] was classified as an essential amino acid binding pocket with 0.96 confidence. Figure 4D shows a valid structure alignment constructed by PocketAlign between this pocket (purple) and a known amino acid binding pocket in L-arginine:glycine amidinotransferase (AT, yellow) complexed with arginine (PDB ID:4jdw) [73]. This alignment has a Cα-RMSD of 1.5 Å calculated over 14 equivalent residues indicating that the binding site in LsrF is structurally related to arginine binding pockets. As a matter of fact, alkyl phosphates and amino acids are connected through common biochemical pathways, e.g., phosphoenol pyruvate is an important citric acid cycle intermediate that produces alpha-ketoglutarate, ultimately leading to the synthesis of amino acid arginine [74,75]. This may explain the classification result by GraphSite of the binding site in LsrF.

Colchicine is an anti-inflammatory agent primarily used to treat gout [76]. A colchicine binding site in human bromodomain-containing protein 4 (BRD4, PDB-ID: 6ajz) [77] was classified by GraphSite as a nucleotide binding site with a confidence score of 0.93. Interestingly, BRD4 is homologous to the murine mitotic chromosome-associated protein [78] and the human RING3 protein [79], both annotated with kinase activity. Colchicine is also effective against acute coronary syndrome by inhibiting a nucleotide-binding domain (NOD)-like receptor protein 3 inflammasome protein complex [80]. The colchicine binding site in BRD4 was aligned to a known ATP binding site in FGAM synthase II with PocketAlign. The resulting alignment shown in Figure 4E has a low Cα-RMSD of 1.7 Å over 9 equivalent residues (BRD4 is purple and FGAM is yellow). This result indicates that both pockets are structurally similar, explaining the classification by GraphSite of the pocket in BRD4 as nucleotide binding.

A few pockets binding essential amino/citric/tartaric acids belonging to class 6 were classified by GraphSite as binding sites for lipids (Figure 3). An example is a pocket in maltose O-acetyltransferase from *E. coli* binding tromethamine (MAT, PDB-ID: 6ag8) [81] assigned by GraphSite to class 5 with a confidence score of 0.98. MAT catalyzes the CoA-dependent transfer of an acetyl group to maltose and other sugars [82]. The fatty acid or lipid biosynthesis pathway produces acetyl CoA that enters the citric acid cycle to produce citrate [83]. According to results by PocketAlign shown in Figure 4F, the binding site in MAT is structurally similar to a pocket in putative endonuclease/exonuclease/phosphatase family protein binding di(hydroxyethyl)ether (BtR318A, PDB-ID: 3mpr) [84] with an RMSD of 1.5 Å over 8 equivalent residues (MAT is purple and BtR318A is yellow). This high similarity to a lipid-binding site gives a reason for the misclassification of a pocket in MAT by GraphSite.

3.3. Performance on Unseen Data

Next, the performance of GraphSite is evaluated against a small dataset of "unseen" pockets. All data in this set were published later than the benchmarking dataset; thus, these pockets have not been used to train the machine learning model. In addition, the unseen dataset comprises only those proteins having low homology to benchmarking proteins. Encouragingly, using GraphSite yields the weighted recall, precision, and F1-score against the unseen dataset of 68.9%, 75.5%, and 70.7%, respectively. Although these values are somewhat lower than those reported in Table 2, the performance of GraphSite is still satisfactory considering that the unseen dataset is smaller and much more challenging than the benchmarking dataset. GraphSite is expected to achieve such performance in real-world applications employing new data.

3.4. Classification of the Negative Dataset

Lastly, GraphSite was applied to the negative dataset of surface pockets having characteristics of binding sites yet not binding any ligands [46]. Figure 5 shows that the distribution of the classification confidence is diametrically different from that obtained for the benchmarking dataset. A purple violin plot on the left shows the distribution of the probability of the top-ranked class predicted by GraphSite for the benchmarking dataset. The median probability of 0.93 indicates that the model produced not only accurate but also highly confident predictions for the benchmarking dataset. Note that this performance was obtained employing a proper cross-validation protocol. In contrast, predictions for the negative dataset are clearly less confident, with a median probability of only 0.67. These results demonstrate that even though non-binding sites were classified into 14 classes as GraphSite was designed for, unconfident predictions indicate that these surface pockets do not fit well any ligand class the model was trained against.

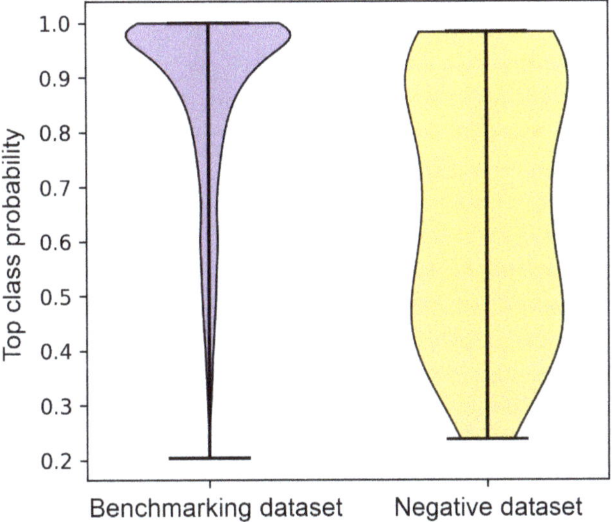

Figure 5. Distribution of the classification confidence for benchmarking and negative datasets. The classification confidence corresponds to a probability of the top-ranked ligand binding class predicted by GraphSite.

3.5. *Siamese-GraphSite Extension*

In addition to the classifier model, we extended GraphSite by adding a Siamese model for metric learning. This model generates two graph embeddings for a pair of input graphs, which are then used to calculate the contrastive loss (CL) [85]:

$$CL(W, y, \mathbf{x}_1, \mathbf{x}_2) = \frac{1}{2}(1-y)(d_W)^2 + \frac{1}{2}(y)(max(0, m - d_W))^2 \quad (5)$$

where y is the label of the pair of input graphs $\mathbf{x}_1$ and $\mathbf{x}_2$ (either 1—similar or 0—dissimilar), W parameterizes the embedding network, d_W is the Euclidean distance between graph embeddings, and $m > 0$ is a distance margin for the input pair to contribute to the loss function. Intuitively, using the contrastive loss in model training results in embeddings from the same class being close to one another in the Euclidean space and far away from each other for embeddings belonging to different classes.

As shown in Figure 6, embedding networks with shared parameters require a pair of graphs representing binding pockets as the input to generate two graph embeddings. These embeddings can subsequently be used in various machine learning applications, such as the visualization of the binding pocket conformational space. As this architecture optimizes the relative distances of the data in the Euclidean space, embeddings generated by Siamese-GraphSite are well suited for distance-based analyses, including, for instance, t-distributed stochastic neighbor embedding (t-SNE) visualization [86] and k-nearest neighbor clustering [87].

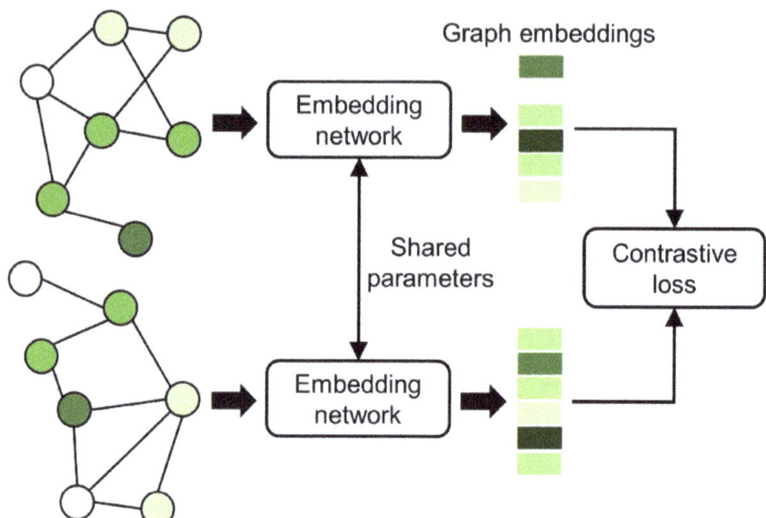

Figure 6. Architecture of Siamese-GraphSite. This model requires a pair of graph-structured data as the input for two embedding networks sharing their parameters and utilizes the contrastive loss function.

To test the distance metric learning on weakly supervised data, we trained Siamese-GraphSite against 8 clusters in the original dataset prior to the manual curation. Figure 7 shows the t-SNE visualization of the clusters from the validation subset (10%) after the model was trained on the remaining subset (90%). Overall, similar pockets are grouped together, while dissimilar pockets are located away from one another. Interestingly, clusters 0 (green dots in Figure 7) and 3 (orange dots in Figure 7) come together according to the t-SNE analysis. The former cluster contains ADP and ANP, whereas the latter is composed of UMP and TMP. Because of the functional similarity of pockets belonging to these clusters, both groups were merged during the manual curation of the dataset into a single class 0 comprising nucleotides (Table 1). Similarly, clusters 3 (red dots in Figure 7) and 8 (yellow dots in Figure 7) are grouped together. These clusters containing glucose and fructose ligands were also manually curated into a single class 2 composed of carbohydrates (Table 1). These observations indicate that the Siamese model effectively learns embeddings to represent functional relations among binding pockets in line with the human expert knowledge.

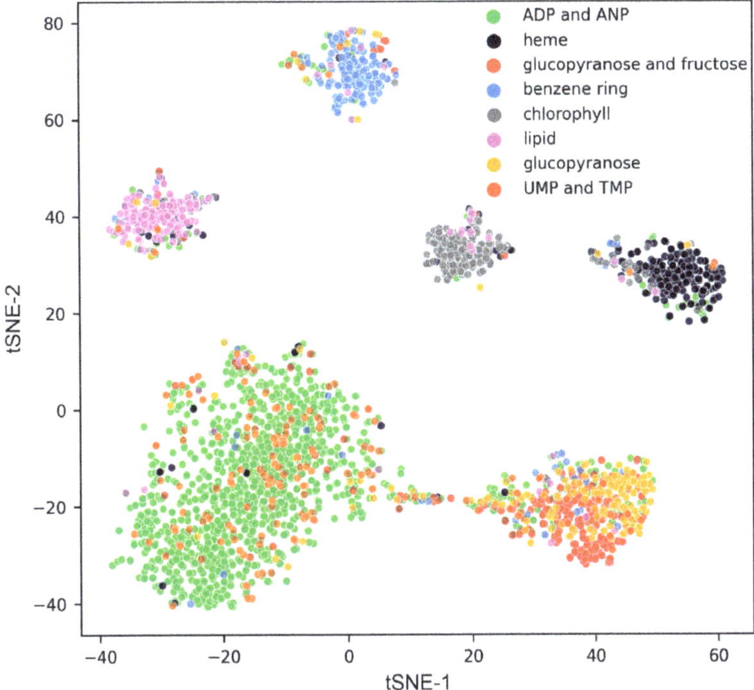

Figure 7. t-SNE visualization of embeddings generated by Siamese-GraphSite. Each dot represents one pocket colored by the cluster assignment.

4. Discussion

In this communication, we describe GraphSite, a method to classify ligand binding sites, represented as graphs, with a graph deep learning model. Comprehensive benchmarking calculations demonstrate that the trained classifier extracts informative features of binding pockets yielding state-of-the-art classification performance. Importantly, GraphSite successfully classifies binding sites without any information on their ligands. It has the desired capability to generalize to unseen data, as shown for an independent dataset of pockets taken from proteins having low homology and solved posterior to training structures. Moreover, calculations conducted for the negative dataset of surface pockets not binding any ligands demonstrate that GraphSite does not overpredict; therefore, the false positive rate in real applications should be low.

GraphSite can further be extended in several directions. Utilizing larger datasets comprising more classes will not only help train a more powerful and accurate classifier, but it will also increase the performance of metric learning by the Siamese model presented here as an example of the extension of GraphSite. However, this plan of action would require employing various data augmentation techniques [31] to account for fewer structures currently available for certain pocket classes. We also expect that exploring additional node features of binding site graphs may also improve the classification performance. GraphSite is a versatile approach that can be useful in other deep learning-based applications involving the analysis of ligand binding sites. For example, it is possible to train a graph autoencoder to generate latent embeddings of binding sites for subsequent use in machine learning. Another potential application is to build a model to predict drug-target interactions where the GNN layers of GraphSite can be used as the feature extractor for input binding sites. These new directions to improve and extend GraphSite will be explored in the future.

Author Contributions: Conceptualization, W.S., L.P. and M.B.; data curation, W.S., M.S., G.S. and M.B.; methodology, W.S. and L.P.; software implementation, W.S.; case studies, M.S.; funding acquisition, J.R. and M.B.; supervision, M.B. and J.R.; manuscript draft, W.S. and M.S.; final manuscript, M.B. All authors have read and agreed to the published version of the manuscript.

Funding: This work has been supported in part by the National Institute of General Medical Sciences of the National Institutes of Health award R35GM119524, the US National Science Foundation award CCF1619303, the Louisiana Board of Regents contract LEQSF(2016-19)-RD-B03, and by the Center for Computation and Technology at Louisiana State University.

Data Availability Statement: GraphSite is available at https://github.com/shiwentao00/Graphsite-classifier and datasets are available at https://osf.io/svwkb/, accessed on 18 July 2022.

Acknowledgments: Portions of this research were conducted with high-performance computational resources provided by Louisiana State University.

Conflicts of Interest: The authors declare no conflict of interests.

References

1. Armstrong, J.D.; Hubbard, R.E.; Farrell, T.; Maiguashca, B. (Eds.) *Structure-Based Drug Discovery: An Overview*; The Royal Society of Chemistry: Cambridge, UK, 2006.
2. Roche, D.B.; Brackenridge, D.A.; McGuffin, L.J. Proteins and Their Interacting Partners: An Introduction to Protein–Ligand Binding Site Prediction Methods. *Int. J. Mol. Sci.* **2015**, *16*, 29829–29842. [CrossRef] [PubMed]
3. Vos, T.; Lim, S.S.; Abbafati, C.; Abbas, K.M.; Abbasi, M.; Abbasifard, M.; Abbasi-Kangevari, M.; Abbastabar, H.; Abd-Allah, F.; Abdelalim, A.; et al. Global burden of 369 diseases and injuries in 204 countries and territories, 1990–2019: A systematic analysis for the Global Burden of Disease Study 2019. *Lancet* **2020**, *396*, 1204–1222. [CrossRef]
4. Govindaraj, R.G.; Naderi, M.; Singha, M.; Lemoine, J.; Brylinski, M. Large-scale computational drug repositioning to find treatments for rare diseases. *npj Syst. Biol. Appl.* **2018**, *4*, 13. [CrossRef] [PubMed]
5. Hendlich, M.; Rippmann, F.; Barnickel, G. LIGSITE: Automatic and efficient detection of potential small molecule-binding sites in proteins. *J. Mol. Graph. Model.* **1997**, *15*, 359–363. [CrossRef]
6. Ngan, C.-H.; Hall, D.R.; Zerbe, B.S.; Grove, L.E.; Kozakov, D.; Vajda, S. FTSite: High accuracy detection of ligand binding sites on unbound protein structures. *Bioinformatics* **2011**, *28*, 286–287. [CrossRef]
7. Brylinski, M.; Feinstein, W.P. eFindSite: Improved prediction of ligand binding sites in protein models using meta-threading, machine learning and auxiliary ligands. *J. Comput. Mol. Des.* **2013**, *27*, 551–567. [CrossRef]
8. Le Guilloux, V.; Schmidtke, P.; Tuffery, P. Fpocket: An open source platform for ligand pocket detection. *BMC Bioinform.* **2009**, *10*, 168. [CrossRef]
9. Lin, Y.; Yoo, S.; Sanchez, R. SiteComp: A server for ligand binding site analysis in protein structures. *Bioinformatics* **2012**, *28*, 1172–1173. [CrossRef]
10. Araki, M.; Iwata, H.; Ma, B.; Fujita, A.; Terayama, K.; Sagae, Y.; Ono, F.; Tsuda, K.; Kamiya, N.; Okuno, Y. Improving the Accuracy of Protein-Ligand Binding Mode Prediction Using a Molecular Dynamics-Based Pocket Generation Approach. *J. Comput. Chem.* **2018**, *39*, 2679–2689. [CrossRef]
11. Kana, O.; Brylinski, M. Elucidating the druggability of the human proteome with eFindSite. *J. Comput. Mol. Des.* **2019**, *33*, 509–519. [CrossRef]
12. Ngan, C.H.; Bohnuud, T.; Mottarella, S.E.; Beglov, D.; Villar, E.A.; Hall, D.R.; Kozakov, D.; Vajda, S. FTMAP: Extended protein mapping with user-selected probe molecules. *Nucleic Acids Res.* **2012**, *40*, W271–W275. [CrossRef]
13. Skolnick, J.; Gao, M.; Roy, A.; Srinivasan, B.; Zhou, H. Implications of the small number of distinct ligand binding pockets in proteins for drug discovery, evolution and biochemical function. *Bioorganic Med. Chem. Lett.* **2015**, *25*, 1163–1170. [CrossRef]
14. Brylinski, M. Local Alignment of Ligand Binding Sites in Proteins for Polypharmacology and Drug Repositioning. In *Protein Function Prediction*; Humana Press: New York, NY, USA, 2017; Volume 1611, pp. 109–122. [CrossRef]
15. Ehrt, C.; Brinkjost, T.; Koch, O. Impact of Binding Site Comparisons on Medicinal Chemistry and Rational Molecular Design. *J. Med. Chem.* **2016**, *59*, 4121–4151. [CrossRef]
16. Naderi, M.; Lemoine, J.M.; Govindaraj, R.G.; Kana, O.Z.; Feinstein, W.P.; Brylinski, M. Binding site matching in rational drug design: Algorithms and applications. *Briefings Bioinform.* **2018**, *20*, 2167–2184. [CrossRef]
17. Pu, L.; Govindaraj, R.G.; Lemoine, J.M.; Wu, H.-C.; Brylinski, M. DeepDrug3D: Classification of ligand-binding pockets in proteins with a convolutional neural network. *PLOS Comput. Biol.* **2019**, *15*, e1006718. [CrossRef]
18. Konc, J.; Janežič, D. ProBiS-ligands: A web server for prediction of ligands by examination of protein binding sites. *Nucleic Acids Res.* **2014**, *42*, W215–W220. [CrossRef]
19. Panwar, B.; Gupta, S.; Raghava, G.P.S. Prediction of vitamin interacting residues in a vitamin binding protein using evolutionary information. *BMC Bioinform.* **2013**, *14*, 44. [CrossRef]
20. Chen, K.; Mizianty, M.J.; Kurgan, L. Prediction and analysis of nucleotide-binding residues using sequence and sequence-derived structural descriptors. *Bioinformatics* **2011**, *28*, 331–341. [CrossRef]

21. Liu, R.; Hu, J. HemeBIND: A novel method for heme binding residue prediction by combining structural and sequence information. *BMC Bioinform.* **2011**, *12*, 207. [CrossRef]
22. Mishra, N.K.; Raghava, G.P. Prediction of FAD interacting residues in a protein from its primary sequence using evolutionary information. *BMC Bioinform.* **2010**, *11*, S48. [CrossRef]
23. Chauhan, J.S.; Mishra, N.K.; Raghava, G.P. Prediction of GTP interacting residues, dipeptides and tripeptides in a protein from its evolutionary information. *BMC Bioinform.* **2010**, *11*, 301. [CrossRef]
24. Ansari, H.R.; Raghava, G.P. Identification of NAD interacting residues in proteins. *BMC Bioinform.* **2010**, *11*, 160. [CrossRef]
25. Horst, J.A.; Samudrala, R. A protein sequence meta-functional signature for calcium binding residue prediction. *Pattern Recognit. Lett.* **2010**, *31*, 2103–2112. [CrossRef]
26. Shu, N.; Zhou, T.; Hovmöller, S. Prediction of zinc-binding sites in proteins from sequence. *Bioinformatics* **2008**, *24*, 775–782. [CrossRef] [PubMed]
27. He, K.; Gkioxari, G.; Dollár, P.; Girshick, R. Mask r-cnn. In Proceedings of the IEEE International Conference on Computer Vision, Venice, Italy, 22–29 October 2017.
28. Schroff, F.; Kalenichenko, D.; Philbin, J. Facenet: A unified embedding for face recognition and clustering. In Proceedings of the IEEE Conference on Computer Vision and Pattern Recognition, Boston, MA, USA, 7–12 June 2015.
29. Wei, S.E.; Ramakrishna, V.; Kanade, T.; Sheikh, Y. Convolutional pose machines. In Proceedings of the IEEE Conference on Computer Vision and Pattern Recognition, Las Vegas, NV, USA, 27–30 June 2016.
30. Bronstein, M.M.; Bruna, J.; LeCun, Y.; Szlam, A.; Vandergheynst, P. Geometric Deep Learning: Going beyond Euclidean data. *IEEE Signal Process. Mag.* **2017**, *34*, 18–42. [CrossRef]
31. Shi, W.; Lemoine, J.M.; Shawky, A.-E.; Singha, M.; Pu, L.; Yang, S.; Ramanujam, J.; Brylinski, M. BionoiNet: Ligand-binding site classification with off-the-shelf deep neural network. *Bioinformatics* **2020**, *36*, 3077–3083. [CrossRef] [PubMed]
32. Simonovsky, M.; Meyers, J. DeeplyTough: Learning Structural Comparison of Protein Binding Sites. *J. Chem. Inf. Model.* **2020**, *60*, 2356–2366. [CrossRef]
33. Jiménez, J.; Doerr, S.; Martínez-Rosell, G.; Rose, A.; De Fabritiis, G. DeepSite: Protein-binding site predictor using 3D-convolutional neural networks. *Bioinformatics* **2017**, *33*, 3036–3042. [CrossRef]
34. Kipf, T.N.; Welling, M. Semi-supervised classification with graph convolutional networks. *arXiv* **2016**, arXiv:1609.02907.
35. Hamilton, W.; Ying, Z.; Leskovec, J. Inductive representation learning on large graphs. In Proceedings of the 31st Conference on Neural Information Processing Systems, Long Beach, CA, USA, 4–9 December 2017.
36. Li, Y.; Yu, R.; Shahabi, C.; Liu, Y. Diffusion convolutional recurrent neural network: Data-driven traffic forecasting. *arXiv* **2017**, arXiv:1707.01926.
37. Sanchez-Gonzalez, A.; Godwin, J.; Pfaff, T.; Ying, R.; Leskovec, J.; Battaglia, P.W. Learning to simulate complex physics with graph networks. In *International Conference on Machine Learning*; PMLR: London, UK, 2020.
38. Gilmer, J.; Schoenholz, S.S.; Riley, P.F.; Vinyals, O.; Dahl, G.E. Neural message passing for quantum chemistry. In Proceedings of the 34th International Conference on Machine Learning, Sydney, Australia, 6–11 August 2017; pp. 1263–1272.
39. Duvenaud, D.K.; Maclaurin, D.; Iparraguirre, J.; Bombarell, R.; Hirzel, T.; Aspuru-Guzik, A.; Adams, R.P. Convolutional networks on graphs for learning molecular fingerprints. In *Advances in Neural Information Processing Systems*; Curran Associates, Inc.: New York, NY, USA, 2015.
40. Fout, A.; Byrd, J.; Shariat, B.; Ben-Hur, A. Protein interface prediction using graph convolutional networks. In Proceedings of the 31st Conference on Neural Information Processing Systems, Long Beach, CA, USA, 4–9 December 2017.
41. Lim, J.; Ryu, S.; Park, K.; Choe, Y.J.; Ham, J.; Kim, W.Y. Predicting Drug–Target Interaction Using a Novel Graph Neural Network with 3D Structure-Embedded Graph Representation. *J. Chem. Inf. Model.* **2019**, *59*, 3981–3988. [CrossRef]
42. Yuan, Q.; Chen, S.; Rao, J.; Zheng, S.; Zhao, H.; Yang, Y. AlphaFold2-aware protein–DNA binding site prediction using graph transformer. *Briefings Bioinform.* **2022**, *23*, bbab564. [CrossRef]
43. Feinstein, W.P.; Brylinski, M. eFindSite: Enhanced Fingerprint-Based Virtual Screening Against Predicted Ligand Binding Sites in Protein Models. *Mol. Inform.* **2014**, *33*, 135–150. [CrossRef]
44. Voigt, J.H.; Bienfait, B.; Wang, S.; Nicklaus, M.C. Comparison of the NCI Open Database with Seven Large Chemical Structural Databases. *J. Chem. Inf. Comput. Sci.* **2001**, *41*, 702–712. [CrossRef]
45. Berman, H.M.; Westbrook, J.; Feng, Z.; Gilliland, G.; Bhat, T.N.; Weissig, H.; Shindyalov, I.N.; Bourne, P.E. The Protein Data Bank. *Nucleic Acids Res.* **2000**, *28*, 235–242. [CrossRef]
46. Santos, J.C.A.; Nassif, H.; Page, D.; Muggleton, S.H.; Sternberg, M.J.E. Automated identification of protein-ligand interaction features using Inductive Logic Programming: A hexose binding case study. *BMC Bioinform.* **2012**, *13*, 162. [CrossRef]
47. Sobolev, V.; Sorokin, A.; Prilusky, J.; Abola, E.E.; Edelman, M. Automated analysis of interatomic contacts in proteins. *Bioinformatics* **1999**, *15*, 327–332. [CrossRef]
48. Feinstein, J.; Shi, W.; Ramanujam, J.; Brylinski, M. Bionoi: A Voronoi Diagram-Based Representation of Ligand-Binding Sites in Proteins for Machine Learning Applications. *Methods Mol. Biol.* **2021**, *2266*, 299–312. [CrossRef]
49. Xu, K.; Li, C.; Tian, Y.; Sonobe, T.; Kawarabayashi, K.-I.; Jegelka, S. Representation learning on graphs with jumping knowledge networks. *arXiv* **2018**, arXiv:1806.03536.
50. Fey, M.; Lenssen, J.E. Fast graph representation learning with PyTorch Geometric. *arXiv* **2019**, arXiv:1903.02428.
51. Xu, K.; Hu, W.; Leskovec, J.; Jegelka, S. How powerful are graph neural networks? *arXiv* **2018**, arXiv:1810.00826.

52. Corso, G.; Cavalleri, L.; Beaini, D.; Liò, P.; Veličković, P. Principal neighbourhood aggregation for graph nets. *arXiv* **2020**, arXiv:2004.05718.
53. Vinyals, O.; Bengio, S.; Kudlur, M. Order matters: Sequence to sequence for sets. *arXiv* **2015**, arXiv:1511.06391.
54. Hochreiter, S.; Schmidhuber, J. Long short-term memory. *Neural Comput.* **1997**, *9*, 1735–1780. [CrossRef]
55. Lin, T.-Y.; Goyal, P.; Girshick, R.; He, K.; Dollár, P. Focal loss for dense object detection. In Proceedings of the IEEE International Conference on Computer Vision, 23–27 July 2018.
56. Koes, D.R.; Baumgartner, M.; Camacho, C.J. Lessons Learned in Empirical Scoring with smina from the CSAR 2011 Benchmarking Exercise. *J. Chem. Inf. Model.* **2013**, *53*, 1893–1904. [CrossRef]
57. Lee, H.S.; Im, W. G-LoSA: An efficient computational tool for local structure-centric biological studies and drug design. *Protein Sci.* **2016**, *25*, 865–876. [CrossRef]
58. Trott, O.; Olson, A.J. AutoDock Vina: Improving the speed and accuracy of docking with a new scoring function, efficient optimization, and multithreading. *J. Comput. Chem.* **2010**, *31*, 455–461. [CrossRef]
59. Good, N.E.; Winget, G.D.; Winter, W.; Connolly, T.N.; Izawa, S.; Singh, R.M.M. Hydrogen Ion Buffers for Biological Research. *Biochemistry* **1966**, *5*, 467–477. [CrossRef]
60. Garcia-Saez, I.; Yen, T.; Wade, R.H.; Kozielski, F. Crystal Structure of the Motor Domain of the Human Kinetochore Protein CENP-E. *J. Mol. Biol.* **2004**, *340*, 1107–1116. [CrossRef]
61. Velankar, S.S.; Best, C.; Beuth, B.; Boutselakis, C.H.; Cobley, N.; da Silva, A.W.S.; Dimitropoulos, D.; Golovin, A.; Hirshberg, M.; John, M.; et al. PDBe: Protein Data Bank in Europe. *Nucleic Acids Res.* **2009**, *38*, D308–D317. [CrossRef] [PubMed]
62. Yeturu, K.; Chandra, N. PocketAlign A Novel Algorithm for Aligning Binding Sites in Protein Structures. *J. Chem. Inf. Model.* **2011**, *51*, 1725–1736. [CrossRef] [PubMed]
63. Parkin, A.; Oswald, I.D.; Parsons, S. Structures of piperazine, piperidine and morpholine. *Acta. Cryst. B* **2004**, *60*, 219–227. [CrossRef] [PubMed]
64. Zhu, R.; Song, Y.; Liu, H.; Yang, Y.; Wang, S.; Yi, C.; Chen, P.R. Allosteric histidine switch for regulation of intracellular zinc(II) fluctuation. *Proc. Natl. Acad. Sci. USA* **2017**, *114*, 13661–13666. [CrossRef]
65. Summerton, J.; Weller, D. Morpholino Antisense Oligomers: Design, Preparation, and Properties. *Antisense Nucleic Acid Drug Dev.* **1997**, *7*, 187–195. [CrossRef]
66. Moulton, J.D. Morpholino Antisense Oligos. Available online: https://www.gene-tools.com/morpholino_antisense_oligos (accessed on 28 February 2022).
67. Zask, A.; Kaplan, J.; Verheijen, J.C.; Richard, D.J.; Curran, K.; Brooijmans, N.; Bennett, E.M.; Toral-Barza, L.; Hollander, I.; Ayral-Kaloustian, S.; et al. Morpholine Derivatives Greatly Enhance the Selectivity of Mammalian Target of Rapamycin (mTOR) Inhibitors. *J. Med. Chem.* **2009**, *52*, 7942–7945. [CrossRef]
68. Avendaño, C.; Menendez, J.C. Drugs That Inhibit Signalling Pathways for Tumor Cell Growth and Proliferation. *Med. Chem. Anticancer. Drugs* **2008**, 251–305. [CrossRef]
69. Liu, Y.; Wan, W.-Z.; Li, Y.; Zhou, G.-L.; Liu, X.-G. Recent development of ATP-competitive small molecule phosphatidylinostitol-3-kinase inhibitors as anticancer agents. *Oncotarget* **2016**, *8*, 7181–7200. [CrossRef]
70. Wu, Y.J. Chapter 1—Heterocycles and Medicine: A Survey of the Heterocyclic Drugs Approved by the U.S. FDA from 2000 to Present. In *Progress in Heterocyclic Chemistry*; Gordon, W.G., John, A.J., Eds.; Elsevier: Amsterdam, The Netherlands, 2012; Volume 24, pp. 1–53.
71. Wilson, C.; Agafonov, R.V.; Hoemberger, M.; Kutter, S.; Zorba, A.; Halpin, J.; Buosi, V.; Otten, R.; Waterman, D.; Theobald, D.L.; et al. Using ancient protein kinases to unravel a modern cancer drug's mechanism. *Science* **2015**, *347*, 882–886. [CrossRef]
72. Marques, J.C.; Oh, I.K.; Ly, D.C.; Lamosa, P.; Ventura, M.R.; Miller, S.T.; Xavier, K.B. LsrF, a coenzyme A-dependent thiolase, catalyzes the terminal step in processing the quorum sensing signal autoinducer-2. *Proc. Natl. Acad. Sci. USA* **2014**, *111*, 14235–14240. [CrossRef]
73. Humm, A.E.A. Crystal structure and mechanism of human L-arginine:glycine amidinotransferase: A mitochondrial enzyme involved in creatine biosynthesis. *EMBO J.* **1997**, *16*, 3373–3385. [CrossRef]
74. Berg, J.M.; Stryer, L. Amino acids are made from intermediates of the citric acid cycle and other major pathways. In *Biochemistry*, 5th ed.; W. H. Freeman: New York, NY, USA, 2002.
75. Berg, J.M.; Tymoczko, J.L.; Stryer, L. Purine bases can be synthesized de novo or recycled by salvage pathways. In *Biochemistry*; W. H. Freeman: New York, NY, USA, 2002.
76. Probenecid and Colchicine Tablets. USP Rx Only. Available online: https://dailymed.nlm.nih.gov/dailymed/fda/fdaDrugXsl.cfm?setid=842dd93d-54e6-43b8-8bd5-d135fc5a3400&type=display (accessed on 28 February 2022).
77. Yokoyama, T.; Matsumoto, K.; Ostermann, A.; Schrader, T.E.; Nabeshima, Y.; Mizuguchi, M. Structural and thermodynamic characterization of the binding of isoliquiritigenin to the first bromodomain of BRD4. *FEBS J.* **2018**, *286*, 1656–1667. [CrossRef]
78. Dey, A.; Ellenberg, J.; Farina, A.; Coleman, A.E.; Maruyama, T.; Sciortino, S.; Lippincott-Schwartz, J.; Ozato, K. A bromodomain protein, MCAP, associates with mitotic chromosomes and affects G(2)-to-M transition. *Mol. Cell. Biol.* **2000**, *20*, 6537–6549. [CrossRef]
79. Denis, G.V.; Vaziri, C.; Guo, N.; Faller, U.V. RING3 kinase transactivates promoters of cell cycle regulatory genes through E2F. *Cell Growth Differ. Mol. Boil. J. Am. Assoc. Cancer Res.* **2000**, *11*, 417–424.

80. McLoughlin, E.C.; O'Boyle, N.M. Colchicine-Binding Site Inhibitors from Chemistry to Clinic: A Review. *Pharmaceuticals* **2020**, *13*, 8. [CrossRef] [PubMed]
81. Zada, B.; Joo, S.; Wang, C.; Tseten, T.; Jeong, S.-H.; Seo, H.; Sohn, J.-H.; Kim, K.-J.; Kim, S.-W. Metabolic engineering of Escherichia coli for production of non-natural acetins from glycerol. *Green Chem.* **2020**, *22*, 7788–7802. [CrossRef]
82. Brand, B.; Boos, W. Maltose transacetylase of Escherichia coli. Mapping and cloning of its structural, gene, mac, and characterization of the enzyme as a dimer of identical polypeptides with a molecular weight of 20,000. *J. Biol. Chem.* **1991**, *266*, 14113–14118. [CrossRef]
83. Kelly, D.J.; Hughes, N.J. The citric acid cycle and fatty acid biosynthesis. In *Helicobacter Pylori: Physiology and Genetics*; ASM Press: Washington, DC, USA, 2001.
84. Kuzin, A.; Su, M.; Seetharaman, J.; Mao, M.; Xiao, R.; Ciccosanti, C.; Lee, D.; Everett, J.K.; Nair, R.; Acton, T.B.; et al. *Northeast Structural Genomics Consortium Target BtR318A*; Northeast Structural Genomics Consortium (NESG), National Institutes of Health (NIH): Bethesda, MD, USA, 2010.
85. Hadsell, R.; Chopra, S.; LeCun, Y. Dimensionality reduction by learning an invariant mapping. In Proceedings of the 2006 IEEE Computer Society Conference on Computer Vision and Pattern Recognition (CVPR'06), New York, NY, USA, 17–22 June 2006.
86. Van der Maaten, L.; Hinton, G. Visualizing data using t-SNE. *J. Mach. Learn. Res.* **2008**, *9*, 2579–2605.
87. Fix, E.; Hodges, J.L. Discriminatory analysis. Nonparametric discrimination: Consistency properties. *Int. Stat. Rev./Rev. Int. Stat.* **1989**, *57*, 238–247. [CrossRef]

 *biomolecules* 

Review

Protein Data Bank: A Comprehensive Review of 3D Structure Holdings and Worldwide Utilization by Researchers, Educators, and Students

Stephen K. Burley [1,2,3,4,5,*], Helen M. Berman [1,2,5], Jose M. Duarte [4], Zukang Feng [1,2], Justin W. Flatt [1,2], Brian P. Hudson [1,2], Robert Lowe [1,2], Ezra Peisach [1,2], Dennis W. Piehl [1,2], Yana Rose [4], Andrej Sali [6], Monica Sekharan [1,2], Chenghua Shao [1,2], Brinda Vallat [1,2,3], Maria Voigt [1,2], John D. Westbrook [1,2,3,†], Jasmine Y. Young [1,2] and Christine Zardecki [1,2]

1 Research Collaboratory for Structural Bioinformatics Protein Data Bank, Rutgers, The State University of New Jersey, Piscataway, NJ 08854, USA
2 Institute for Quantitative Biomedicine, Rutgers, The State University of New Jersey, Piscataway, NJ 08854, USA
3 Cancer Institute of New Jersey, Rutgers, The State University of New Jersey, New Brunswick, NJ 08901, USA
4 Research Collaboratory for Structural Bioinformatics Protein Data Bank, San Diego Supercomputer Center, University of California San Diego, La Jolla, CA 92093, USA
5 Department of Chemistry and Chemical Biology, Rutgers, The State University of New Jersey, Piscataway, NJ 08854, USA
6 Research Collaboratory for Structural Bioinformatics Protein Data Bank, Department of Bioengineering and Therapeutic Sciences, Department of Pharmaceutical Chemistry, Quantitative Biosciences Institute, University of California San Francisco, San Francisco, CA 94158, USA
* Correspondence: stephen.burley@rcsb.org
† Deceased.

Citation: Burley, S.K.; Berman, H.M.; Duarte, J.M.; Feng, Z.; Flatt, J.W.; Hudson, B.P.; Lowe, R.; Peisach, E.; Piehl, D.W.; Rose, Y.; et al. Protein Data Bank: A Comprehensive Review of 3D Structure Holdings and Worldwide Utilization by Researchers, Educators, and Students. *Biomolecules* **2022**, *12*, 1425. https://doi.org/10.3390/biom12101425

Academic Editors: Cameron Mura and Lei Xie

Received: 30 August 2022
Accepted: 26 September 2022
Published: 4 October 2022

Abstract: The Research Collaboratory for Structural Bioinformatics Protein Data Bank (RCSB PDB), funded by the United States National Science Foundation, National Institutes of Health, and Department of Energy, supports structural biologists and Protein Data Bank (PDB) data users around the world. The RCSB PDB, a founding member of the Worldwide Protein Data Bank (wwPDB) partnership, serves as the US data center for the global PDB archive housing experimentally-determined three-dimensional (3D) structure data for biological macromolecules. As the wwPDB-designated Archive Keeper, RCSB PDB is also responsible for the security of PDB data and weekly update of the archive. RCSB PDB serves tens of thousands of data depositors (using macromolecular crystallography, nuclear magnetic resonance spectroscopy, electron microscopy, and micro-electron diffraction) annually working on all permanently inhabited continents. RCSB PDB makes PDB data available from its research-focused web portal at no charge and without usage restrictions to many millions of PDB data consumers around the globe. It also provides educators, students, and the general public with an introduction to the PDB and related training materials through its outreach and education-focused web portal. This review article describes growth of the PDB, examines evolution of experimental methods for structure determination viewed through the lens of the PDB archive, and provides a detailed accounting of PDB archival holdings and their utilization by researchers, educators, and students worldwide.

Keywords: Protein Data Bank; Open Access; Worldwide Protein Data Bank; macromolecular crystallography; cryogenic electron microscopy; cryogenic electron tomography; electron crystallography; micro-electron diffraction; nuclear magnetic resonance spectroscopy; biological macromolecules; proteins; nucleic acids; DNA; RNA; carbohydrates; small-molecule ligands

1. Introduction

The Protein Data Bank (PDB) is now in its 51st year of continuous operations. As the first open-access digital data resource in biology, it was established in 1971 with just seven protein structures [1]. At the time of writing, PDB holdings numbered nearly 200,000 experimentally-determined three-dimensional (3D) structures of proteins and nucleic acids (DNA and RNA) and their complexes with one another and small-molecule ligands (e.g., enzyme co-factors, drugs, investigational agents). Since 2003, the PDB archive has been jointly managed by the Worldwide Protein Data Bank (wwPDB, wwpdb.org, accessed on 28 August 2022) partnership [2,3]. wwPDB Full Members include the US-funded Research Collaboratory for Structural Bioinformatics Protein Data Bank (RCSB PDB, RCSB.org, [4–7]); Protein Data Bank in Europe (PDBe, PDBe.org, [8]); Protein Data Bank Japan (PDBj, PDBj.org, accessed on 28 August 2022 [9]); the Electron Microscopy Data Bank (EMDB, emdb-empiar.org, accessed on 28 August 2022 [10,11]); and the Biological Magnetic Resonance Bank (BMRB, bmrb.io, accessed on 28 August 2022 [12,13]). The activities of the wwPDB are governed by a charter, which was last renewed in 2021 on the occasion of the accession of EMDB (www.wwpdb.org/about/agreement, accessed on 28 August 2022). The RCSB PDB is headquartered at Rutgers, The State University of New Jersey with smaller teams based at the University of California San Diego (UCSD) and the University of California San Francisco (UCSF). Within the wwPDB, RCSB PDB serves as the designated Archive Keeper for the PDB, responsible for safeguarding both digital information and a physical archive of correspondence. A conservative estimate of USD 100,000 for the average replacement cost of each individual PDB structure translates to a replacement cost of the structures in the entire archive of nearly USD 20 billion (as of mid-2022).

wwPDB partners are committed to the FAIR (Findability, Accessibility, Interoperability, and Reusability [14]) and FACT (Fairness, Accuracy, Confidentiality, and Transparency [15]) Principles emblematic of responsible data stewardship in the modern era. The PDB archive has been accredited by CoreTrustSeal (coretrustseal.org accessed on 28 August 2022). Since its inception, the PDB has been regarded as a pioneer in the open-access data movement. More than 60,000 structural biologists working on every inhabited continent have generously deposited 3D structure information (atomic coordinates, experimental data, and related metadata) to the archive over more than fifty years. Today, many millions of PDB data consumers worldwide working in fundamental biology, biomedicine, bioengineering, biotechnology, and energy sciences enjoy no-cost access to 3D biostructure information with no limitations on data usage. Many scientific research areas have been profoundly impacted by the creation and availability of the PDB archive [16–42].

This review article is published in a Special Issue of *Biomolecules* honoring Professor Phil Bourne, who served as Associate Director of the RCSB PDB from 1998–2014. Phil led the UCSD site, where he focused on database development, integration with the scientific literature, and PDB search and data visualization tools. Bourne and Helge Weissig played critical roles in developing the inaugural version of the RCSB PDB data-delivery web portal at RCSB.org [4,43]. Access to PDB data and development of tools for query, visualization, and analysis as supported by the wwPDB partnership have helped drive the growth of structural and computational biology. PDB data and its usage by researchers, educators, and students over more than five decades is presented to highlight the evolution of these scientific fields and inform the next fifty years of successful PDB operations.

2. Results

2.1. PDB Data Metrics and Trends

Since 1971, PDB structures have been contributed freely by more than sixty thousand structural biologists (depositors) working on every permanently inhabited continent (Figure 1). Structural biologists in 53 countries, territories, etc. recognized by the United Nations deposited data to PDB during 2021. All used the wwPDB OneDep software system (deposit.wwpdb.org) that enables complete structure data deposition [44], rigorous validation [45,46], and expert biocuration [47]. OneDep currently supports 3D macromolecular structures determined using the following experimental methods: macromolecular crystallography (MX), 3D electron microscopy (3DEM), nuclear magnetic resonance (NMR) spectroscopy, electron crystallography (EC), and micro-electron diffraction (microED). Currently, newly deposited structures are processed at RCSB PDB (Americas, Oceania), PDBe (Europe, Africa), or PDBj (Asia, Middle East), allocated based on the depositor's IP address location.

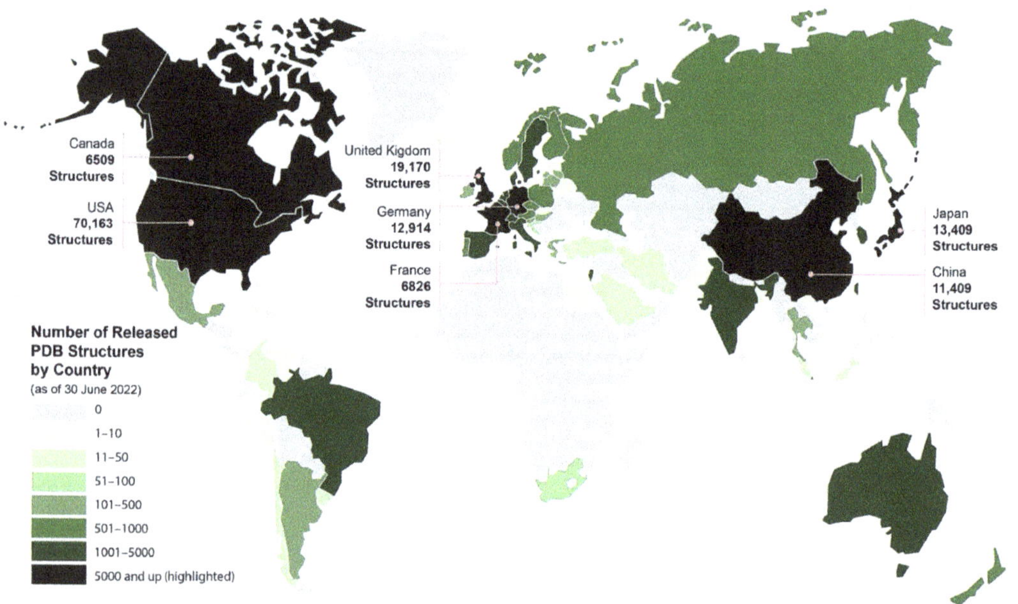

Figure 1. Geographic distribution of PDB depositions from 1971 to mid-2022.

Figure 2A illustrates growth of the PDB archive over the past 50+ years. Since the first X-ray crystal structure of a protein (sperm whale myoglobin) was determined by Sir John Kendrew and his colleagues [48], the discipline has become central to molecular and cellular biology. Figure 2B documents the impact of MX, 3DEM, and NMR on annual PDB data releases. Since 2016, annual releases of PDB MX structures have plateaued at ~10,000, with the exception of substantial spike in 2020 driven by the pandemic lockdown and various MX-based fragment screening campaigns against SARS-CoV-2 proteins thought to represent good drug discovery targets. During the same period, NMR structure releases declined, and 3DEM structure releases grew exponentially (increasing ~6-fold in only 4 years). As of mid-2022, the archive contained 166,894 MX structures, 11,294 3DEM structures, and 13,738 NMR structures. Given current deposition metrics, aggregate 3DEM structure holdings are expected to surpass those of NMR in late 2022 or early 2023. Of immediate importance to those working to combat the COVID-19 pandemic, the PDB archive currently holds >2600 SARS-CoV-2 related structures (~800 released in 2020, and ~900 released in 2021). Figure 2C shows the number of PDB MX and 3DEM structures broken down as a function of resolution (median value ~2.0 Å). While nearly all PDB structures determined at better than 2.5 Å resolution came from MX (~99.6%), 3DEM is now capable of delivering structures to nearly 1Å resolution (e.g., 1.15 Å resolution structure of apoferritin, PDB ID 7a6a [49]).

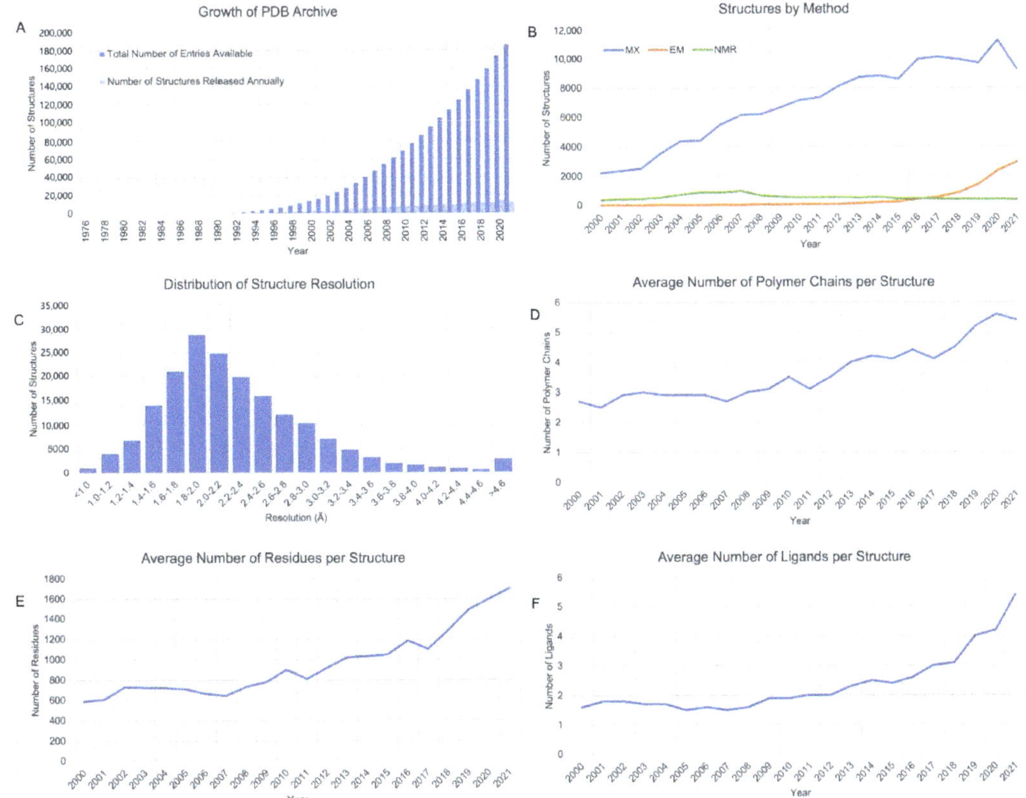

Figure 2. PDB archive metrics. (**A**). Growth 1976–2021. (**B**). New MX, 3DEM, and NMR structures released annually (2000–2021). (**C**). MX and 3DEM structure counts vs. resolution (Å). (**D**). Average number of residues per structure for structures released annually (2000–2021). (**E**). Average number of polymer chains per structure for structures released annually (2000–2021). (**F**). Average number of non-polymer ligands per structure for structures released annually (2000–2021).

While the total number of PDB structures continues to grow, their complexity is increasing year-on-year. Figure 2D illustrates structure complexity as a function of time as judged by the average number of amino acid and/or nucleotide residues per PDB ID. As of mid-2022, the total number of residues (proteins and nucleic acid) in the archive exceeded 200 million and the total number of atoms exceeded 1.5 billion. Figure 2E,F show similar trends for the average number of polymer chains per PDB ID and average number of ligands per PDB ID (excluding bound water molecules, other solvents, salts, ions, common buffers, crystallization and cryoprotection agents as specified in Shao et al. [50]), respectively.

2.2. Evolution of Structural Biology Methods Viewed through the Lens of the PDB

As evidenced in Figure 2A, growth of the PDB has been much faster than linear. This section examines the evolution of structural biology as a discipline viewed through the lens of PDB archival holdings. Technical innovations in MX, 3DEM, and NMR are discussed in some detail, followed by a brief account of the emergence of microED as an exciting new diffraction method for structure determination of biological macromolecules.

2.3. Macromolecular Crystallography (MX)

Structures determined using the MX method were the first to be deposited into the PDB. All of these early structures were determined using isomorphous replacement (IR) [51] to solve the crystallographic phase problem. Slow but steady growth of the PDB archive during the 1980s combined with development of the molecular replacement (MR) method for structure determination by Michael Rossmann [52] helped to accelerate MX. In 2001, after PDB first began systematic collection of phasing method information, it was already apparent that most 3D structures being deposited to the archive were determined using MR. Figure 3 also shows that by 2001 IR had been largely abandoned as a de novo structure determination method in favor of multiple-wavelength anomalous dispersion (MAD, to be supplanted by single-wavelength anomalous dispersion or SAD) for new structure determinations for which MR was not feasible. Analyses across the entire archive revealed that MR was used to determine ~85% of all PDB MX structures as of mid-2022. This method depends critically on the parsimony of macromolecular evolution. Protein domain folds (3D structures) are reused repeatedly within biomolecules carrying out similar biochemical or biological functions. According to generally accepted estimates, ~10,000 distinct polypeptide chain folds account for the vast majority of naturally occurring proteins.

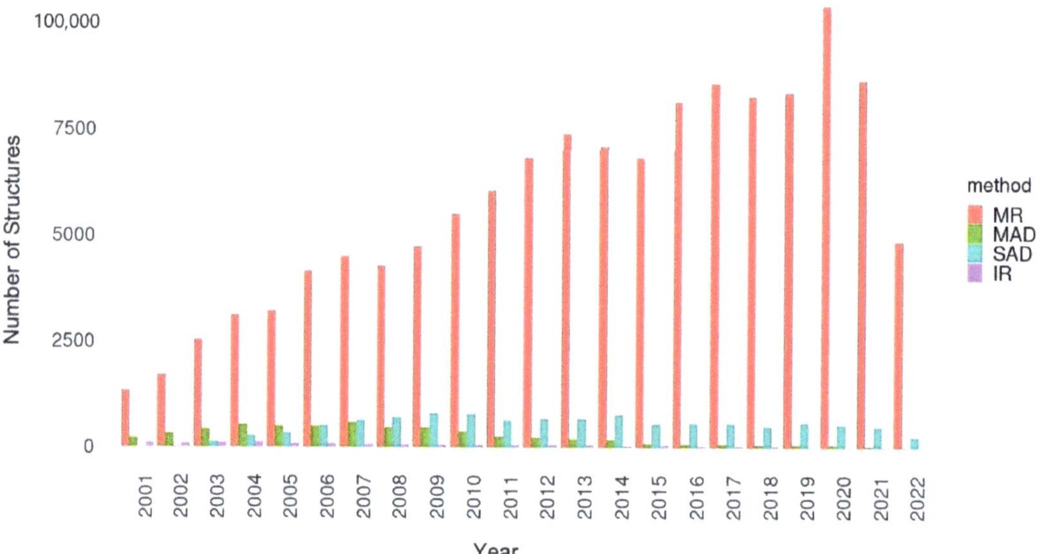

Figure 3. PDB MX structure phasing method trends vs. year of structure release from 2001–2021 (MR: molecular replacement; MAD: multi-wavelength anomalous dispersion; SAD: single-wavelength anomalous dispersion; IR: isomorphous replacement).

The other important trend in MX structure determination practices evident from historical PDB data concerns X-ray sources. Widespread availability of MX beamlines at synchrotron radiation sources transformed how protein crystallographers work. As of mid-2022, ~85% of PDB MX structures relied on diffraction data collected at synchrotrons vs. ~15% that used home X-ray sources. Before 2000, most PDB MX structures released annually were the products of home sources. In contrast, only ~7% of new PDB MX structures came from home sources during the period of 2017 through 2021. Among global synchrotron sources worldwide, the top five contributors of PDB MX structures in rank order as of mid-2022 were the Advanced Photon Source (APS, ~21% of all PDB MX structures), the European Synchrotron Research Facility (ESRF, ~12%), Diamond (~9%), the Advanced

Light Source (ALS, ~7%), and the National Synchrotron Light Source (NSLS, ~6%). Three of these top five biostructure-producing synchrotrons (APS, ALS, and NSLS) and others operated by the US Department of Energy contributed ~41% of all PDB MX structures worldwide as of mid-2022.

Given the critical roles played by synchrotron radiation sources in MX structure studies, one could reasonably expect that bright X-ray sources combined with cryogenic data collection would have contributed to ongoing improvements in structure resolution throughout the history of the PDB. Figure 4 tells an entirely different story. As of 1990, well before access to synchrotron beamlines and cryo-cooling of protein crystals became routine, median resolution of new MX structures released by the PDB annually plateaued at ~2.0 Å. Since then, median resolution of PDB MX structures has not changed appreciably. This reality almost certainly reflects limitations due to the degree of order (or disorder) typical of crystalline preparations of biological macromolecules. Absent new crystallization strategies that markedly increase the order of protein crystals or modeling methods that deconvolute this disorder into multiple structural states, it appears unlikely that median resolution of MX structures in PDB will improve substantially, if at all. Fortunately for most PDB data consumers, 2 Å resolution usually suffices to reveal features of macromolecules relevant for understanding biological phenomena in 3D. In contrast, higher resolution studies may be required to understand fully biochemical functions of proteins and nucleic acids (e.g., reactions catalyzed by protein enzymes and ribozymes).

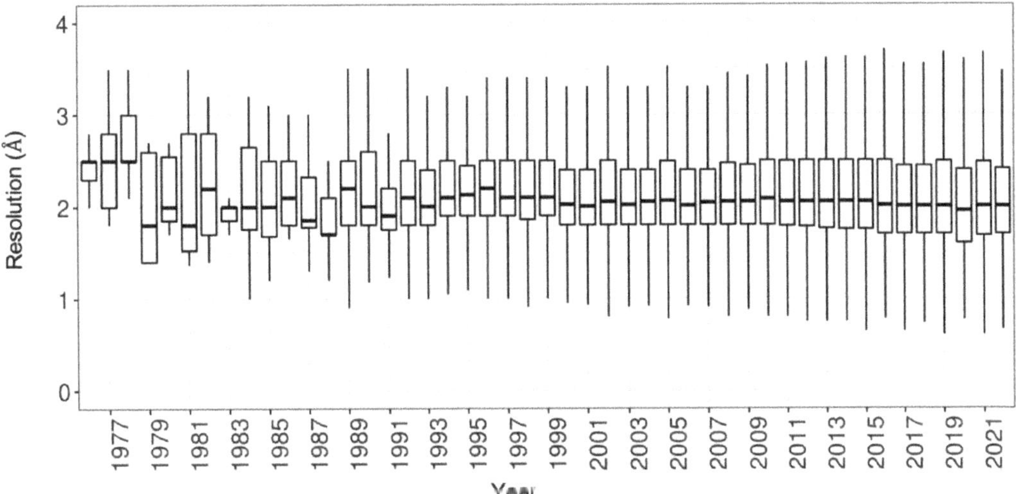

Figure 4. Box plot display of PDB MX structure resolution vs. time. The bold solid bar within each box corresponds to the median value for structures publicly released that year. (N.B.: Small numbers of extreme outliers with resolution > 4 Å were excluded from this analysis for clarity).

Geometric validation of atomic coordinates deposited to the PDB was introduced in the 1990s. Validation of 3D structures vs. experimental structure factors was not routinely performed until 2008, when deposition of experimental structure factor data became mandatory at the behest of the MX community. Stakeholder recommendations regarding some additional means of validating MX structures were subsequently provided in 2011 by the wwPDB X-ray Validation Task Force [53] and implemented in wwPDB legacy deposition systems in 2013 before the wwPDB global OneDep system was launched in 2014 [44]. Availability of experimental data has enabled systematic validation of atomic structures and contributed to development of better validation tools [45] and improved quality of the archived data [54].

Notwithstanding numerous aspects of 3D structure validation initially implemented within the wwPDB OneDep software system validation module, ligand validation was somewhat limited at the outset. The 2016 wwPDB/CCDC/D3R Ligand Validation Workshop recommended best practices for validation of MX co-crystal structures [55]. These recommendations were subsequently incorporated into the OneDep validation module to provide "Buster-like" 2D geometry quality and 3D electron density graphical overlays with small-molecule ligands [46]. Validation of PDB MX structures was further enhanced with introduction of uniform representation for carbohydrates [56].

Arguably, one of the most exciting new methods for measuring diffraction data at the time of writing is serial crystallography [57–59]. This approach is being used to probe dynamic properties of proteins and nucleic acids and visualize progress of chemical reactions in 3D (e.g., *M. tuberculosis* β-lactamase (BlaC) inactivating the β-lactam antibiotic ceftriaxone: PDB IDs 6b5x, 6b5y, 6b6a-6b6f, 6b68, and 6b69 [60]). Both X-ray free-electron lasers (XFELs) and 3rd generation synchrotron sources are being used to conduct such experiments. As of mid-2022, PDB archival holdings included 587 serial crystallography structures, with 343 (~58%) coming from XFELs and 244 (~42%) based on data collected from synchrotrons. Additionally, 217 PDB MX structures were determined using XFEL data without recourse to serial methods (e.g., PDB ID 3pcq [61]).

2.4. 3D Electron Microscopy (3DEM)

Over the last decade, resolution of 3DEM PDB structures has improved dramatically. Since 2013, average resolution of a 3DEM PDB structure has improved from worse than 14 Å to better than ~4 Å (Figure 5A). These overall statistics, however, obscure some of the most impressive recent developments in 3DEM. Between the beginning of 2019 and mid-2022, 40 3DEM structures with resolution better than 2.0 Å were publicly released by the PDB.

Technical breakthroughs in four critical areas were responsible for this "Resolution Revolution" [62,63]. First, improvements in electron optics, driven by the needs of materials scientists and the semiconductor industry, ensure that state-of-the-art transmission electron microscopes (TEM, e.g., Thermo-Fisher Titan Krios, Waltham, MA, USA) preserve phase information at atomic resolution. Second, vitrification of biological samples and imaging under cryogenic conditions is now routine [64]. Third, direct electron detectors (DEDs) have revolutionized how we collect TEM data for single particles arrayed on EM grids. The move away from charge-coupled device (CCD) detectors to DEDs has been nothing short of a stampede. Figure 5B illustrates the trend. In 2013, only ~5% of new 3DEM PDB structures relied on DEDs. By 2017, the fraction relying on DEDs exceeded 90%, and in 2021 the fraction was ~99%. In aggregate, DEDs have been used to collect data for 10,406 3DEM PDB structures released as of mid-2022 (vs. 11,309 total 3DEM PDB structures). Finally, the other key contributor to the rapid rise of 3DEM has been advances made in data processing software. Key software engineering developments include beam-induced motion correction [65–67] and use of Bayesian maximum-likelihood statistics [68]. Figure 5C shows that the most popular 3DEM reconstruction software package at the time of writing is RELION [69], which has been used for determination of more than 4000 3DEM PDB structures since 2013.

Year-on-year growth of 3DEM PDB structure depositions evident in Figure 3B was driven by the single-particle method, which is revealing structures of ever more complex macromolecular assemblies and illuminating important areas of biology (e.g., ion channels, transcription–translation expressome complexes, nuclear pore complexes). Arguably even more exciting advances are yet to be made using cryo-electron tomography (cryo-ET) combined with sub-tomogram averaging [70]. One of the earliest cryo-ET structures in the archive is PDB ID 4bzj (40 Å resolution COPII Transport-Vesicle Coat Assembled on Membranes [71]). As of mid-2022, the highest resolution cryo-ET structure in the archive was PDB ID 7zbt (3.3 Å resolution RuBisCO visualized within native *Halothiobacillus neapolitanus* carboxysomes [72]). At better than 3.5 Å resolution, both α-helix and β-strand secondary structural elements and bulky amino acid sidechains are discernible in experimental 3DEM density maps (deposited to EMDB) revealing molecular details in 3D important for understanding biochemical and biological function.

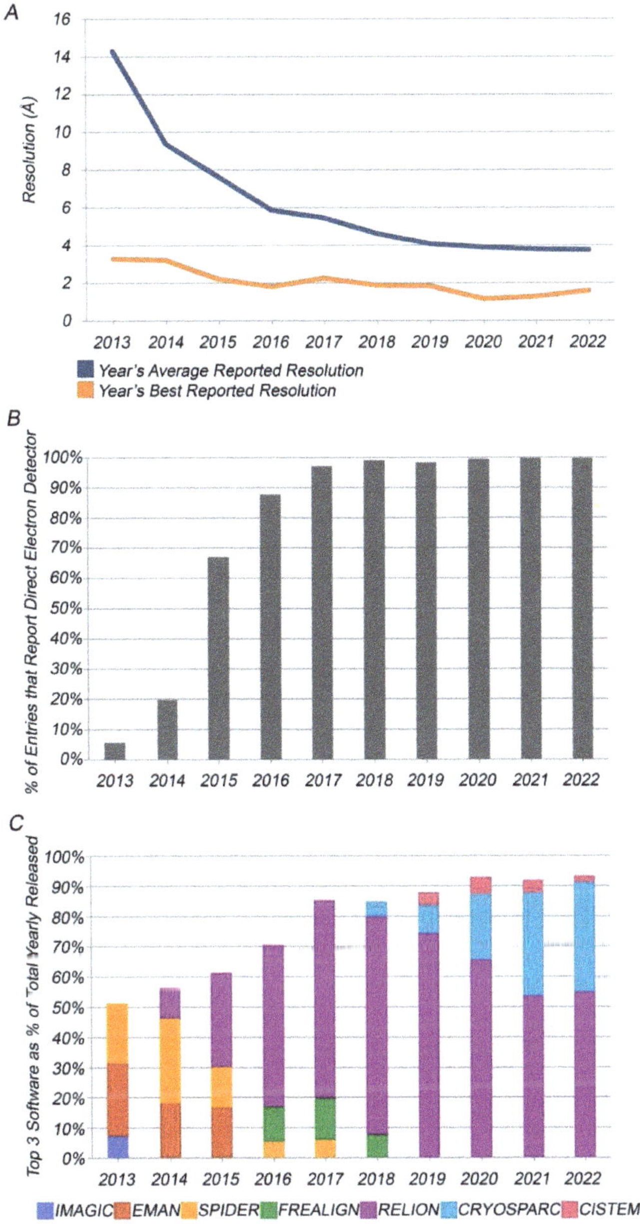

Figure 5. (**A**). Annual average reported resolution (blue) and annual best reported resolution (orange) for 3DEM PDB structures released 2013–2022. (**B**). Percentage of 3DEM PDB structures released per year reporting use of direct electron detectors. (**C**). Top-three reported image reconstruction software packages per year shown as a percentage of 3DEM PDB structures reporting reconstruction software.

The *H. neapolitanus* RuBisCO cryo-ET structure employed a relatively new sample preparation technique that relies on cryogenic dual-beam focused ion beam/scanning electron microscopes (cryo-FIB/SEM) to generate 10–20 nm thickness *lamellae* of vitrified samples using the focused ion beam to "mill" away unwanted parts of the sample. This tool allows researchers to isolate thin wafer-like volumes from inside frozen cells for subsequent cryo-ET imaging and sub-tomogram averaging. Immediate-term prospects for cryo-ET plus cryo-FIB/SEM milling with sub-tomogram averaging brightened considerable with the advent of AlphaFold2 [73–75] and RoseTTAFold [76]. For example, in 2021, computed structure models of human nuclear pore complex (NPC) proteins from AlphaFoldDB were combined with cellular cryo-ET and molecular dynamics simulations, to generate composite 3DEM density maps of the human NPC in both dilated and constricted conformations (PDB IDs 7r5k, 7tbl, 7tbm, 7tbj, 7tbk, and 7tbi [77]). Combining cryo-FIB/SEM with correlative light microscopy prior to cryo-ET imaging of *lamellae* holds the promise of improving the efficiency of the method by maximizing the number of molecular assemblies of interest present in a given wafer-like sample for imaging and subsequent sub-tomogram averaging [78].

At the time of writing, wwPDB validation reports for 3DEM structures included: (a) assessment of model geometry similar to that used for all MX and NMR structures (ClashScore, Ramachandran outliers, Sidechain outliers, nucleic acid polymer backbone); (b) orthogonal projections of map and map-model overlays; (c) half-map FSC plot based on mandatory half-maps collected at deposition; (d) voxel-value distribution and volume-estimation graph; (e) evaluation of map-model fit via atom-inclusion plot and residue inclusion analysis; and (f) finer evaluation of map-model fit incorporating both overall and per residue Q-scores [79]. EMDB also provides 3DEM density map and structure quality assessments on its website, including Q-scores [80]. (For more details regarding the history of 3DEM validation in the PDB, see [81]).

2.5. Nuclear Magnetic Resonance (NMR) Spectroscopy

Solution nuclear magnetic resonance (NMR) spectroscopy can be used to determine 3D structures of biomolecules (e.g., [82,83]). The first NMR structure of a protein was deposited to the PDB in 1988 and released publicly in 1989 (PDB ID 1bds [84]). By the end of the 1980s, solution NMR structures of 10 proteins had been determined, for which no crystallographic data were previously available [85]. At the same time, heteronuclear 3D and 4D NMR experiments were introduced to overcome limitations of spectral complexity and increased molecular weight (polypeptide chains longer than 150 amino acid residues, hereafter residues) [86]. At the beginning of the 1990s, the first NMR data file that included NMR restraints used to determine the 3D structure of Interleukin-8 (IL-8/NAP) was deposited to the archive (PDB ID 1il8 [87]). At the end of the 1990s, the first chemical shift file (containing a total of 179 chemical shifts) was deposited as part of PDB ID 1qlo [88]. Upon the recommendation of the wwPDB NMR Validation Task Force (NMR-VTF), NMR PDB structure depositions were required to include NMR restraint data and chemical shift data, in 2008 and 2010, respectively [89].

The number of new NMR structures released to the public annually from the PDB peaked in 2007 at 965, when NMR structures accounted for ~17% of the entire archive. Annual depositions have been trending downward ever since (362 NMR structures released publicly in 2021), and NMR structures now account for only ~7% of PDB holdings. As of mid-2022, the archive housed 13,733 NMR structures, 13,602 solution plus 131 solid-state. Figure 6 provides a breakdown of NMR PDB structures as a function of biomolecule sample type.

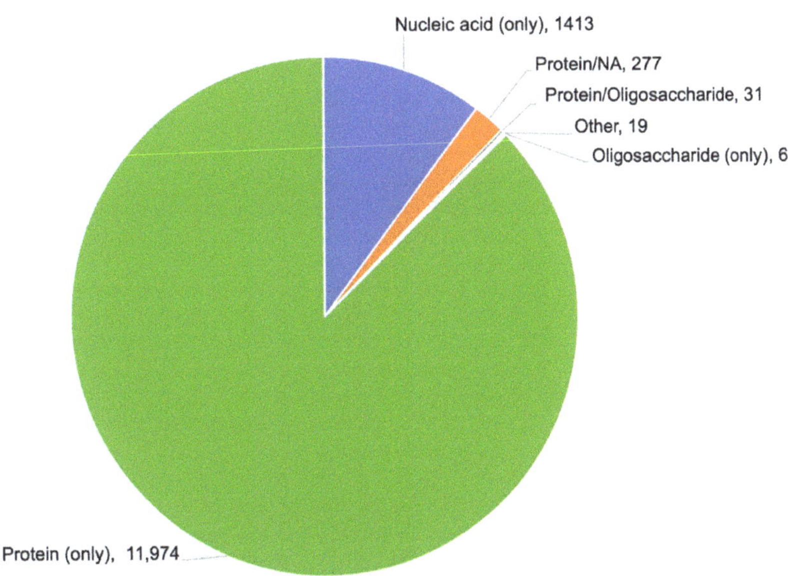

Figure 6. Breakdown of NMR PDB structure holdings by sample type.

Historically, NMR structural studies of biomolecules were size-limited. Most NMR PDB structures are those of smaller proteins or isolated protein domains (polymer entities < 8.5 kDa). Both solution and solid-state NMR (SSNMR) can, however, be used to study larger, more complex structures. SSNMR has been utilized to overcome some of the obstacles restricting the purview of solution NMR (e.g., relatively insoluble proteins). Both techniques can be deployed in tandem to overcome respective limitations. As of mid-2022, the PDB archive housed at least six structures determined using a combination of solution and SSNMR (e.g., *O. cuniculus* phosphorylated phospholamban homopentamer PDB ID 2m3b [90]).

Advances in technology for both solution and SSNMR have allowed for larger structures to be determined. For example, the largest solution NMR structure in the archive (as judged by total number of residues) is the Box C/D enzyme, a multimeric complex consisting of four instances of three unique proteins totaling 3044 residues (PDB ID 4by9 [91]). Additionally, use of magic angle spinning (MAS) SSNMR has enabled determination of structures with no inherent molecular size limitation, overcoming obstacles faced by solution NMR and MX. Exploiting these capabilities, SSNMR has been used to elucidate structures of complex assemblies similar in size to those studied by cryo-EM while in their native state, without the need for cryogenic preservation. As of mid-2022, the largest macromolecular structure determined by MAS SSNMR is the HIV-1 Capsid Tube, containing 378 repeats of a 231-residue subunit for a total of 87,318 residues (PDB ID 6x63 [92]). Larger structures have also been determined using integrative or hybrid methods, including that of a 484.61 kDa, 24mer αB-crystallin oligomer (4200 residues), incorporating experimental data from solution NMR, solution scattering, and 3DEM (PDB ID 3j07 [93]), and that of the 470.42 kDa tetrahedral aminopeptidase TET2 (4236 residues total), incorporating data from SSNMR and 3DEM (PDB ID 6r8n [94]).

With use of membrane-mimicking systems (e.g., micelles, bicelles, and nanodiscs), it is possible to study integral membrane proteins in their near-native environments using NMR [95]. A structure of the 7.77 kDa transmembrane domain of bacterioopsin (residues 1–71) was determined using solution NMR by solubilizing the protein in methanol/chloroform and SDS micelles, and deposited into PDB in 1993 (PDB IDs 1bha and 1bhb [96]). At the time of writing, the largest membrane protein structure determined via solution NMR deposited to the PDB is that of 149.16 kDa, 1360 residue human α7 nicotinic acetylcholine receptor, determined by a combination of solution NMR, electron spin resonance spectroscopy, and Rosetta calculations (PDB ID 7rpm [97]). As of mid-2022, the largest membrane protein structure determined by SSNMR in the PDB is that of 183.51 kDa, 1750 residue M13 bacteriophage capsid (PDB ID 2mjz [98]).

In addition to the study of 3D structures of biological macromolecules, examination of dynamics is often important for understanding function. Insights into a biomolecule's local dynamic behavior can be used to identify parts of structures important for ligand binding, protein–protein or protein–nucleic acid interactions, allostery, or conformational changes (e.g., integral membrane proteins). NMR spectroscopy is uniquely capable of studying macromolecular movement because of its ability to study samples spanning a wide range of solvent/solute conditions at atomic resolution over relevant timescales (i.e., picoseconds to seconds). Such studies are also possible using MAS SSNMR, which can be used to interrogate dynamics of the protein backbone atoms and sidechains (both globally and locally). As of mid-2022, the PDB archive housed results of dynamics studies of both small proteins (e.g., 8.58 kDa ubiquitin, PDB ID 2k39 [99]) and large biological nanomachines (e.g., 181.87 kDa proteasome subunit alpha heptamer, PDB ID 2ku1 [100]).

As is the case for MX and 3DEM, validation standards for NMR structures archived in the PDB are being developed collaboratively by the wwPDB and independent experts. Following implementation of chemical shift validation in 2015 at the behest of community stakeholders, the NMR Data Exchange Format (NEF) Working Group, which includes developers of NMR structure determination and refinement software packages, recommended use of a common exchange format to represent NMR chemical shifts, restraints, and related metadata [101]). NMR structure validation utilizing this unified exchange format was incorporated within the wwPDB OneDep software system and wwPDB validation reports in 2020. At the time of writing, archive-wide regeneration of extant NMR structure validation reports to enable restraint validation was underway. Completion of this remediation project and public release of regenerated wwPDB validation reports for all NMR structures archived in the PDB is anticipated in 2023. Additional improvements in wwPDB validation of NMR structures is expected to encompass data representation and validation of multiple conformers (e.g., pro-islet amyloid polypeptide open conformer (PDB ID 6ucj) and pro-islet amyloid polypeptide bent conformer (PDB ID 6uck [102]) and validation of structures determined using NMR combined with other experimental methods (e.g., PDB ID 3j07 [93]).

2.6. *Electron Crystallography (EC) and Micro-Electron Diffraction (microED)*

Electron diffraction or electron crystallography (EC) has also been used to determine 3D structures of biological macromolecules. The method employs 2D crystals, beginning with those of bacteriorhodopsin, the first integral membrane protein structure to be deposited into the archive (PDB ID 1brd [103], resolution 3.5 Å). Prior to 2013, a total of 37 biostructures determined using EC were deposited to PDB. With the advent of modern electron microscopes, a new electron diffraction method using miniscule 3D crystals (micro-electron diffraction or microED) has been developed [104]. The first microED structure of a globular protein (hen egg white lysozyme, PDB ID 3j4g [105], resolution 2.9 Å) was deposited to the PDB in late 2013. As of mid-2022, the PDB housed 137 microED structures of biomolecules, the largest two of which are human adenosine receptor A2a/cytochrome b562 chimeric protein (PDB ID 7rm5, 50 kDa, resolution 2.8Å [106]) and bovine catalase (PDB ID 3j7b, 60 kDa, resolution 3.2Å [107]). Unlike most EC structures archived in PDB, microED structures are typically determined at very high resolution. As of mid-2022, the highest resolution microED structure in PDB was that of hen egg white lysozyme (PDB ID 7skw [108], resolution 0.87 Å).

2.7. PDB Archive Management and Weekly Update/Release

The PDB data standard is defined by the PDBx/mmCIF dictionary [109–111]. It is the macromolecular extension of an earlier community data standard, the Crystallization Information Framework (cif.iucr.org, accessed on 28 August 2022), developed for small molecules by the International Union of Crystallography [112]. The macromolecular data standard is maintained by the wwPDB partnership together with the wwPDB PDBx/mmCIF Working Group (wwpdb.org/task/mmcif, accessed on 28 August 2022) [111]. wwPDB partners and the Working Group collaborate on developing terminologies for new and rapidly evolving methodologies and remediating (or enhancing) representations for existing data content.

In its role as wwPDB-designated PDB Archive Keeper, RCSB PDB is responsible for safeguarding >100 TB of digital information and a physical archive that includes correspondence and other archive-related artifacts dating back to the early 1970s. Snapshots of the digital information are preserved annually and following large-scale archive-wide data remediation campaigns, the most recent of which involved standardizing atom naming, etc. for >14,000 carbohydrate-containing structures in the PDB [56]. The size of the 2021 digital snapshot was ~1 TB, which does not include ~4.5 TB of 3DEM density map information archived in EMDB (also jointly managed by the wwPDB partnership).

In its role as wwPDB-designated Archive Keeper, RCSB PDB is responsible for weekly updates of the PDB archive using the following two-stage process:

Stage One releases sequence(s) for each distinct polymer (amino acid or nucleotide) in the structure; InChI string(s) for each distinct ligand; and crystallization *p*H value(s), where appropriate, on the wwPDB web portal (see www.wwpdb.org/ftp/pdb-ftp-sites, accessed on 28 August 2022) every Saturday by 03:00 Universal Time Coordinated (UTC). This first stage in the process supports weekly blind challenges for in silico prediction of protein structure (CAMEO, cameo3d.org, accessed on 28 August 2022 [113]) and small-molecule docking (CELPP, drugdesigndata.org/about/celpp, accessed on 28 August 2022 [114]).

Stage Two completes the weekly process every Wednesday at 00:00 UTC by releasing the updated PDB archive in full (currently adding ~300 new structures/week, updating previously released structures with literature citation information, etc., and on occasion removing obsolete structures).

PDB data are freely distributed online, providing universal open access to the archival information in two forms (latest archive, files.wwpdb.org/pub/pdb/data, accessed on 28 August 2022; and latest and prior versions of archive, files-versioned.wwpdb.org, accessed on 28 August 2022). Hypertext Transfer Protocol (HTTP) and remote sync (rsync) are recommended for access; File Transfer Protocol (FTP) access will be retired in late 2024. PDB data are also made available without storage fees or egress charges by Amazon Web Services (AWS) through its Open Data Sponsorship Program (registry.opendata.aws/pdb-3d-structural-biology-data/, accessed on 28 August 2022).

Global PDB archive data downloads in 2021 reached a record high of 2,364,150,827 structure data files, which represents an ~80% increase vs. the previous record of 1,323,213,832 set in 2020. Approximately 70% of global structure data file downloads in 2021 originated from the FTP archive. The remainder were accessed by users of wwPDB member web portals.

2.8. All Three Kingdoms of Life Are Represented in the PDB Archive

As of mid-2022, MX, 3DEM, NMR, EC, and microED had been used collectively to determine >190,000 3D biostructures housed in the PDB archive, which encompasses proteins from organisms representing all living kingdoms (Figure 7). Archaebacterial proteins were the least numerous (totaling 5664 structures), followed by bacteria (65,967 structures). PDB holdings of eukaryotic protein structures exceeded 105,000, with more than half being human in origin. There is limited PDB coverage across the so-called model organisms, with mouse proteins being most numerous at >8000 structures.

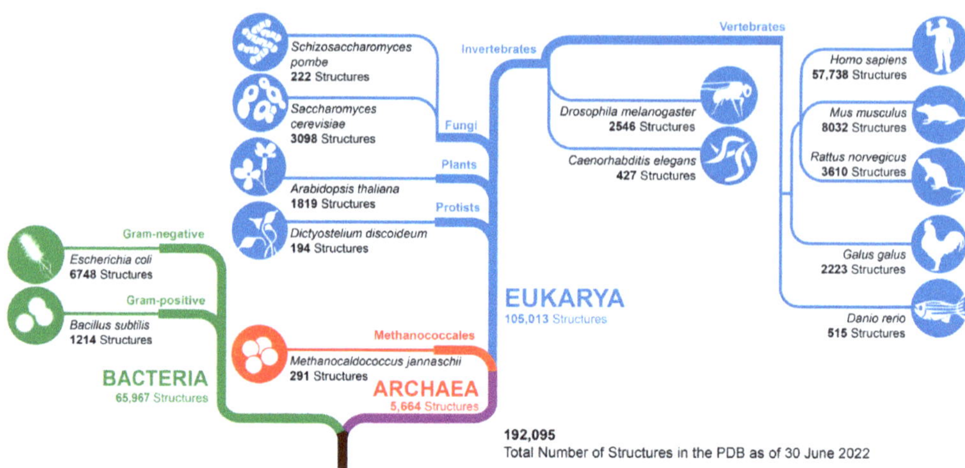

Figure 7. Phylogenetic Tree showing PDB holdings (as of mid-2022). Within each of the three branches, PDB structure totals are provided for selected organisms. N.B.: The PDB also houses 3D structures that solely contain nucleic acids (DNA, RNA, DNA-RNA hybrids, etc.) and/or viral proteins or human-designed proteins, which collectively accounted for ~8% of archival holdings as of mid-2022.

2.9. PDB Data Delivery/Usage Metrics

Most RCSB PDB users access the archive through our RCSB.org research-focused web portal, which makes PDB data available at no cost with no limitations on usage via the Creative Commons CC0 1.0 Universal license (creativecommons.org/publicdomain/zero/1.0/, accessed on 28 August 2022). In 2021, 6,845,233 unique internet protocol (IP) addresses from more than 240 countries and territories recognized by the United Nations (Figure 8A) were used to access RCSB.org (exceeding the 2020 pandemic lock-down record of 6,677,853). Figure 8B ranks RCSB.org utilization for the top ten user countries for 2019–2021. Not surprisingly, the US–RCSB PDB's host country–has the largest percentage of users, followed by the world's two most populous nations, India and the People's Republic of China.

We estimate that ~99% of PDB data consumers are not experts in structural biology. Their research interests are extremely broad, encompassing fundamental biology, biomedicine, energy sciences, bioengineering, and biotechnology [115,116]. Beyond the natural, physical, mathematical, and engineering sciences, there is also use of PDB data by social scientists (e.g., economists, [117,118]).

The RCSB.org web portal provides added value to PDB users that goes well beyond the content of the archive itself. On a weekly basis, RCSB PDB integrates PDB data with information from ~50 trusted external resources (Table 1). Integrating individual PDB structures with information from trusted external resources ensures that the RCSB.org web portal operates as a "living data resource." Scholarly journal articles describing PDB structures are static documents, reflecting what was known about the biomolecule(s) at the time of publication. Thereafter, it is not uncommon for new biological or biochemical functions of a macromolecule to come to light, or new disease-causing mutations to be identified. Such new findings are integrated with PDB data every week, thereby ensuring that RCSB.org users have access to the most current information pertaining to every 3D biostructure in the public domain.

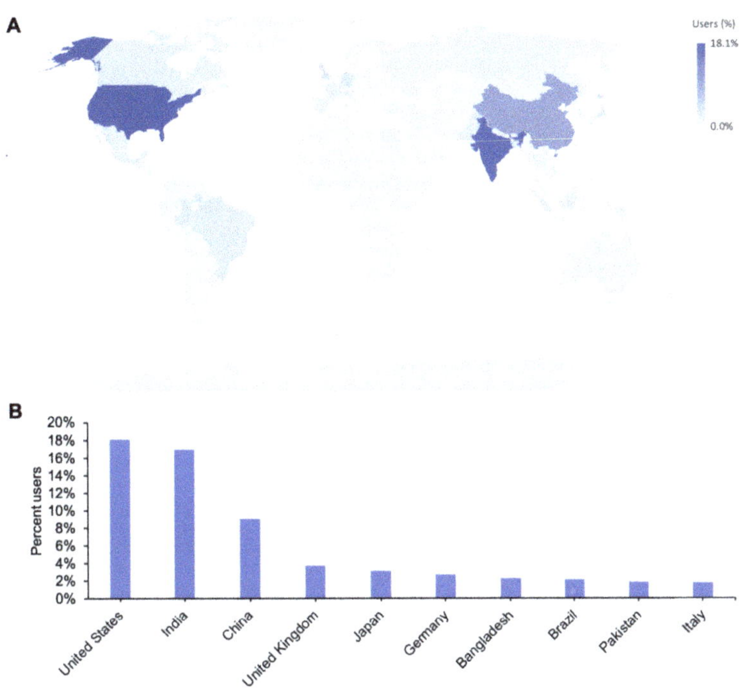

Figure 8. (**A**). Geographic distribution of RCSB.org users by country. (**B**). Top 10 countries with the highest percentage of users from 2019–2021. Data from Google Analytics.

Table 1. Trusted external resources/data content integrated weekly with PDB archival data by RCSB PDB from rcsb.org/docs/general-help/data-from-external-resources-integrated-into-rcsb-pdb (accessed on 28 August 2022). (N.B.: In response to community input, RCSB PDB continues to integrate new external data resources.).

Resource	Description
AlphaFold DB [73,74]	Computed Structure Models by AlphaFold2
ATC	Anatomical Therapeutic Chemical (ATC) Classification System from World Health Organization
Binding MOAD [119]	Binding affinities
BindingDB [120]	Binding affinities
BMRB [13]	BMRB-to-PDB mappings
CATH [121]	Protein structure classification
CCDC [122]	Cambridge Structural Database (CSD)
ChEBI [123]	Chemical entities of biological interest
ChEMBL [124]	Manually curated database of bioactive molecules with drug-like properties
DrugBank [125]	Drug and drug target data
ECOD [126]	Evolutionary Classification of Protein Domains
EMDB [11]	3DEM density maps and associated metadata
ExplorEnz [127]	IUBMB Enzyme nomenclature and classification

Table 1. *Cont.*

Resource	Description
Gencode [128]	Gene structure data
Gene Ontology [129]	Gene structure data
Genotype-Tissue Expression (GTEx) [130]	Tissue-specific gene expression data
GlyCosmos [131]	Web portal integrating the glycosciences with the life sciences
GlyGen [132]	Data integration and dissemination resource for carbohydrates and glycoconjugates
GlyTouCan [133]	Glycan structure repository
Human Gene Nomenclature Committee (genenames.org, accessed on 28 August 2022)	Human gene name nomenclature and genomic information
IMGT [134]	International ImMunoGeneTics information system
Immune Epitope Database [135]	Antibody and T cell epitopes
International Mouse Phenotyping Consortium (mousephenotype.org, accessed on 28 August 2022)	Mouse gene phenotype data
InterPro [136]	Classification of Protein Families
MemProtMD [137]	Database of Membrane Proteins Embedded in Lipid Bilayers
ModelArchive (modelarchive.org accessed on 28 August 2022)	Computed Structure Models (e.g., by RoseTTAFold)
Mpstruc [138]	Classification of transmembrane protein structures
NCBI Gene [139]	Gene info, reference sequences, etc.
NCBI Taxonomy [139]	Organism classification
NDB [140]	Experimentally determined nucleic acids and complex assemblies
OPM [141]	Orientations of Proteins in Membranes database; Classification of transmembrane protein structures and membrane segments
PDBbind-CN [142]	Binding affinities
PDBflex [143]	Protein structure flexibility
PDBTM [144]	Protein Data Bank of Transmembrane Proteins
Pharos [145]	Drug targets and diseases
ProteinDiffraction.org (proteindiffraction.org, accessed on 28 August 2022)	Diffraction images
PubChem [146]	Chemical information
PubMed [139]	Citation information
PubMedCentral [139]	Open access literature
RECOORD [147]	NMR structure ensembles
RESID [148]	Protein modifications
SAbDab [149]	The Structural Antibody Database
Thera-SAbDab [150]	Therapeutic Structural Antibody Database
SBGrid [151]	Structural Biology Data Grid/diffraction images
SCOP [152]	Structural Classification of Proteins
SCOPe [153]	Structural Classification of Proteins—extended
SIFTS [154]	Structure, function, taxonomy, sequence
UniProt [155]	Protein sequences and annotations

PDB data utilization worldwide is also mediated by third parties that repackage and reuse the archival information. While the RCSB PDB is unable to assess utilization of the archive via third parties, review of the Nucleic Acids Research Online Molecular Biology Database Collection [156], which comprises databases from *Nucleic Acids Research* annual Database Issues, identified 460 external data resources that distribute repackaged PDB data (Supplementary Materials Table S1). Additional utilization of PDB data occurs within all major biopharmaceutical companies and many smaller biotechnology companies that maintain copies of the archive inside company firewalls. They frequently use PDB data alongside proprietary MX structures determined by company structural biologists or their contractors. Most, if not all, global biopharmaceutical companies (e.g., Pfizer, Novartis, Eli Lilly and Company) rely on structure-guided drug discovery of small-molecule, orally bioavailable therapeutic agents, which typically begins with scanning of PDB archival holdings for a public domain structure of the target protein to begin the discovery process [25,157,158]. They also make use of PDB structures when engineering new biologic agents (monoclonal antibodies, cytokines, etc.) for use as injectables [159].

Literature searching provides another means of assessing utilization and impact of PDB data. As of mid-2022, 162,262 (~84%) of PDB structures are described in 75,497 unique primary publications, the vast majority of which appeared in peer-reviewed journals. Citation analyses carried out using EuropePMC revealed that in 2021, the PDB was mentioned by name in 23,030 publications. It further documented that PDB IDs were mentioned in 585,903 publications in 2021. An RCSB PDB study published in 2018 [160] documented that citations of PDB data spanned the sciences, literally from Agriculture to Zoology. Not surprisingly, nearly 90% of published PDB structures analyzed in 2018 were cited by journals in the area of Biochemistry and Molecular Biology. High impact within other areas of biomedicine (Cell Biology, Pharmacology and Pharmacy, Microbiology, Genetics and Heredity) was, as expected, also documented. Further RCSB PDB analyses on this topic highlighted PDB structure publications that were frequently cited in scientific journals focused on Materials Science, Physics, Computer Science, Chemistry, Engineering, and Mathematics [116].

Searching of the patent literature in August 2022 also documented substantial impact of PDB data. Directed searches for PDB mentions using the US Patent and Trademark Office website (uspto.gov, accessed on 28 August 2022) identified nearly 19,000 in-process patent applications and ~10,000 issued US patents (vs. ~20,000 in process applications and ~6500 issued patents in June 2017 [160]). Analogous searches of global patent literature using PatSeer (patseer.com) documented ~90,000 issued patents and patent applications in process worldwide that include PDB mentions (vs. ~50,000 in June 2017 [160]).

Finally, RCSB PDB also operates a second web portal focused on outreach and education (PDB101.RCSB.org, with PDB-101 denoting an introductory course) [161]. PDB-101 was launched in 2011 to support PDB archive exploration and training by university faculty, postdoctoral researchers, undergraduate and graduate students, school teachers and their pupils, and the general public. It was established to help train the next generation of PDB users and promote structural biology and protein science to non-experts. Regularly published features include the highly popular *Molecule of the Month* series [162], 3D biostructure-related activities, molecular animations and videos, and educational curricula, many of which are organized around a public health topic [163]. The *Guide to Understanding PDB Data* covers key topics, including file format information and explanations of the types of data included with a PDB entry. Materials are organized into various categories (Health and Disease, Molecules of Life, Biotech and Nanotech, and Structures and Structure Determination) and searchable by keyword (e.g., cancer, checkpoint therapy, antibody). Although it is not as intensively accessed as our RCSB.org research-focused web portal, there is substantial utilization of PDB101.RCSB.org by users from around the world (Figure 9).

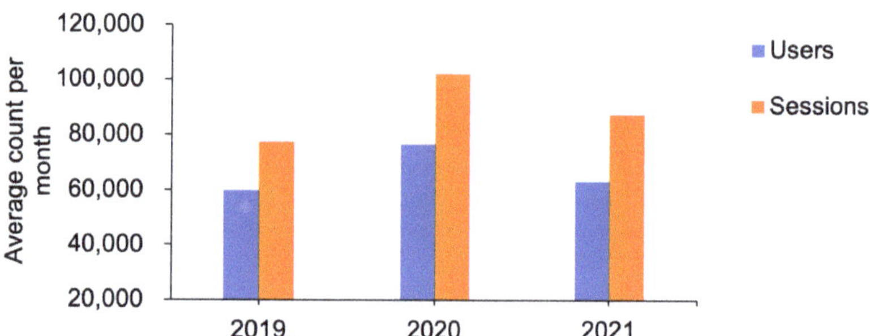

Figure 9. Average monthly usage of PDB-101 (PDB101.RCSB.org, accessed on 28 August 2022) from 2019–2021. Data from Google Analytics.

2.10. Impact of PDB Data on Computational Structure Modeling

Use of PDB data to compute 3D structure information for other proteins is well-established. For many years, publicly available computational services (e.g., Modeller/ModBase [164–166] and ProMod3/SWISS-MODEL, [167,168] and Rosetta [169]) used comparative or homology modeling to predict protein structures. This approach depends on finding an experimentally-determined protein structure in the PDB with an amino acid sequence similar to that of the target protein to use as a modeling template or scaffold. Homology modeling typically succeeds when a structural template with >40% sequence identity is available. Like MR, homology modeling is often useful because of the parsimony of macromolecular evolution.

As the PDB archive grew, template-free computational structure modeling became possible for very small globular proteins. Continuous advances in both homology modeling and template-free protein structure prediction were fostered by two community-led blind challenges (i.e., CASP [170], and the weekly Continuous Automated Model EvaluatiOn (or CAMEO) online challenge [113]). Both CASP and CAMEO rely on coordination with structural biologists and the wwPDB to ensure relevant structure data are not publicly released before each challenge concludes.

Google DeepMind emerged as the top performer in the 2020 CASP challenge [170]. Its AlphaFold2 software uses artificial intelligence/machine learning (AI/ML) to predict 3D structures of smaller globular proteins with accuracies comparable to that of low-resolution experimental methods [74]. It was rightly heralded as a major breakthrough in de novo protein structure prediction. Subsequently, the Rosetta team led by David A. Baker (University of Washington/Howard Hughes Medical Institute) released RoseTTAFold [76] and then RoseTTAFold2, which also use AI/ML methods to generate computed structure models (CSMs) of proteins with reported accuracies comparable to that of AlphaFold2. Figure 10 contrasts experimental structure determination with computed structure model calculation. At the time of writing, CSMs for nearly every protein sequence represented in UniProt [155] generated by DeepMind using AlphaFold2 were publicly available from AlphaFold DB [73–75]. Some of the CSMs generated by computational biologists operating independently of DeepMind (using RoseTTAFold, AlphaFold2, etc.) are available from the open access ModelArchive (modelarchive.org, accessed on 28 August 2022).

Of particular importance when evaluating CSMs for use in research are pLDDT (predicted local distance difference test) scores or confidence estimates generated by AlphaFold2 [74,171]. pLDDT scores (scaled between 0 and 100) denote polypeptide chain segments as very high confidence (pLDDT $\geq$ 90), confident (90 > pLDDT $\geq$ 70), low confidence (70 > pLDDT $\geq$ 50), and very low confidence (pLDDT < 50). We do not yet know how much enhanced AI/ML methods will improve prediction accuracy and expand the scope thereof to larger, multidomain proteins, but history shows us that continued growth of the PDB should only help in this regard.

It is no exaggeration to say that neither AlphaFold2 nor RoseTTAFold2 would exist today without open access to complete, rigorously validated, expertly biocurated 3D biostructure data from the PDB [172]. Looking ahead, use of AI/ML methods for accurate prediction of structures of macromolecular assemblies and, perhaps even more challenging, transient intermolecular interactions that underpin complex regulatory processes in biology will depend critically on continued growth in the number of 3DEM structures of large molecular machines deposited to the PDB. Successful application of AI/ML methods for predicting small-molecule ligand binding to protein targets may not be possible in the near term given current PDB data deposition trends. The number of co-crystal structures of small molecules binding to proteins in the PDB is dwarfed by 3D structure data collectively held as trade secrets across the biopharmaceutical industry. Contributions of significantly more co-crystal structure data from industry would almost certainly fuel advances in prediction of small-molecule binding to proteins. With sufficient data placed in the public domain, we can reasonably expect that AI/ML methods would accelerate drug discovery and development efforts in both academe and industry for the greater good [172].

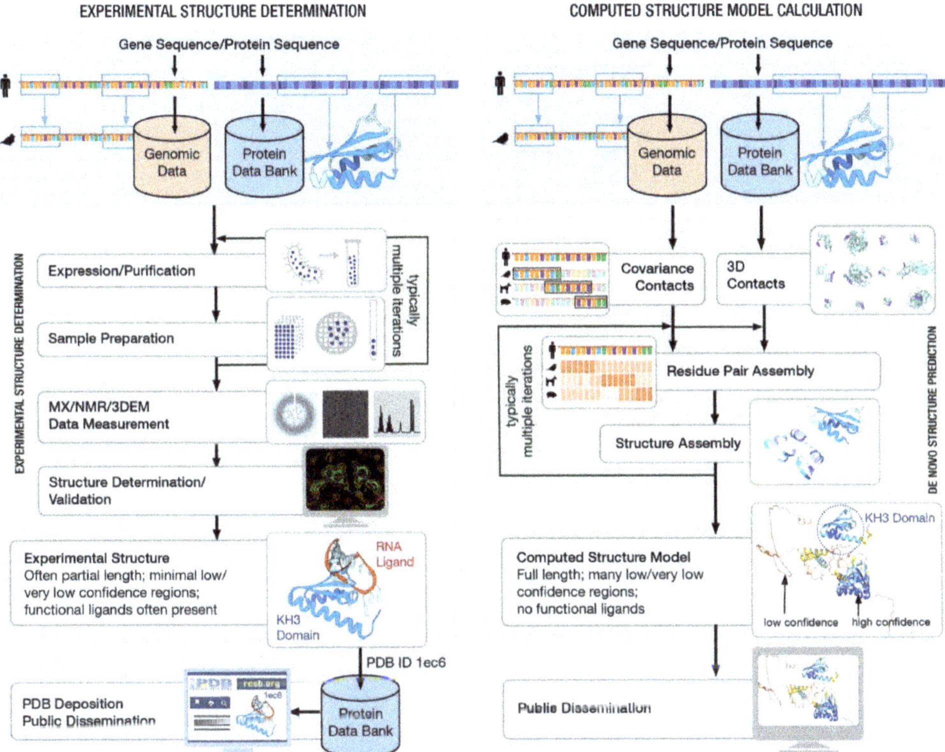

Figure 10. Experimental approaches for determination of protein structures and computational methods for predicting structures both rely on open access to genomic and 3D structure data. Here, methods for determining the structure of the RNA-binding protein Nova-2 are shown. The MX structure (**left**) was determined for an isolated domain of the protein bound to its RNA target. The computed structure (**right**) includes the entire polypeptide chain, which is predicted to include three well-folded domains (blue/cyan) connected by apparently unstructured linkers (yellow/orange). Image adapted from *New England Journal of Medicine*, Stephen K. Burley, Wadih Arap, Renata Pasqualini, Predicting Proteome-Scale Protein Structure with Artificial Intelligence, 385, 2191–2194 [173].Copyright © 2022 Massachusetts Medical Society. Reprinted with permission.

2.11. Future Directions

The futures of structural biologists and the PDB appear even brighter, contrary to post-AlphaFold2 rumors to the effect that experimental structural biology is on the verge of precipitous decline. Depositions of structures to the PDB in 2022 are on track to exceed those in all previous years. Experimentally determined 3D biostructures are highly prized accomplishments. Medium-to-high resolution experimental structures (e.g., MX structures better than 3.5 Å resolution) are more accurate than CSMs [174]. Moreover, they frequently contain bound small-molecule ligands of biological or biomedical importance. They may also include more than one macromolecule, providing information regarding homo- and hetero-meric assemblies that underpin the workings of complex molecular machines.

CSMs generated with AI/ML methods are of considerable interest to experimental structural biologists. Many are taking a "glass half full" approach to this information. They often rely on CSMs of large multi-domain eukaryotic proteins for designing protein expression constructs by excluding low confidence and very low confidence regions when generating truncations suitable for MX, NMR, or 3DEM studies. (N.B.: CSMs are not eligible for archiving in the PDB, because they do not involve measurements from a sample of the biological macromolecule for which the structure is determined.)

The future of experimental structural biology is also looking bright. Researchers are tackling ever larger and more complex macromolecular machines using so-called integrative or hybrid methods that combine experimental measurements from more than one biophysical technique. Anticipating this trend, a wwPDB Integrative/Hybrid Methods (IHM) Task Force was assembled to make recommendations regarding data archiving and structure validation [175,176]. As an interim measure, the wwPDB established PDB-Dev as a standalone prototype system [177–179] for archiving and publicly disseminating integrative structures and associated data. Integrative structure determination entails making measurements using complementary experimental methods (e.g., 3DEM and chemical cross-linking) and converting the results into spatial restraints that are applied to with known starting structures of molecular components to determine the structures of complex macromolecular assemblies.

The PDB-Dev software system supports data collection, processing, curation, validation, archiving, and distribution of integrative biostructures. It is underpinned by ModelCIF (github.com/ihmwg/ModelCIF, accessed on September 29 2022), an expanded set of data standards based on the PDBx/mmCIF data standard (above) for representing integrative structures and associated experimental restraints; a software library that supports the new data standards; a data harvesting system for collecting heterogeneous data from diverse experimental techniques, methods for curating, validating and visualizing integrative structures; and web services for distributing archived data. The PDB-Dev prototype system has allowed structural biologists to make their integrative structures publicly available, including but by no means limited to those involved in transport of proteins and nucleic acids across the nuclear envelope (nuclear pore complex [180]), regulation of gene expression (expressome complex [181]), cellular vesicle trafficking (exocyst complex [182]), and regulation of genomic architecture (BAF complex [183]). Importantly, the PDB-Dev data standard was designed to interoperate with PDBx/mmCIF and the PDB, so that integrative structures and related metadata can eventually be archived in the PDB.

In parallel with building PDB-Dev, wwPDB partners are working to establish a federated network of interoperating structural biology data resources, as recommended by the IHM Task Force [176]. This effort involves collaboration with other experimental data repositories (e.g., SASBDB [184] and PRIDE [185]). Tools are being created to support automated data exchange between PDB-Dev and these and other biodata repositories (e.g., BioImage Archive, www.ebi.ac.uk/bioimage-archive, accessed on 28 August 2022 [186]). The overarching goal of the wwPDB partnership is to foster federation of structural biology data resources across length scales ranging from atoms to individual proteins to macromolecular machines to organelles to cells and eventually tissues to maximize the impact that atomic level 3D biostructures will have on research and education across basic and applied biological, biomedical and energy sciences.

Supplementary Materials: The following supporting information can be downloaded at: https://www.mdpi.com/article/10.3390/biom12101425/s1, Table S1 Cumulative list of external data resources identified as repackaging and redistributing PDB data.

Author Contributions: Conceptualization, S.K.B. and H.M.B.; software, J.M.D., Z.F., R.L., E.P., D.W.P., Y.R., C.S., M.V. and J.D.W.; writing—original draft preparation, S.K.B.; writing—review and editing, S.K.B., H.M.B., J.M.D., Z.F., J.W.F., B.P.H., R.L., E.P., D.W.P., Y.R., A.S., M.S., C.S., B.V., M.V., J.Y.Y. and C.Z.; visualization, C.S., J.W.F., B.P.H., D.W.P., C.S. and M.V.; supervision, J.M.D., R.L., Y.R., J.Y.Y. and C.Z.; funding acquisition, S.K.B., A.S. and B.V. All authors have read and agreed to the published version of the manuscript.

Funding: RCSB PDB core operations are jointly funded by the National Science Foundation (NSF; DBI-1832184, PI: S.K.B.), the US Department of Energy (DE-SC0019749, PI: S.K.B.), and the National Cancer Institute, the National Institute of Allergy and Infectious Diseases, and the National Institute of General Medical Sciences of the National Institutes of Health (R01GM133198, PI: S.K.B.). Other funding awards to RCSB PDB by the NSF and to PDBe by the UK Biotechnology and Biological Research Council are jointly supporting development of a Next Generation PDB archive (DBI-2019297, PI: S.K.B.; BB/V004247/1, PI: Sameer Velankar) and new Mol* features (DBI-2129634, PI: S.K.B.; BB/W017970/1, PI: Sameer Velankar). PDB-dev development supported NSF awards DBI-1756248, DBI-2112966 (PI: B.V.) and DBI-1756250, DBI-2112967 (A.S.). The content is solely the responsibility of the authors and does not necessarily represent the official views of the National Institutes of Health.

Institutional Review Board Statement: Not applicable.

Informed Consent Statement: Not applicable.

Data Availability Statement: PDB data are made freely available by the wwPDB (wwPDB.org, accessed on 28 August 2022).

Acknowledgments: The authors thank the tens of thousands of structural biologists worldwide who deposited structures to the PDB since 1971 and the many millions of researchers, educators, and students around the world who consume PDB data. We also gratefully acknowledge contributions to the success of the PDB archive made by past members of RCSB PDB and our Worldwide Protein Data Bank partners (PDBe, PDBj, EMDB, and BMRB).

Conflicts of Interest: The authors declare no conflict of interest.

References

1. Protein Data Bank. Crystallography: Protein Data Bank. *Nat. New Biol.* **1971**, *233*, 223. [CrossRef]
2. Berman, H.M.; Henrick, K.; Nakamura, H. Announcing the worldwide Protein Data Bank. *Nat. Struct. Biol.* **2003**, *10*, 980. [CrossRef]
3. wwPDB consortium. Protein Data Bank: The single global archive for 3D macromolecular structure data. *Nucleic Acids Res.* **2019**, *47*, D520–D528. [CrossRef]
4. Berman, H.M.; Westbrook, J.; Feng, Z.; Gilliland, G.; Bhat, T.N.; Weissig, H.; Shindyalov, I.N.; Bourne, P.E. The Protein Data Bank. *Nucleic Acids Res.* **2000**, *28*, 235–242. [CrossRef]
5. Burley, S.K.; Bhikadiya, C.; Bi, C.; Bittrich, S.; Chen, L.; Crichlow, G.; Christie, C.H.; Dalenberg, K.; Costanzo, L.D.; Duarte, J.M.; et al. RCSB Protein Data Bank: Powerful new tools for exploring 3D structures of biological macromolecules for basic and applied research and education in fundamental biology, biomedicine, biotechnology, bioengineering, and energy sciences. *Nucleic Acid Res.* **2021**, *49*, D437–D451. [CrossRef] [PubMed]
6. Burley, S.K.; Bhikadiya, C.; Bi, C.; Bittrich, S.; Chen, L.; Crichlow, G.V.; Duarte, J.M.; Dutta, S.; Fayazi, M.; Feng, Z.; et al. RCSB Protein Data Bank: Celebrating 50 years of the PDB with new tools for understanding and visualizing biological macromolecules in 3D. *Protein Sci.* **2022**, *31*, 187–208. [CrossRef]
7. Burley, S.K.; Bhikadiya, C.; Bi, C.; Bittrich, S.; Chao, H.; Chen, L.; Craig, P.A.; Crichlow, G.V.; Dalenberg, K.; Duarte, J.M.; et al. RCSB Protein Data Bank: Tools for visualizing and understanding biological macromolecules in 3D. *Protein Sci.* 2022; *submitted*.
8. Armstrong, D.R.; Berrisford, J.M.; Conroy, M.J.; Gutmanas, A.; Anyango, S.; Choudhary, P.; Clark, A.R.; Dana, J.M.; Deshpande, M.; Dunlop, R.; et al. PDBe: Improved findability of macromolecular structure data in the PDB. *Nucleic Acids Res.* **2020**, *48*, D335–D343. [CrossRef]

9. Bekker, G.J.; Yokochi, M.; Suzuki, H.; Ikegawa, Y.; Iwata, T.; Kudou, T.; Yura, K.; Fujiwara, T.; Kawabata, T.; Kurisu, G. Protein Data Bank Japan: Celebrating our 20th anniversary during a global pandemic as the Asian hub of three dimensional macromolecular structural data. *Protein Sci.* **2022**, *31*, 173–186. [CrossRef] [PubMed]
10. Tagari, M.; Newman, R.; Chagoyen, M.; Carazo, J.M.; Henrick, K. New electron microscopy database and deposition system. *Trends Biochem. Sci.* **2002**, *27*, 589. [CrossRef]
11. Lawson, C.L.; Patwardhan, A.; Baker, M.L.; Hryc, C.; Garcia, E.S.; Hudson, B.P.; Lagerstedt, I.; Ludtke, S.J.; Pintilie, G.; Sala, R.; et al. EMDataBank unified data resource for 3DEM. *Nucleic Acids Res.* **2016**, *44*, D396–D403. [CrossRef] [PubMed]
12. Ulrich, E.L.; Akutsu, H.; Doreleijers, J.F.; Harano, Y.; Ioannidis, Y.E.; Lin, J.; Livny, M.; Mading, S.; Maziuk, D.; Miller, Z.; et al. BioMagResBank. *Nucleic Acids Res.* **2008**, *36*, D402–D408. [CrossRef]
13. Romero, P.R.; Kobayashi, N.; Wedell, J.R.; Baskaran, K.; Iwata, T.; Yokochi, M.; Maziuk, D.; Yao, H.; Fujiwara, T.; Kurusu, G.; et al. BioMagResBank (BMRB) as a Resource for Structural Biology. *Methods Mol. Biol.* **2020**, *2112*, 187–218. [CrossRef] [PubMed]
14. Wilkinson, M.D.; Dumontier, M.; Aalbersberg, I.J.; Appleton, G.; Axton, M.; Baak, A.; Blomberg, N.; Boiten, J.W.; da Silva Santos, L.B.; Bourne, P.E.; et al. The FAIR Guiding Principles for scientific data management and stewardship. *Sci. Data* **2016**, *3*, 1–9. [CrossRef] [PubMed]
15. van der Aalst, W.M.P.; Bichler, M.; Heinzl, A. Responsible Data Science. *Bus. Inf. Syst. Eng.* **2017**, *59*, 311–313. [CrossRef]
16. Moore, P.B. The PDB and the ribosome. *J. Biol. Chem.* **2021**, *296*, 100561. [CrossRef] [PubMed]
17. Johnson, J.E.; Olson, A.J. Icosahedral virus structures and the protein data bank. *J. Biol. Chem.* **2021**, *296*, 100554. [CrossRef]
18. Neidle, S. Beyond the double helix: DNA structural diversity and the PDB. *J. Biol. Chem.* **2021**, *296*, 100553. [CrossRef] [PubMed]
19. Westhof, E.; Leontis, N.B. An RNA-centric historical narrative around the Protein Data Bank. *J. Biol. Chem.* **2021**, *296*, 100555. [CrossRef] [PubMed]
20. Prestegard, J.H. A perspective on the PDB's impact on the field of glycobiology. *J. Biol. Chem.* **2021**, *296*, 100556. [CrossRef] [PubMed]
21. Li, F.; Egea, P.F.; Vecchio, A.J.; Asial, I.; Gupta, M.; Paulino, J.; Bajaj, R.; Dickinson, M.S.; Ferguson-Miller, S.; Monk, B.C.; et al. Highlighting membrane protein structure and function: A celebration of the Protein Data Bank. *J. Biol. Chem.* **2021**, *296*, 100557. [CrossRef]
22. Chiu, W.; Schmid, M.F.; Pintilie, G.D.; Lawson, C.L. Evolution of standardization and dissemination of cryo-EM structures and data jointly by the community, PDB, and EMDB. *J. Biol. Chem.* **2021**, *296*, 100560. [CrossRef] [PubMed]
23. Pan, X.; Kortemme, T. Recent advances in de novo protein design: Principles, methods, and applications. *J. Biol. Chem.* **2021**, *296*, 100558. [CrossRef]
24. Murray, D.; Petrey, D.; Honig, B. Integrating 3D structural information into systems biology. *J. Biol. Chem.* **2021**, *296*, 100562. [CrossRef]
25. Burley, S.K. Impact of structural biologists and the Protein Data Bank on small-molecule drug discovery and development. *J. Biol. Chem.* **2021**, *296*, 100559. [CrossRef]
26. Taylor, S.S.; Wu, J.; Bruystens, J.G.H.; Del Rio, J.C.; Lu, T.W.; Kornev, A.P.; Ten Eyck, L.F. From structure to the dynamic regulation of a molecular switch: A journey over 3 decades. *J. Biol. Chem.* **2021**, *296*, 100746. [CrossRef]
27. Wolberger, C. How structural biology transformed studies of transcription regulation. *J. Biol. Chem.* **2021**, *296*, 100741. [CrossRef]
28. Wilson, I.A.; Stanfield, R.L. 50 Years of structural immunology. *J. Biol. Chem.* **2021**, *296*, 100745. [CrossRef]
29. Saibil, H.R. The PDB and protein homeostasis: From chaperones to degradation and disaggregase machines. *J. Biol. Chem.* **2021**, *296*, 100744. [CrossRef]
30. Michalska, K.; Joachimiak, A. Structural genomics and the Protein Data Bank. *J. Biol. Chem.* **2021**, *296*, 100747. [CrossRef]
31. Sali, A. From integrative structural biology to cell biology. *J. Biol. Chem.* **2021**, *296*, 100743. [CrossRef]
32. Miller, M.D.; Phillips, G.N., Jr. Moving beyond static snapshots: Protein dynamics and the Protein Data Bank. *J. Biol. Chem.* **2021**, *296*, 100749. [CrossRef]
33. Richardson, J.S.; Richardson, D.C.; Goodsell, D.S. Seeing the PDB. *J. Biol. Chem.* **2021**, *296*, 100742. [CrossRef] [PubMed]
34. Cohen, A.E. A new era of synchrotron-enabled macromolecular crystallography. *Nat. Methods* **2021**, *18*, 433–434. [CrossRef]
35. Kern, D. From structure to mechanism: Skiing the energy landscape. *Nat. Methods* **2021**, *18*, 435–436. [CrossRef] [PubMed]
36. Vinothkumar, K.R. Expanding capabilities and infrastructure for cryo-EM. *Nat. Methods* **2021**, *18*, 437–438. [CrossRef] [PubMed]
37. Das, R. RNA structure: A renaissance begins? *Nat. Methods* **2021**, *18*, 439. [CrossRef]
38. Li, X. Cryo-electron tomography: Observing the cell at the atomic level. *Nat. Methods* **2021**, *18*, 440–441. [CrossRef] [PubMed]
39. Wozny, M.R.; Kukulski, W. Molecular visualization of cellular complexity. *Nat. Methods* **2021**, *18*, 442–443. [CrossRef]
40. Narykov, O.; Srinivasan, S.; Korkin, D. Computational protein modeling and the next viral pandemic. *Nat. Methods* **2021**, *18*, 444–445. [CrossRef] [PubMed]
41. Luthey-Schulten, Z. Integrating experiments, theory and simulations into whole-cell models. *Nat. Methods* **2021**, *18*, 446–447. [CrossRef]
42. Bonvin, A. 50 years of PDB: A catalyst in structural biology. *Nat. Methods* **2021**, *18*, 448–449. [CrossRef]
43. Bourne, P.E.; Addess, K.J.; Bluhm, W.F.; Chen, L.; Deshpande, N.; Feng, Z.; Fleri, W.; Green, R.; Merino-Ott, J.C.; Townsend-Merino, W.; et al. The distribution and query systems of the RCSB Protein Data Bank. *Nucleic Acids Res.* **2004**, *32*, D223–D225. [CrossRef] [PubMed]

44. Young, J.Y.; Westbrook, J.D.; Feng, Z.; Sala, R.; Peisach, E.; Oldfield, T.J.; Sen, S.; Gutmanas, A.; Armstrong, D.R.; Berrisford, J.M.; et al. OneDep: Unified wwPDB System for Deposition, Biocuration, and Validation of Macromolecular Structures in the PDB Archive. *Structure* **2017**, *25*, 536–545. [CrossRef] [PubMed]
45. Gore, S.; Sanz Garcia, E.; Hendrickx, P.M.S.; Gutmanas, A.; Westbrook, J.D.; Yang, H.; Feng, Z.; Baskaran, K.; Berrisford, J.M.; Hudson, B.P.; et al. Validation of Structures in the Protein Data Bank. *Structure* **2017**, *25*, 1916–1927. [CrossRef]
46. Feng, Z.; Westbrook, J.D.; Sala, R.; Smart, O.S.; Bricogne, G.; Matsubara, M.; Yamada, I.; Tsuchiya, S.; Aoki-Kinoshita, K.F.; Hoch, J.C.; et al. Enhanced validation of small-molecule ligands and carbohydrates in the protein databank. *Structure* **2021**, *29*, 393–400.e391. [CrossRef]
47. Young, J.Y.; Westbrook, J.D.; Feng, Z.; Peisach, E.; Persikova, I.; Sala, R.; Sen, S.; Berrisford, J.M.; Swaminathan, G.J.; Oldfield, T.J.; et al. Worldwide Protein Data Bank biocuration supporting open access to high-quality 3D structural biology data. *Database* **2018**, *2018*, bay002. [CrossRef]
48. Kendrew, J.C.; Dickerson, R.E.; Strandberg, B.E.; Hart, R.G.; Davies, D.R.; Phillips, D.C.; Shore, V.C. Structure of myoglobin: A three-dimensional Fourier synthesis at 2 A. resolution. *Nature* **1960**, *185*, 422–427. [CrossRef]
49. Yip, K.M.; Fischer, N.; Paknia, E.; Chari, A.; Stark, H. Atomic-resolution protein structure determination by cryo-EM. *Nature* **2020**, *587*, 157–161. [CrossRef] [PubMed]
50. Shao, C.; Westbrook, J.D.; Lu, C.; Bhikadiya, C.; Peisach, E.; Young, J.Y.; Duarte, J.M.; Lowe, R.; Wang, S.; Rose, Y.; et al. Simplified Quality Assessment for Small-molecule Ligands in the PDB Archive. *Structure* **2022**, *30*, 252–262. [CrossRef] [PubMed]
51. Blundell, T.L.; Johnson, L.N. *Protein Crystallography*; Academic Press: New York, NY, USA, 1976.
52. Rossmann, M.G. The molecular replacement method. *Acta Cryst. A* **1990**, *46 Pt 2*, 73–82. [CrossRef]
53. Read, R.J.; Adams, P.D.; Arendall, W.B., 3rd; Brunger, A.T.; Emsley, P.; Joosten, R.P.; Kleywegt, G.J.; Krissinel, E.B.; Lutteke, T.; Otwinowski, Z.; et al. A new generation of crystallographic validation tools for the protein data bank. *Structure* **2011**, *19*, 1395–1412. [CrossRef]
54. Shao, C.; Yang, H.; Westbrook, J.D.; Young, J.Y.; Zardecki, C.; Burley, S.K. Multivariate Analyses of Quality Metrics for Crystal Structures in the PDB Archive. *Structure* **2017**, *25*, 458–468. [CrossRef]
55. Adams, P.D.; Aertgeerts, K.; Bauer, C.; Bell, J.A.; Berman, H.M.; Bhat, T.N.; Blaney, J.M.; Bolton, E.; Bricogne, G.; Brown, D.; et al. Outcome of the First wwPDB/CCDC/D3R Ligand Validation Workshop. *Structure* **2016**, *24*, 502–508. [CrossRef] [PubMed]
56. Shao, C.; Feng, Z.; Westbrook, J.D.; Peisach, E.; Berrisford, J.; Ikegawa, Y.; Kurisu, G.; Velankar, S.; Burley, S.K.; Young, J.Y. Modernized Uniform Representation of Carbohydrate Molecules in the Protein Data Bank. *Glycobiology* **2021**, *31*, 1204–1218. [CrossRef]
57. Barends, T.R.M.; Stauch, B.; Cherezov, V.; Schlichting, I. Serial femtosecond crystallography. *Nat. Rev. Methods Prim.* **2022**, *2*, 59. [CrossRef]
58. Pearson, A.R.; Mehrabi, P. Serial synchrotron crystallography for time-resolved structural biology. *Curr. Opin. Struct. Biol.* **2020**, *65*, 168–174. [CrossRef]
59. Schmidt, M. Macromolecular movies, storybooks written by nature. *Biophys. Rev.* **2021**, *13*, 1191–1197. [CrossRef] [PubMed]
60. Olmos, J.L., Jr.; Pandey, S.; Martin-Garcia, J.M.; Calvey, G.; Katz, A.; Knoska, J.; Kupitz, C.; Hunter, M.S.; Liang, M.; Oberthuer, D.; et al. Enzyme intermediates captured "on the fly" by mix-and-inject serial crystallography. *BMC Biol.* **2018**, *16*, 59. [CrossRef] [PubMed]
61. Chapman, H.N.; Fromme, P.; Barty, A.; White, T.A.; Kirian, R.A.; Aquila, A.; Hunter, M.S.; Schulz, J.; DePonte, D.P.; Weierstall, U.; et al. Femtosecond X-ray protein nanocrystallography. *Nature* **2011**, *470*, 73–77. [CrossRef] [PubMed]
62. Kuhlbrandt, W. Biochemistry. The resolution revolution. *Science* **2014**, *343*, 1443–1444. [CrossRef] [PubMed]
63. Herzik, M.A., Jr. Cryo-electron microscopy reaches atomic resolution. *Nature* **2020**, *587*, 39–40. [CrossRef]
64. Passmore, L.A.; Russo, C.J. Specimen Preparation for High-Resolution Cryo-EM. *Methods Enzym.* **2016**, *579*, 51–86. [CrossRef]
65. Brilot, A.F.; Chen, J.Z.; Cheng, A.; Pan, J.; Harrison, S.C.; Potter, C.S.; Carragher, B.; Henderson, R.; Grigorieff, N. Beam-induced motion of vitrified specimen on holey carbon film. *J. Struct. Biol.* **2012**, *177*, 630–637. [CrossRef]
66. Li, X.; Mooney, P.; Zheng, S.; Booth, C.R.; Braunfeld, M.B.; Gubbens, S.; Agard, D.A.; Cheng, Y. Electron counting and beam-induced motion correction enable near-atomic-resolution single-particle cryo-EM. *Nat. Methods* **2013**, *10*, 584–590. [CrossRef]
67. Bai, X.C.; Fernandez, I.S.; McMullan, G.; Scheres, S.H. Ribosome structures to near-atomic resolution from thirty thousand cryo-EM particles. *eLife* **2013**, *2*, e00461. [CrossRef]
68. Scheres, S.H. A Bayesian view on cryo-EM structure determination. *J. Mol. Biol.* **2012**, *415*, 406–418. [CrossRef]
69. Scheres, S.H.W. RELION: Implementation of a Bayesian approach to cryo-EM structure determination. *J. Struct. Biol.* **2012**, *180*, 519–530. [CrossRef]
70. Zhang, P. Advances in cryo-electron tomography and subtomogram averaging and classification. *Curr. Opin. Struct. Biol.* **2019**, *58*, 249–258. [CrossRef]
71. Zanetti, G.; Prinz, S.; Daum, S.; Meister, A.; Schekman, R.; Bacia, K.; Briggs, J.A. The structure of the COPII transport-vesicle coat assembled on membranes. *eLife* **2013**, *2*, e00951. [CrossRef]
72. Ni, T.; Sun, Y.; Seaton-Burn, W.; Al-Hazeem, M.M.J.; Zhu, Y.; Yu, X.; Liu, L.-N.; Zhang, P. Tales of Two α-Carboxysomes: The Structure and Assembly of Cargo Rubisco. *bioRxiv* **2022**. [CrossRef]

73. Varadi, M.; Anyango, S.; Deshpande, M.; Nair, S.; Natassia, C.; Yordanova, G.; Yuan, D.; Stroe, O.; Wood, G.; Laydon, A.; et al. AlphaFold Protein Structure Database: Massively expanding the structural coverage of protein-sequence space with high-accuracy models. *Nucleic Acids Res.* **2022**, *50*, D439–D444. [CrossRef]
74. Jumper, J.; Evans, R.; Pritzel, A.; Green, T.; Figurnov, M.; Ronneberger, O.; Tunyasuvunakool, K.; Bates, R.; Zidek, A.; Potapenko, A.; et al. Highly accurate protein structure prediction with AlphaFold. *Nature* **2021**, *596*, 583–589. [CrossRef]
75. Senior, A.W.; Evans, R.; Jumper, J.; Kirkpatrick, J.; Sifre, L.; Green, T.; Qin, C.; Zidek, A.; Nelson, A.W.R.; Bridgland, A.; et al. Improved protein structure prediction using potentials from deep learning. *Nature* **2020**, *577*, 706–710. [CrossRef]
76. Baek, M.; DiMaio, F.; Anishchenko, I.; Dauparas, J.; Ovchinnikov, S.; Lee, G.R.; Wang, J.; Cong, Q.; Kinch, L.N.; Schaeffer, R.D.; et al. Accurate prediction of protein structures and interactions using a three-track neural network. *Science* **2021**, *373*, 871–876. [CrossRef]
77. Mosalaganti, S.; Obarska-Kosinska, A.; Siggel, M.; Taniguchi, R.; Turonova, B.; Zimmerli, C.E.; Buczak, K.; Schmidt, F.H.; Margiotta, E.; Mackmull, M.T.; et al. AI-based structure prediction empowers integrative structural analysis of human nuclear pores. *Science* **2022**, *376*, eabm9506. [CrossRef]
78. Turk, M.; Baumeister, W. The promise and the challenges of cryo-electron tomography. *FEBS Lett.* **2020**, *594*, 3243–3261. [CrossRef]
79. Pintilie, G.; Zhang, K.; Su, Z.; Li, S.; Schmid, M.F.; Chiu, W. Measurement of atom resolvability in cryo-EM maps with Q-scores. *Nat. Methods* **2020**, *17*, 328–334. [CrossRef] [PubMed]
80. Wang, Z.; Patwardhan, A.; Kleywegt, G.J. Validation analysis of EMDB entries. *Acta Crystallogr. Sect. D Struct. Biol.* **2022**, *78*, 542–552. [CrossRef]
81. Burley, S.K.; Berman, H.M.; Chiu, W.; Dai, W.; Flatt, J.W.; Hudson, B.P.; Kaelber, J.; Khare, S.; Kulczyk, A.; Lawson, C.L.; et al. Electron Microscopy Holdings of the Protein Data Bank: Impact of the Resolution Revolution and Implications for the Future. *Biophys Rev.* 2022; *submitted*.
82. Williamson, M.P.; Havel, T.F.; Wuthrich, K. Solution conformation of proteinase inhibitor IIA from bull seminal plasma by 1H nuclear magnetic resonance and distance geometry. *J. Mol. Biol.* **1985**, *182*, 295–315. [CrossRef]
83. Kaptein, R.; Zuiderweg, E.R.; Scheek, R.M.; Boelens, R.; van Gunsteren, W.F. A protein structure from nuclear magnetic resonance data. lac repressor headpiece. *J. Mol. Biol.* **1985**, *182*, 179–182. [CrossRef]
84. Driscoll, P.C.; Gronenborn, A.M.; Beress, L.; Clore, G.M. Determination of the three-dimensional solution structure of the antihypertensive and antiviral protein BDS-I from the sea anemone Anemonia sulcata: A study using nuclear magnetic resonance and hybrid distance geometry-dynamical simulated annealing. *Biochemistry* **1989**, *28*, 2188–2198. [CrossRef] [PubMed]
85. Kaptein, R.; Boelens, R.; Scheek, R.M.; van Gunsteren, W.F. Protein structures from NMR. *Biochemistry* **1988**, *27*, 5389–5395. [CrossRef] [PubMed]
86. Gronenborn, A.M.; Bax, A.; Wingfield, P.T.; Clore, G.M. A powerful method of sequential proton resonance assignment in proteins using relayed 15N-1H multiple quantum coherence spectroscopy. *FEBS Lett.* **1989**, *243*, 93–98. [CrossRef]
87. Clore, G.M.; Appella, E.; Yamada, M.; Matsushima, K.; Gronenborn, A.M. Three-dimensional structure of interleukin 8 in solution. *Biochemistry* **1990**, *29*, 1689–1696. [CrossRef] [PubMed]
88. Pfander, R.; Neumann, L.; Zweckstetter, M.; Seger, C.; Holak, T.A.; Tampe, R. Structure of the active domain of the herpes simplex virus protein ICP47 in water/sodium dodecyl sulfate solution determined by nuclear magnetic resonance spectroscopy. *Biochemistry* **1999**, *38*, 13692–13698. [CrossRef]
89. Montelione, G.T.; Nilges, M.; Bax, A.; Guntert, P.; Herrmann, T.; Richardson, J.S.; Schwieters, C.D.; Vranken, W.F.; Vuister, G.W.; Wishart, D.S.; et al. Recommendations of the wwPDB NMR Validation Task Force. *Structure* **2013**, *21*, 1563–1570. [CrossRef]
90. Vostrikov, V.V.; Mote, K.R.; Verardi, R.; Veglia, G. Structural dynamics and topology of phosphorylated phospholamban homopentamer reveal its role in the regulation of calcium transport. *Structure* **2013**, *21*, 2119–2130. [CrossRef] [PubMed]
91. Lapinaite, A.; Simon, B.; Skjaerven, L.; Rakwalska-Bange, M.; Gabel, F.; Carlomagno, T. The structure of the box C/D enzyme reveals regulation of RNA methylation. *Nature* **2013**, *502*, 519–523. [CrossRef]
92. Lu, M.; Russell, R.W.; Bryer, A.J.; Quinn, C.M.; Hou, G.; Zhang, H.; Schwieters, C.D.; Perilla, J.R.; Gronenborn, A.M.; Polenova, T. Atomic-resolution structure of HIV-1 capsid tubes by magic-angle spinning NMR. *Nat. Struct. Mol. Biol.* **2020**, *27*, 863–869. [CrossRef]
93. Jehle, S.; Vollmar, B.S.; Bardiaux, B.; Dove, K.K.; Rajagopal, P.; Gonen, T.; Oschkinat, H.; Klevit, R.E. N-terminal domain of alphaB-crystallin provides a conformational switch for multimerization and structural heterogeneity. *Proc. Natl. Acad. Sci. USA* **2011**, *108*, 6409–6414. [CrossRef]
94. Gauto, D.F.; Estrozi, L.F.; Schwieters, C.D.; Effantin, G.; Macek, P.; Sounier, R.; Sivertsen, A.C.; Schmidt, E.; Kerfah, R.; Mas, G.; et al. Integrated NMR and cryo-EM atomic-resolution structure determination of a half-megadalton enzyme complex. *Nat. Commun.* **2019**, *10*, 2697. [CrossRef]
95. Puthenveetil, R.; Vinogradova, O. Solution NMR: A powerful tool for structural and functional studies of membrane proteins in reconstituted environments. *J. Biol. Chem.* **2019**, *294*, 15914–15931. [CrossRef]
96. Pervushin, K.V.; Orekhov, V.; Popov, A.I.; Musina, L.; Arseniev, A.S. Three-dimensional structure of (1-71)bacterioopsin solubilized in methanol/chloroform and SDS micelles determined by 15N-1H heteronuclear NMR spectroscopy. *Eur. J. Biochem.* **1994**, *219*, 571–583. [CrossRef]
97. Bondarenko, V.; Wells, M.M.; Chen, Q.; Tillman, T.S.; Singewald, K.; Lawless, M.J.; Caporoso, J.; Brandon, N.; Coleman, J.A.; Saxena, S.; et al. Structures of highly flexible intracellular domain of human alpha7 nicotinic acetylcholine receptor. *Nat. Commun.* **2022**, *13*, 793. [CrossRef]

98. Morag, O.; Sgourakis, N.G.; Baker, D.; Goldbourt, A. The NMR-Rosetta capsid model of M13 bacteriophage reveals a quadrupled hydrophobic packing epitope. *Proc. Natl. Acad. Sci. USA* **2015**, *112*, 971–976. [CrossRef]
99. Lange, O.F.; Lakomek, N.A.; Fares, C.; Schroder, G.F.; Walter, K.F.; Becker, S.; Meiler, J.; Grubmuller, H.; Griesinger, C.; de Groot, B.L. Recognition dynamics up to microseconds revealed from an RDC-derived ubiquitin ensemble in solution. *Science* **2008**, *320*, 1471–1475. [CrossRef] [PubMed]
100. Religa, T.L.; Sprangers, R.; Kay, L.E. Dynamic regulation of archaeal proteasome gate opening as studied by TROSY NMR. *Science* **2010**, *328*, 98–102. [CrossRef] [PubMed]
101. Gutmanas, A.; Adams, P.D.; Bardiaux, B.; Berman, H.M.; Case, D.A.; Fogh, R.H.; Guntert, P.; Hendrickx, P.M.; Herrmann, T.; Kleywegt, G.J.; et al. NMR Exchange Format: A unified and open standard for representation of NMR restraint data. *Nat. Struct. Mol. Biol.* **2015**, *22*, 433–434. [CrossRef] [PubMed]
102. DeLisle, C.F.; Malooley, A.L.; Banerjee, I.; Lorieau, J.L. Pro-islet amyloid polypeptide in micelles contains a helical prohormone segment. *FEBS J.* **2020**, *287*, 4440–4457. [CrossRef] [PubMed]
103. Henderson, R.; Baldwin, J.M.; Ceska, T.A.; Zemlin, F.; Beckmann, E.; Downing, K.H. Model for the structure of bacteriorhodopsin based on high-resolution electron cryo-microscopy. *J. Mol. Biol.* **1990**, *213*, 899–929. [CrossRef]
104. Nannenga, B.L.; Gonen, T. The cryo-EM method microcrystal electron diffraction (MicroED). *Nat. Methods* **2019**, *16*, 369–379. [CrossRef]
105. Shi, D.; Nannenga, B.L.; Iadanza, M.G.; Gonen, T. Three-dimensional electron crystallography of protein microcrystals. *eLife* **2013**, *2*, e01345. [CrossRef] [PubMed]
106. Martynowycz, M.W.; Shiriaeva, A.; Ge, X.; Hattne, J.; Nannenga, B.L.; Cherezov, V.; Gonen, T. MicroED structure of the human adenosine receptor determined from a single nanocrystal in LCP. *Proc. Natl. Acad. Sci. USA* **2021**, *118*, e2106041118. [CrossRef]
107. Nannenga, B.L.; Shi, D.; Hattne, J.; Reyes, F.E.; Gonen, T. Structure of catalase determined by MicroED. *eLife* **2014**, *3*, e03600. [CrossRef]
108. Martynowycz, M.W.; Clabbers, M.T.B.; Hattne, J.; Gonen, T. Ab initio phasing macromolecular structures using electron-counted MicroED data. *Nat. Methods* **2022**, *19*, 724–729. [CrossRef]
109. Westbrook, J.; Bourne, P.E. STAR/mmCIF: An extensive ontology for macromolecular structure and beyond. *Bioinformatics* **2000**, *16*, 159–168. [CrossRef]
110. Fitzgerald, P.M.D.; Westbrook, J.D.; Bourne, P.E.; McMahon, B.; Watenpaugh, K.D.; Berman, H.M. 4.5 Macromolecular dictionary (mmCIF). In *International Tables for Crystallography G. Definition and Exchange of Crystallographic Data*; Hall, S.R., McMahon, B., Eds.; Springer: Dordrecht, The Netherlands, 2005; pp. 295–443.
111. Westbrook, J.D.; Young, J.Y.; Shao, C.; Feng, Z.; Guranovic, V.; Lawson, C.; Vallat, B.; Adams, P.D.; Berrisford, J.M.; Bricogne, G.; et al. PDBx/mmCIF Ecosystem: Foundational semantic tools for structural biology. *J. Mol. Biol.* **2022**, *434*, 167599. [CrossRef]
112. Hall, S.R.; Allen, F.H.; Brown, I.D. The crystallographic information file (CIF): A new standard archive file for crystallography. *Acta Crystallogr. Sect. A Found. Crystallogr.* **1991**, *47*, 655–685. [CrossRef]
113. Haas, J.; Barbato, A.; Behringer, D.; Studer, G.; Roth, S.; Bertoni, M.; Mostaguir, K.; Gumienny, R.; Schwede, T. Continuous Automated Model EvaluatiOn (CAMEO) complementing the critical assessment of structure prediction in CASP12. *Proteins Struct. Funct. Genet.* **2018**, *86* (Suppl. S1), 387–398. [CrossRef]
114. Wagner, J.R.; Churas, C.P.; Liu, S.; Swift, R.V.; Chiu, M.; Shao, C.; Feher, V.A.; Burley, S.K.; Gilson, M.K.; Amaro, R.E. Continuous Evaluation of Ligand Protein Predictions: A Weekly Community Challenge for Drug Docking. *Structure* **2019**, *27*, 1326–1335. [CrossRef]
115. Markosian, C.; Di Costanzo, L.; Sekharan, M.; Shao, C.; Burley, S.K.; Zardecki, C. Analysis of impact metrics for the Protein Data Bank. *Sci. Data* **2018**, *5*, 180212. [CrossRef] [PubMed]
116. Feng, Z.; Verdiguel, N.; Di Costanzo, L.; Goodsell, D.S.; Westbrook, J.D.; Burley, S.K.; Zardecki, C. Impact of the Protein Data Bank Across Scientific Disciplines. *Data Sci. J.* **2020**, *19*, 1–14. [CrossRef]
117. Sullivan, K.P.; Brennan-Tonetta, P.; Marxen, L.J. Economic Impacts of the Research Collaboratory for Structural Bioinformatics (RCSB) Protein Data Bank. 2017. Available online: https://doi.org/10.2210/rcsb_pdb/pdb-econ-imp-2017 (accessed on 28 August 2022).
118. Hill, R.; Stein, C. *Scooped! Estimating Rewards for Priority in Science*; Working Paper; Massachusetts Institute of Technology: Cambridge, MA, USA, 2019.
119. Ahmed, A.; Smith, R.D.; Clark, J.J.; Dunbar, J.B., Jr.; Carlson, H.A. Recent improvements to Binding MOAD: A resource for protein-ligand binding affinities and structures. *Nucleic Acids Res.* **2015**, *43*, D465–D469. [CrossRef]
120. Gilson, M.K.; Liu, T.; Baitaluk, M.; Nicola, G.; Hwang, L.; Chong, J. BindingDB in 2015: A public database for medicinal chemistry, computational chemistry and systems pharmacology. *Nucleic Acids Res.* **2016**, *44*, D1045–D1053. [CrossRef] [PubMed]
121. Sillitoe, I.; Bordin, N.; Dawson, N.; Waman, V.P.; Ashford, P.; Scholes, H.M.; Pang, C.S.M.; Woodridge, L.; Rauer, C.; Sen, N.; et al. CATH: Increased structural coverage of functional space. *Nucleic Acids Res.* **2021**, *49*, D266–D273. [CrossRef] [PubMed]
122. Groom, C.R.; Bruno, I.J.; Lightfoot, M.P.; Ward, S.C. The Cambridge Structural Database. *Acta Cryst. B Struct. Sci. Cryst. Eng. Mater.* **2016**, *72*, 171–179. [CrossRef] [PubMed]
123. Hastings, J.; Owen, G.; Dekker, A.; Ennis, M.; Kale, N.; Muthukrishnan, V.; Turner, S.; Swainston, N.; Mendes, P.; Steinbeck, C. ChEBI in 2016: Improved services and an expanding collection of metabolites. *Nucleic Acids Res.* **2016**, *44*, D1214–D1219. [CrossRef]

124. Gaulton, A.; Hersey, A.; Nowotka, M.; Bento, A.P.; Chambers, J.; Mendez, D.; Mutowo, P.; Atkinson, F.; Bellis, L.J.; Cibrian-Uhalte, E.; et al. The ChEMBL database in 2017. *Nucleic Acids Res.* **2017**, *45*, D945–D954. [CrossRef]
125. Wishart, D.S.; Feunang, Y.D.; Guo, A.C.; Lo, E.J.; Marcu, A.; Grant, J.R.; Sajed, T.; Johnson, D.; Li, C.; Sayeeda, Z.; et al. DrugBank 5.0: A major update to the DrugBank database for 2018. *Nucleic Acids Res.* **2018**, *46*, D1074–D1082. [CrossRef]
126. Cheng, H.; Liao, Y.; Schaeffer, R.D.; Grishin, N.V. Manual classification strategies in the ECOD database. *Proteins Struct. Funct. Genet.* **2015**, *83*, 1238–1251. [CrossRef]
127. McDonald, A.G.; Boyce, S.; Tipton, K.F. ExplorEnz: The primary source of the IUBMB enzyme list. *Nucleic Acids Res.* **2009**, *37*, D593–D597. [CrossRef] [PubMed]
128. Harrow, J.; Frankish, A.; Gonzalez, J.M.; Tapanari, E.; Diekhans, M.; Kokocinski, F.; Aken, B.L.; Barrell, D.; Zadissa, A.; Searle, S.; et al. GENCODE: The reference human genome annotation for The ENCODE Project. *Genome Res.* **2012**, *22*, 1760–1774. [CrossRef]
129. Gene Ontology Consortium. The Gene Ontology resource: Enriching a GOld mine. *Nucleic Acids Res.* **2021**, *49*, D325–D334. [CrossRef]
130. GTEx Consortium. The GTEx Consortium atlas of genetic regulatory effects across human tissues. *Science* **2020**, *369*, 1318–1330. [CrossRef]
131. Yamada, I.; Shiota, M.; Shinmachi, D.; Ono, T.; Tsuchiya, S.; Hosoda, M.; Fujita, A.; Aoki, N.P.; Watanabe, Y.; Fujita, N.; et al. The GlyCosmos Portal: A unified and comprehensive web resource for the glycosciences. *Nat. Methods* **2020**, *17*, 649–650. [CrossRef]
132. York, W.S.; Mazumder, R.; Ranzinger, R.; Edwards, N.; Kahsay, R.; Aoki-Kinoshita, K.F.; Campbell, M.P.; Cummings, R.D.; Feizi, T.; Martin, M.; et al. GlyGen: Computational and Informatics Resources for Glycoscience. *Glycobiology* **2020**, *30*, 72–73. [CrossRef] [PubMed]
133. Tiemeyer, M.; Aoki, K.; Paulson, J.; Cummings, R.D.; York, W.S.; Karlsson, N.G.; Lisacek, F.; Packer, N.H.; Campbell, M.P.; Aoki, N.P.; et al. GlyTouCan: An accessible glycan structure repository. *Glycobiology* **2017**, *27*, 915–919. [CrossRef]
134. Lefranc, M.P.; Giudicelli, V.; Duroux, P.; Jabado-Michaloud, J.; Folch, G.; Aouinti, S.; Carillon, E.; Duvergey, H.; Houles, A.; Paysan-Lafosse, T.; et al. IMGT(R), the international ImMunoGeneTics information system(R) 25 years on. *Nucleic Acids Res.* **2015**, *43*, D413–D422. [CrossRef]
135. Vita, R.; Overton, J.A.; Greenbaum, J.A.; Ponomarenko, J.; Clark, J.D.; Cantrell, J.R.; Wheeler, D.K.; Gabbard, J.L.; Hix, D.; Sette, A.; et al. The immune epitope database (IEDB) 3.0. *Nucleic Acids Res.* **2015**, *43*, D405–D412. [CrossRef]
136. Blum, M.; Chang, H.Y.; Chuguransky, S.; Grego, T.; Kandasaamy, S.; Mitchell, A.; Nuka, G.; Paysan-Lafosse, T.; Qureshi, M.; Raj, S.; et al. The InterPro protein families and domains database: 20 years on. *Nucleic Acids Res.* **2021**, *49*, D344–D354. [CrossRef]
137. Newport, T.D.; Sansom, M.S.P.; Stansfeld, P.J. The MemProtMD database: A resource for membrane-embedded protein structures and their lipid interactions. *Nucleic Acids Res.* **2019**, *47*, D390–D397. [CrossRef] [PubMed]
138. White, S.H.; Snider, C. Membrane Proteins of Known 3D Structure (MPStruc). Available online: http://blanco.biomol.uci.edu/mpstruc/ (accessed on 28 August 2022).
139. Sayers, E.W.; Bolton, E.E.; Brister, J.R.; Canese, K.; Chan, J.; Comeau, D.C.; Connor, R.; Funk, K.; Kelly, C.; Kim, S.; et al. Database resources of the national center for biotechnology information. *Nucleic Acids Res.* **2022**, *50*, D20–D26. [CrossRef]
140. Berman, H.M.; Olson, W.K.; Beveridge, D.L.; Westbrook, J.; Gelbin, A.; Demeny, T.; Hsieh, S.H.; Srinivasan, A.R.; Schneider, B. The nucleic acid database. A comprehensive relational database of three-dimensional structures of nucleic acids. *Biophys. J.* **1992**, *63*, 751–759. [CrossRef]
141. Lomize, M.A.; Lomize, A.L.; Pogozheva, I.D.; Mosberg, H.I. OPM: Orientations of proteins in membranes database. *Bioinformatics* **2006**, *22*, 623–625. [CrossRef] [PubMed]
142. Su, M.; Yang, Q.; Du, Y.; Feng, G.; Liu, Z.; Li, Y.; Wang, R. Comparative Assessment of Scoring Functions: The CASF-2016 Update. *J. Chem. Inf. Model.* **2019**, *59*, 895–913. [CrossRef]
143. Hrabe, T.; Li, Z.; Sedova, M.; Rotkiewicz, P.; Jaroszewski, L.; Godzik, A. PDBFlex: Exploring flexibility in protein structures. *Nucleic Acids Res.* **2016**, *44*, D423–D428. [CrossRef]
144. Tusnady, G.E.; Dosztanyi, Z.; Simon, I. Transmembrane proteins in the Protein Data Bank: Identification and classification. *Bioinformatics* **2004**, *20*, 2964–2972. [CrossRef]
145. Nguyen, D.T.; Mathias, S.; Bologa, C.; Brunak, S.; Fernandez, N.; Gaulton, A.; Hersey, A.; Holmes, J.; Jensen, L.J.; Karlsson, A.; et al. Pharos: Collating protein information to shed light on the druggable genome. *Nucleic Acids Res.* **2017**, *45*, D995–D1002. [CrossRef]
146. Kim, S.; Chen, J.; Cheng, T.; Gindulyte, A.; He, J.; He, S.; Li, Q.; Shoemaker, B.A.; Thiessen, P.A.; Yu, B.; et al. PubChem in 2021: New data content and improved web interfaces. *Nucleic Acids Res.* **2021**, *49*, D1388–D1395. [CrossRef]
147. Nederveen, A.J.; Doreleijers, J.F.; Vranken, W.; Miller, Z.; Spronk, C.A.; Nabuurs, S.B.; Guntert, P.; Livny, M.; Markley, J.L.; Nilges, M.; et al. RECOORD: A recalculated coordinate database of 500+ proteins from the PDB using restraints from the BioMagResBank. *Proteins Struct. Funct. Genet.* **2005**, *59*, 662–672. [CrossRef]
148. Garavelli, J.S. The RESID Database of Protein Modifications as a resource and annotation tool. *Proteomics* **2004**, *4*, 1527–1533. [CrossRef]
149. Dunbar, J.; Krawczyk, K.; Leem, J.; Baker, T.; Fuchs, A.; Georges, G.; Shi, J.; Deane, C.M. SAbDab: The structural antibody database. *Nucleic Acids Res.* **2014**, *42*, D1140–D1146. [CrossRef]

150. Raybould, M.I.J.; Marks, C.; Lewis, A.P.; Shi, J.; Bujotzek, A.; Taddese, B.; Deane, C.M. Thera-SAbDab: The Therapeutic Structural Antibody Database. *Nucleic Acids Res.* **2020**, *48*, D383–D388. [CrossRef] [PubMed]
151. Morin, A.; Eisenbraun, B.; Key, J.; Sanschagrin, P.C.; Timony, M.A.; Ottaviano, M.; Sliz, P. Collaboration gets the most out of software. *eLife* **2013**, *2*, e01456. [CrossRef]
152. Andreeva, A.; Kulesha, E.; Gough, J.; Murzin, A.G. The SCOP database in 2020: Expanded classification of representative family and superfamily domains of known protein structures. *Nucleic Acids Res.* **2020**, *48*, D376–D382. [CrossRef]
153. Chandonia, J.M.; Fox, N.K.; Brenner, S.E. SCOPe: Classification of large macromolecular structures in the structural classification of proteins-extended database. *Nucleic Acids Res.* **2019**, *47*, D475–D481. [CrossRef]
154. Dana, J.M.; Gutmanas, A.; Tyagi, N.; Qi, G.; O'Donovan, C.; Martin, M.; Velankar, S. SIFTS: Updated Structure Integration with Function, Taxonomy and Sequences resource allows 40-fold increase in coverage of structure-based annotations for proteins. *Nucleic Acids Res.* **2019**, *47*, D482–D489. [CrossRef]
155. UniProt Consortium. UniProt: The universal protein knowledgebase in 2021. *Nucleic Acids Res.* **2021**, *49*, D480–D489. [CrossRef]
156. Rigden, D.J.; Fernandez, X.M. The 2022 Nucleic Acids Research database issue and the online molecular biology database collection. *Nucleic Acids Res.* **2022**, *50*, D1–D10. [CrossRef]
157. Westbrook, J.D.; Burley, S.K. How Structural Biologists and the Protein Data Bank Contributed to Recent FDA New Drug Approvals. *Structure* **2019**, *27*, 211–217. [CrossRef]
158. Westbrook, J.D.; Soskind, R.; Hudson, B.P.; Burley, S.K. Impact of Protein Data Bank on Anti-neoplastic Approvals. *Drug Discov. Today* **2020**, *25*, 837–850. [CrossRef]
159. Chiu, M.L.; Gilliland, G.L. Engineering antibody therapeutics. *Curr. Opin. Struct. Biol.* **2016**, *38*, 163–173. [CrossRef]
160. Burley, S.K.; Berman, H.M.; Christie, C.; Duarte, J.M.; Feng, Z.; Westbrook, J.; Young, J.; Zardecki, C. RCSB Protein Data Bank: Sustaining a living digital data resource that enables breakthroughs in scientific research and biomedical education. *Protein Sci.* **2018**, *27*, 316–330. [CrossRef]
161. Zardecki, C.; Dutta, S.; Goodsell, D.S.; Lowe, R.; Voigt, M.; Burley, S.K. PDB-101: Educational resources supporting molecular explorations through biology and medicine. *Protein Sci.* **2022**, *31*, 129–140. [CrossRef]
162. Goodsell, D.S.; Zardecki, C.; Berman, H.M.; Burley, S.K. Insights from 20 Years of the Molecule of the Month. *Biochem. Mol. Biol. Educ.* **2020**, *48*, 350–355. [CrossRef]
163. Goodsell, D.S.; Dutta, S.; Voigt, M.; Zardecki, C.; Burley, S.K. Molecular explorations of cancer biology and therapeutics at PDB-101. *Oncogene* **2022**, *41*, 4333–4335. [CrossRef]
164. Webb, B.; Sali, A. Comparative Protein Structure Modeling Using MODELLER. *Curr. Protoc. Bioinform.* **2016**, *54*, 5.6.1–5.6.37. [CrossRef] [PubMed]
165. Webb, B.; Sali, A. Protein structure modeling with MODELLER. *Methods Mol. Biol.* **2014**, *1137*, 1–15. [CrossRef] [PubMed]
166. Pieper, U.; Webb, B.M.; Dong, G.Q.; Schneidman-Duhovny, D.; Fan, H.; Kim, S.J.; Khuri, N.; Spill, Y.G.; Weinkam, P.; Hammel, M.; et al. ModBase, a database of annotated comparative protein structure models and associated resources. *Nucleic Acids Res.* **2014**, *42*, D336–D346. [CrossRef] [PubMed]
167. Biasini, M.; Schmidt, T.; Bienert, S.; Mariani, V.; Studer, G.; Haas, J.; Johner, N.; Schenk, A.D.; Philippsen, A.; Schwede, T. OpenStructure: An integrated software framework for computational structural biology. *Acta Crystallogr. Ser. D* **2013**, *69*, 701–709. [CrossRef]
168. Waterhouse, A.; Bertoni, M.; Bienert, S.; Studer, G.; Tauriello, G.; Gumienny, R.; Heer, F.T.; de Beer, T.A.P.; Rempfer, C.; Bordoli, L.; et al. SWISS-MODEL: Homology modelling of protein structures and complexes. *Nucleic Acids Res.* **2018**, *46*, W296–W303. [CrossRef]
169. Alford, R.F.; Leaver-Fay, A.; Jeliazkov, J.R.; O'Meara, M.J.; DiMaio, F.P.; Park, H.; Shapovalov, M.V.; Renfrew, P.D.; Mulligan, V.K.; Kappel, K.; et al. The Rosetta All-Atom Energy Function for Macromolecular Modeling and Design. *J. Chem. Theory Comput.* **2017**, *13*, 3031–3048. [CrossRef] [PubMed]
170. Alexander, L.T.; Lepore, R.; Kryshtafovych, A.; Adamopoulos, A.; Alahuhta, M.; Arvin, A.M.; Bomble, Y.J.; Bottcher, B.; Breyton, C.; Chiarini, V.; et al. Target highlights in CASP14: Analysis of models by structure providers. *Proteins Struct. Funct. Genet.* **2021**, *89*, 1647–1672. [CrossRef]
171. Tunyasuvunakool, K.; Adler, J.; Wu, Z.; Green, T.; Zielinski, M.; Zidek, A.; Bridgland, A.; Cowie, A.; Meyer, C.; Laydon, A.; et al. Highly accurate protein structure prediction for the human proteome. *Nature* **2021**, *596*, 590–596. [CrossRef] [PubMed]
172. Burley, S.K.; Berman, H.M. Open-access data: A cornerstone for artificial intelligence approaches to protein structure prediction. *Structure* **2021**, *29*, 515–520. [CrossRef]
173. Burley, S.K.; Arap, W.; Pasqualini, R. Predicting Proteome-Scale Protein Structure with Artificial Intelligence. *N. Engl. J. Med.* **2021**, *385*, 2191–2194. [CrossRef]
174. Shao, C.; Bittrich, S.; Wang, W.; Burley, S.K. Assessing PDB Macromolecular Crystal Structure Confidence at the Individual Amino Acid Residue Level. *Structure*, 2022; *in press*.
175. Berman, H.M.; Adams, P.D.; Bonvin, A.A.; Burley, S.K.; Carragher, B.; Chiu, W.; DiMaio, F.; Ferrin, T.E.; Gabanyi, M.J.; Goddard, T.D.; et al. Federating Structural Models and Data: Outcomes from A Workshop on Archiving Integrative Structures. *Structure* **2019**, *27*, 1745–1759. [CrossRef] [PubMed]

176. Sali, A.; Berman, H.M.; Schwede, T.; Trewhella, J.; Kleywegt, G.; Burley, S.K.; Markley, J.; Nakamura, H.; Adams, P.; Bonvin, A.M.; et al. Outcome of the First wwPDB Hybrid/Integrative Methods Task Force Workshop. *Structure* **2015**, *23*, 1156–1167. [CrossRef] [PubMed]
177. Vallat, B.; Webb, B.; Westbrook, J.D.; Sali, A.; Berman, H.M. Development of a Prototype System for Archiving Integrative/Hybrid Structure Models of Biological Macromolecules. *Structure* **2018**, *26*, 894–904.e892. [CrossRef] [PubMed]
178. Vallat, B.; Webb, B.; Fayazi, M.; Voinea, S.; Tangmunarunkit, H.; Ganesan, S.J.; Lawson, C.L.; Westbrook, J.D.; Kesselman, C.; Sali, A.; et al. New system for archiving integrative structures. *Acta Crystallogr. Sect. D Struct. Biol.* **2021**, *77*, 1486–1496. [CrossRef]
179. Burley, S.K.; Kurisu, G.; Markley, J.L.; Nakamura, H.; Velankar, S.; Berman, H.M.; Sali, A.; Schwede, T.; Trewhella, J. PDB-Dev: A Prototype System for Depositing Integrative/Hybrid Structural Models. *Structure* **2017**, *25*, 1317–1318. [CrossRef]
180. Kim, S.J.; Fernandez-Martinez, J.; Nudelman, I.; Shi, Y.; Zhang, W.; Raveh, B.; Herricks, T.; Slaughter, B.D.; Hogan, J.A.; Upla, P.; et al. Integrative structure and functional anatomy of a nuclear pore complex. *Nature* **2018**, *555*, 475–482. [CrossRef]
181. O'Reilly, F.J.; Xue, L.; Graziadei, A.; Sinn, L.; Lenz, S.; Tegunov, D.; Blotz, C.; Singh, N.; Hagen, W.J.H.; Cramer, P.; et al. In-cell architecture of an actively transcribing-translating expressome. *Science* **2020**, *369*, 554–557. [CrossRef]
182. Ganesan, S.J.; Feyder, M.J.; Chemmama, I.E.; Fang, F.; Rout, M.P.; Chait, B.T.; Shi, Y.; Munson, M.; Sali, A. Integrative structure and function of the yeast exocyst complex. *Protein Sci.* **2020**, *29*, 1486–1501. [CrossRef]
183. Mashtalir, N.; Suzuki, H.; Farrell, D.P.; Sankar, A.; Luo, J.; Filipovski, M.; D'Avino, A.R.; St Pierre, R.; Valencia, A.M.; Onikubo, T.; et al. A Structural Model of the Endogenous Human BAF Complex Informs Disease Mechanisms. *Cell* **2020**, *183*, 802–817.e824. [CrossRef]
184. Kikhney, A.G.; Borges, C.R.; Molodenskiy, D.S.; Jeffries, C.M.; Svergun, D.I. SASBDB: Towards an automatically curated and validated repository for biological scattering data. *Protein Sci.* **2020**, *29*, 66–75. [CrossRef]
185. Perez-Riverol, Y.; Csordas, A.; Bai, J.; Bernal-Llinares, M.; Hewapathirana, S.; Kundu, D.J.; Inuganti, A.; Griss, J.; Mayer, G.; Eisenacher, M.; et al. The PRIDE database and related tools and resources in 2019: Improving support for quantification data. *Nucleic Acids Res.* **2019**, *47*, D442–D450. [CrossRef] [PubMed]
186. Ellenberg, J.; Swedlow, J.R.; Barlow, M.; Cook, C.E.; Sarkans, U.; Patwardhan, A.; Brazma, A.; Birney, E. A call for public archives for biological image data. *Nat. Methods* **2018**, *15*, 849–854. [CrossRef]

Perspective

From Genes to Geography, from Cells to Community, from Biomolecules to Behaviors: The Importance of Social Determinants of Health

Jaysón Davidson [1,2,*], Rohit Vashisht [2] and Atul J. Butte [2]

1 Pharmaceutical Science and Pharmacogenomics Graduate Program, University of California San Francisco, San Francisco, CA 94143, USA
2 Bakar Computational Health Sciences Institute, University of California San Francisco, San Francisco, CA 94143, USA
* Correspondence: jayso'n.davidson@ucsf.edu

Abstract: Much scientific work over the past few decades has linked health outcomes and disease risk to genomics, to derive a better understanding of disease mechanisms at the genetic and molecular level. However, genomics alone does not quite capture the full picture of one's overall health. Modern computational biomedical research is moving in the direction of including social/environmental factors that ultimately affect quality of life and health outcomes at both the population and individual level. The future of studying disease now lies at the hands of the social determinants of health (SDOH) to answer pressing clinical questions and address healthcare disparities across population groups through its integration into electronic health records (EHRs). In this perspective article, we argue that the SDOH are the future of disease risk and health outcomes studies due to their vast coverage of a patient's overall health. SDOH data availability in EHRs has improved tremendously over the years with EHR toolkits, diagnosis codes, wearable devices, and census tract information to study disease risk. We discuss the availability of SDOH data, challenges in SDOH implementation, its future in real-world evidence studies, and the next steps to report study outcomes in an equitable and actionable way.

Keywords: social determinants of health; electronic health records; real-world evidence; census tract; data science

Citation: Davidson, J.; Vashisht, R.; Butte, A.J. From Genes to Geography, from Cells to Community, from Biomolecules to Behaviors: The Importance of Social Determinants of Health. *Biomolecules* **2022**, *12*, 1449. https://doi.org/10.3390/biom12101449

Academic Editor: Cameron Mura

Received: 26 August 2022
Accepted: 6 October 2022
Published: 9 October 2022

1. Introduction

Understanding disease at the molecular level has dominated the field of genetics, which has been the major basis for studying disease risk over the past two decades. Researchers have presented considerable evidence that disease risk is generally conferred through genetic inheritance, and now more recently, through specific rare and common mutations [1,2]. Using tools in molecular and cellular biology, researchers and medical providers can investigate many diseases and conditions. However, the results of previous investigations have shown that disease risk is too complex to model using genetics or molecules alone. Indeed, genetic, social, and environmental factors including socioeconomic status, geolocation, and age, as well as racial and ethnic background play a role in disease risk across different population groups [3]. Growing evidence increasingly indicates the importance of accounting for the social and environmental factors that are likely to affect health outcomes. While Dr. Phil Bourne, whom this Special Issue honors, is certainly known for his work in computational methodologies and structural biology, he also understood the importance of external influences on health and called for better methods to measure and "describe individuals' activity spaces and exposure to the built, natural, social, and economic environments that influence behaviors and health outcomes" [4].

2. Social Determinants of Health

Social determinants of health (SDOH) are one of the ways to capture, represent, and assess the impact of social and environmental factors in clinical research, thus improving patient care. SDOH are the conditions in which people are born, live, work, play, worship, and age, which affect a wide range of health, functioning, quality of life outcomes, and risks [5]. A patient's SDOH can be used to estimate their access to healthcare and treatments, their positive or negative health outcomes, and to assess comorbidities by using information related to an individual's health including alcohol and tobacco usage, socioeconomic status, insurance status, living situation, access to healthy foods, access to health literacy, and access to quality of care [5]. The main components of the SDOH commonly gathered in medicine are grouped into five domains: economic stability, education access and quality, healthcare access and equity, social and community context, and neighborhood and built environment [5]. Though noted separately, each domain is interconnected to match the complexity of SDOH variables and represent SDOH at both the population and individual levels [6,7].

Population-level SDOH measures are heavily reliant on census tract information derived from the United States Census Bureau (U.S. Census). Census tracts are indicative of geographical areas, which are defined as small, relatively permanent statistical subdivisions of a county providing information on demographic and housing estimates, occupation codes, industry codes, product and service codes, and material/fuel codes [8]. Census tracts have surveys such as the American community survey, decennial census, economic surveys, population estimates, public sector census, and economic censuses that can be leveraged to assess the overall impact of socioeconomic parameters on the health and wellbeing of patients in a given healthcare system at a given geographical location. Census tract information is gathered by assigning each person, household, housing unit, institution, farm, business establishment, or other responding entity to a specific location, and then assigning that location to a zip code tabulation area appropriate to the census or sample survey by way of geocoding [8,9]. The geocoding process ensures that the Census Bureau can provide correct counts for small geographic entities and that both the Census Bureau and data users can accumulate the data for small entities to provide totals for larger geographic entities such as zip code areas. Census tract information has been used to develop indices that directly explain the SDOH of people by using their zip code location to develop the area deprivation index, social vulnerability index, and modified retail food index [10–13]. Indices that use census tract information often categorize data by socioeconomic status, location, and education to calculate the deprivation or vulnerability of people residing in a location.

SDOH are utilized in clinical care and research studies by way of electronic health records (EHRs) which are the primary way to capture real-world data from providers on patient encounters in a health system [14]. EHRs provide a unique opportunity to study the relationship between SDOH and the management and outcomes of clinical diseases through real-world data (RWD). RWD captured in EHRs are used to develop real-world evidence (RWE) studies that analyze data and inform providers about the causes of different treatment strategies, disease risk, quality of life, and outcomes for different patients and populations. RWE studies often contain diverse patient populations that are representative of real patients' health where common SDOH are collected. Prior to EHRs, the SDOH were primarily captured by population-level questionnaires administered by the U.S. Census or through direct questionnaires administered in clinical trials. However, the innovation of EHRs has provided us with patient-derived data to help us understand the social and lifestyle factors of patients. SDOH data coupled with questionnaires and clinical data in EHRs could be used to enable precision medical studies on healthcare access and health outcomes, by linking with data about treatments, disease conditions, drug response, insurance status, and demographics.

Although the classification of SDOH at the individual or patient level is becoming increasingly standardized for operational and clinical research purposes, a current challenge in the wide adoption of SDOH in RWE studies is that of missing data, HIPAA regulations, and quality control issues that severely limit the amount of data available to answer clinical questions with high precision [15–18]. Therefore, the roles that the SDOH play in various chronic illnesses and diseases are ill-defined but have the potential to address population- and person-specific questions in the future. Research shows that public health goals cannot be realized without addressing the underlying SDOH that contribute to disparities and outcomes [19,20]. Therefore, healthcare research should strive to include SDOH in addition to race/ethnicity in RWE studies. A plethora of research reveals numerous socioeconomic parameters potentially accelerating disease risk, especially among minorities [20]. We must improve our understanding of the impact of SDOH on disease risk by investigating the different roles that SDOH play for patients, population groups, healthcare providers, healthcare access, and health outcomes (Figure 1).

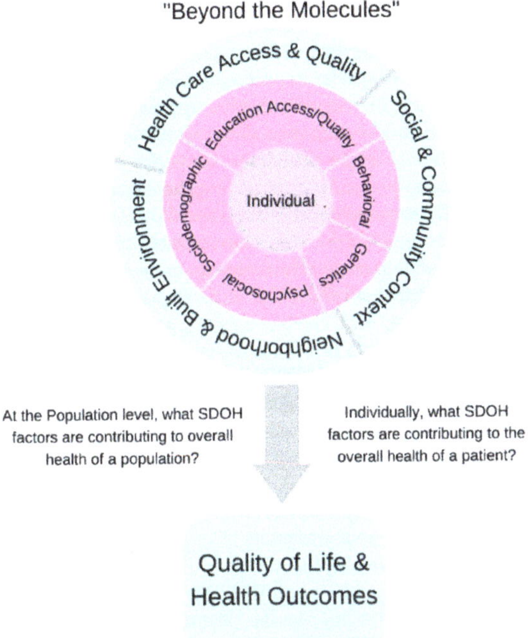

Figure 1. Grouped SDOH factors are categorized at the population and individual levels. At each level, we define the SDOH factors that contribute to the overall health of a population and the overall health of a patient, which mark the difference between a good outcome and a poor outcome.

3. SDOH Integration into Electronic Health Records

In EHR databases across the country, there is incompleteness of SDOH data, which has led previous RWE studies to use only race/ethnicity, sex, and age as measures of SDOH. In theory, those demographics can provide context, but cannot capture the full picture of one's overall health. In an attempt to capture SDOH effectively in EHRs, efforts to map de-identified patients' information to census tracts have been extremely important in providing researchers the ability to use evidence-based SDOH to answer clinical questions. However, the SDOH are often only captured in clinical notes, without structured coding, and we need better methods to obtain SDOH data trapped in notes. Currently, social aspects found in clinical notes vary across EHR databases in the country; however, the Institute of Medicine (IOM) has worked aggressively on identifying SDOH domains to be suggested for use in EHRs for academic research purposes [21]. The type of data suggested

will increase the opportunities to use precision medicine to target clinical diseases. Precision medicine can be used to accurately prescribe patients based on RWE of health outcomes and treatment patterns respective to different population groups. More importantly, it will give clinicians the ability to prescribe patients using patient-centered approaches derived from research. Nevertheless, it is critical to remember that precision medicine means more than just genes, molecules, and cells. The future implementation of SDOH will provide greater precision of treatments based on an array of demographics, lifestyle factors, and environmental factors, all of which are likely to make a greater difference for our patients than any given measured cell or base-pair in their genome.

Author Contributions: Conceptualization, J.D. and A.J.B.; writing—original draft preparation, J.D.; writing—review and editing, J.D., R.V. and A.J.B.; visualization, J.D. All authors have read and agreed to the published version of the manuscript.

Funding: Research reported in this publication was supported by the UCSF Bakar Computational Health Sciences Institute, and the National Center for Advancing Translational Sciences, National Institutes of Health, through UCSF-CTSI grant number UL1 TR001872, along with the Food and Drug Administration, through U01 FD005978 to the UCSF–Stanford Center of Excellence in Regulatory Sciences and Innovation (CERSI). Its contents are solely the responsibility of the authors and do not necessarily represent the official views of the NIH or FDA. None of the study sponsors had any influence over the data interpretation or conclusions of this study.

Conflicts of Interest: Atul Butte is a co-founder and consultant to Personalis and NuMedii; consultant to Mango Tree Corporation, and in the recent past, Samsung, 10x Genomics, Helix, Pathway Genomics, and Verinata (Illumina); has served on paid advisory panels or boards for Geisinger Health, Regenstrief Institute, Gerson Lehman Group, AlphaSights, Covance, Novartis, Genentech, Merck, and Roche; is a shareholder in Personalis and NuMedii; is a minor shareholder in Apple, Meta (Facebook), Alphabet (Google), Microsoft, Amazon, Snap, 10x Genomics, Illumina, Regeneron, Sanofi, Pfizer, Royalty Pharma, Moderna, Sutro, Doximity, BioNtech, Invitae, Pacific Biosciences, Editas Medicine, Nuna Health, Assay Depot, and Vet24seven, and several other non-health-related companies and mutual funds; and has received honoraria and travel reimbursement for invited talks from Johnson and Johnson, Roche, Genentech, Pfizer, Merck, Lilly, Takeda, Varian, Mars, Siemens, Optum, Abbott, Celgene, AstraZeneca, AbbVie, Westat, and many academic institutions, medical or disease-specific foundations and associations, and health systems. Atul Butte receives royalty payments through Stanford University for several patents and other disclosures licensed to NuMedii and Personalis. Atul Butte's research has been funded by NIH, Peraton (as the prime on an NIH contract), Genentech, Johnson and Johnson, FDA, Robert Wood Johnson Foundation, Leon Lowenstein Foundation, Intervalien Foundation, Priscilla Chan and Mark Zuckerberg, the Barbara and Gerson Bakar Foundation, and in the recent past, the March of Dimes, Juvenile Diabetes Research Foundation, California Governor's Office of Planning and Research, California Institute for Regenerative Medicine, L'Oreal, and Progenity.

References

1. Baptista, P.V. Principles in genetic risk assessment. *Ther. Clin. Risk Manag.* **2005**, *1*, 15–20. [CrossRef]
2. Hall, W.D.; Morley, K.I.; Lucke, J.C. The prediction of disease risk in genomic medicine. *EMBO Rep.* **2004**, *5*, S22–S26. [CrossRef] [PubMed]
3. Fraiman, Y.S.; Wojcik, M.H. The influence of social determinants of health on the genetic diagnostic odyssey: Who remains undiagnosed, why, and to what effect? *Pediatr. Res.* **2020**, *89*, 295–300. [CrossRef] [PubMed]
4. Breen, N.; Berrigan, D.; Jackson, J.S.; Wong, D.W.; Wood, F.B.; Denny, J.C.; Zhang, X.; Bourne, P.E. Translational Health Disparities Research in a Data-Rich World. *Health Equity* **2019**, *3*, 588–600. [CrossRef] [PubMed]
5. Healthy People 2030, U.S. Department of Health and Human Services, Office of Disease Prevention and Health Promotion. 2018. Available online: https://health.gov/healthypeople/objectives-and-data/social-determinants-health (accessed on 11 March 2022).
6. Haire-Joshu, D.; Hill-Briggs, F. The Next Generation of Diabetes Translation: A Path to Health Equity. *Annu. Rev. Public Health* **2019**, *40*, 391–410. [CrossRef] [PubMed]
7. Kind, A.J.; Buckingham, W.R. Making Neighborhood-Disadvantage Metrics Accessible—The Neighborhood Atlas. *N. Engl. J. Med.* **2018**, *378*, 2456–2458. [CrossRef]
8. U.S. Census Bureau. 1994. Census Tracts and Block Numbering Areas. Available online: https://www2.census.gov/geo/pdfs/reference/GARM/Ch10GARM.pdf (accessed on 11 November 2021).

9. Kolak, M.; Bhatt, J.; Park, Y.H.; Padrón, N.A.; Molefe, A. Quantification of Neighborhood-Level Social Determinants of Health in the Continental United States. *JAMA Netw. Open* **2020**, *3*, e1919928. [CrossRef]
10. Singh, G.K. Area Deprivation and Widening Inequalities in US Mortality, 1969–1998. *Am. J. Public Health* **2003**, *93*, 1137–1143. [CrossRef]
11. Flanagan, B.E.; Gregory, E.W.; Hallisey, E.J.; Heitgerd, J.L.; Lewis, B. A Social Vulnerability Index for Disaster Management. *J. Homel. Secur. Emerg. Manag.* **2020**, 4–6. [CrossRef]
12. Nagano, I.; Wu, Z.; Asano, K.; Takahashi, Y. Census Tract Level State Maps of the Modified Retail Food Environment Index (mRFEI). *Natl. Cent. Chronic Dis. Prev. Health Promot. Div.* **2011**, *178*, 134–142.
13. Knighton, A.J.; Savitz, L.; Belnap, T.; Stephenson, B.; VanDerslice, J. Introduction of an Area Deprivation Index Measuring Patient Socio-economic Status in an Integrated Health System: Implications for Population Health. *EGEMs (Gener. Evid. Methods Improv. Patient Outcomes)* **2016**, *4*, 9. [CrossRef] [PubMed]
14. Nelson, C.A.; Butte, A.J.; Baranzini, S.E. Integrating biomedical research and electronic health records to create knowledge-based biologically meaningful machine-readable embeddings. *Nat. Commun.* **2019**, *10*, 1–7. [CrossRef] [PubMed]
15. The Office for Civil Rights (OCR); Malin, B. Guidance Regarding Methods for de-identification of protected health information in accordance with the Health Insurance Portability and Accountability Act (HIPAA) Privacy Rule. *Health Inf. Priv.* **2012**, 1–32. Available online: https://privacysecurityacademy.com/wp-content/uploads/2021/03/HHS-OCR-Guidance-on-De-Identification-of-PHI-2012.pdf (accessed on 25 August 2022).
16. Shortreed, S.M.; Cook, A.J.; Coley, R.Y.; Bobb, J.F.; Nelson, J.C. Challenges and opportunities for using big health care data to advance medical science and public health. *Am. J. Epidemiol.* **2019**, *188*, 851–861. [CrossRef] [PubMed]
17. Rudrapatna, V.A.; Butte, A.J. Opportunities and challenges in using real-world data for health care. *J. Clin. Investig.* **2020**, *130*, 565–574. [CrossRef] [PubMed]
18. Cantor, M.N.; Thorpe, L. Integrating data on social determinants of health into electronic health records. *Health Aff.* **2018**, *37*, 585–590. [CrossRef] [PubMed]
19. Marmot, M. Social determinants of health inequalities. *Lancet* **2005**, *365*, 1099–1104. [CrossRef]
20. Thornton, P.L.; Kumanyika, S.K.; Gregg, E.W.; Araneta, M.R.; Baskin, M.L.; Chin, M.H.; Mangione, C.M. New research directions on disparities in obesity and type 2 diabetes. *Ann. N. Y. Acad. Sci.* **2020**, *1461*, 5–24. [CrossRef]
21. IOM (Institute of Medicine). *Capturing Social and Behavioral Domains and Measures in Electronic Health Records: Phase 2*; National Academies Press: Washington, DC, USA, 2015; pp. 1–351. [CrossRef]
22. Hamilton, C.M.; Strader, L.C.; Pratt, J.G.; Maiese, D.; Hendershot, T.; Kwok, R.K.; Hammond, J.A.; Huggins, W.; Jackman, D.; Pan, H.; et al. The PhenX Toolkit: Get the Most From Your Measures. *Am. J. Epidemiol.* **2011**, *174*, 253–260. [CrossRef]
23. Brämer, G.R. *International Statistical Classification of Diseases and Related Health Problems: Tenth Revision*, 2nd ed.; World Health Organization: Geneva, Switzerland, 2004.
24. MIT Critical Data. Secondary Analysis of Electronic Health Records. In *Secondary Analysis of Electronic Health Records*; Springer International Publishing: Berlin/Heidelberg, Germany, 2016; pp. 1–427. [CrossRef]
25. World Health Organization. *mHealth: New Horizons for Health through Mobile Technologies: Second Global Survey on Ehealth*; World Health Organization: Geneva, Switzerland, 2011.
26. Akbari, P.; Sosina, O.A.; Bovijn, J.; Landheer, K.; Nielsen, J.B.; Kim, M.; Aykul, S.; De, T.; Haas, M.E.; Hindy, G.; et al. Multiancestry exome sequencing reveals INHBE mutations associated with favorable fat distribution and protection from diabetes. *Nat. Commun.* **2022**, *13*, 4844. [CrossRef]
27. Hughes, M.M.; Wang, A.; Grossman, M.K.; Pun, E.; Whiteman, A.; Deng, L.; Hallisey, E.; Sharpe, J.D.; Ussery, E.N.; Stokley, S.; et al. County-Level COVID-19 Vaccination Coverage and Social Vulnerability—United States, 14 December 2020–1 March 2021. *MMWR. Morb. Mortal. Wkly. Rep.* **2021**, *70*, 431–436. [CrossRef] [PubMed]
28. Azap, R.A.; Paredes, A.Z.; Diaz, A.; Hyer, J.M.; Pawlik, T.M. The association of neighborhood social vulnerability with surgical textbook outcomes among patients undergoing hepatopancreatic surgery. *Surgery* **2020**, *168*, 868–875. [CrossRef] [PubMed]
29. Ghirimoldi, F.M.; Schmidt, S.; Simon, R.C.; Wang, C.-P.; Wang, Z.; Brimhall, B.B.; Damien, P.; Moffett, E.E.; Manuel, L.S.; Sarwar, Z.U.; et al. Association of Socioeconomic Area Deprivation Index with Hospital Readmissions After Colon and Rectal Surgery. *J. Gastrointest. Surg.* **2020**, *25*, 795–808. [CrossRef] [PubMed]
30. Rosenzweig, M.Q.; Althouse, A.D.; Sabik, L.; Arnold, R.; Chu, E.; Smith, T.J.; Smith, K.; White, D.; Schenker, Y. The Association Between Area Deprivation Index and Patient-Reported Outcomes in Patients with Advanced Cancer. *Health Equity* **2021**, *5*, 8–16. [CrossRef]

 biomolecules

Article

HBcompare: Classifying Ligand Binding Preferences with Hydrogen Bond Topology

Justin Z. Tam, Zhaoming Kong [†], Omar Ahmed [†], Lifang He and Brian Y. Chen *

Department Computer Science and Engineering, Lehigh University, 113 Research Drive, Bethlehem, PA 19004, USA
* Correspondence: chen@cse.lehigh.edu
† These authors contributed equally to this work.

Abstract: This paper presents HBcompare, a method that classifies protein structures according to ligand binding preference categories by analyzing hydrogen bond topology. HBcompare excludes other characteristics of protein structure so that, in the event of accurate classification, it can implicate the involvement of hydrogen bonds in selective binding. This approach contrasts from methods that represent many aspects of protein structure because holistic representations cannot associate classification with just one characteristic. To our knowledge, HBcompare is the first technique with this capability. On five datasets of proteins that catalyze similar reactions with different preferred ligands, HBcompare correctly categorized proteins with similar ligand binding preferences 89.5% of the time. Using only hydrogen bond topology, classification accuracy with HBcompare surpassed standard structure-based comparison algorithms that use atomic coordinates. As a tool for implicating the role of hydrogen bonds in protein function categories, HBcompare represents a first step towards the automatic explanation of biochemical mechanisms.

Keywords: structural bioinformatics; function annotation; specificity annotation

Citation: Tam, J.Z.; Kong, Z.; Ahmed, O.; He, L.; Chen, B.Y. HBcompare: Classifying Ligand Binding Preferences with Hydrogen Bond Topology. *Biomolecules* **2022**, *12*, 1589. https://doi.org/10.3390/biom12111589

Academic Editors: Lei Xie and Cameron Mura

Received: 18 August 2022
Accepted: 25 October 2022
Published: 28 October 2022

1. Introduction

Exploring the space of protein structures with algorithms that compare molecular shape can reveal structural similarities that point to shared evolutionary origins and biological functions. The nature of these observations is influenced strongly by the way in which molecular structure is represented. Algorithms that represent protein structure as a geometric arrangement of secondary structure elements [1,2] or as a collection of alpha carbon coordinates [3,4] can reveal relationships between families of protein folds [5,6]. Comparisons of binding sites, represented as collections of atomic coordinates [7,8], molecular surface patches [9,10] or volumetric constructs [11], can identify proteins with similar catalytic functions and different overall folds [12]. Representing binding site geometry or electrostatic isopotentials as geometric solids can reveal differences in binding site geometry and charge that identify mechanisms that alter binding specificity [13–16].

Existing representations integrate many aspects of protein structure, but none to our knowledge focus exclusively on the topology of hydrogen bonds. Yet hydrogen bonds play a central role in organizing tertiary structure and in governing the specificity of molecular recognition. For this reason, we hypothesize that the topology of hydrogen bonds, alone, can distinguish proteins with different binding preferences, even if they have the same overall fold. To evaluate this hypothesis, we developed *HBcompare*, a deep learning algorithm for comparing the topology of hydrogen bonds in protein structures.

The specific problem studied here begins with a superfamily of proteins that perform the same catalytic function, which have been classified into subfamilies with different binding preferences. The goal is to classify a new protein into one of these subfamilies based on similarities in hydrogen bond topology. In such cases, the superfamily exhibits the same overall fold, so the topology of their hydrogen bonds is largely conserved. At the same time, critical variations in hydrogen bonding patterns could lead to differences in binding specificity that differentiate subfamilies in terms of preferred binding partners. Correctly classifying a protein into one of the subfamilies requires a look beyond the shared similarities of the superfamily to focus on differences that betray subfamily membership.

HBcompare describes the topology of hydrogen bonds in a protein structure using a *molecular graph*, which we define in detail below. As a representation of protein structures, graphs have been used frequently to describe spatial relationships between atoms, amino acids and secondary structure elements (e.g., [17]) or protein structure prediction (e.g., [18]). Rather than represent more aspects of protein structure, HBcompare is first to use graphs that exclusively represent the topology of hydrogen bonds.

This exclusivity enables a novel capability: Since HBcompare atomistically considers only hydrogen bond topology, the classification of a protein into a subfamily with specific binding preferences is also predicting a role for hydrogen bond topology in the specificity mechanism. That is, since only hydrogen bond topology is considered, it must be at least related to the difference between categories. We call this feature "mechanism prediction", and it cannot be performed with holistic methods. In the holistic case, multiple biophysical mechanisms, such as atomic coordinates and electrostatic potentials, are used together in a weighted fashion to distinguish between specificity categories. In such cases, a single mechanism cannot be said to explain the distinction between categories.

The atomistic approach has useful applications. By suggesting a role for hydrogen bonding, HBcompare generates explanations that a non-computational user can adapt into experimental design. For example, if similarities in hydrogen bond topology justify the classification of a protein structure into a category with well defined binding preferences, then it is logical that experiments that mutate hydrogen bond donors and acceptors may reveal the bonds that play an important role in recognition. Without that observation, a much larger space of experimental redesigns must be considered.

Naturally, HBcompare is only a first step in creating possibilities for automatically explaining binding mechanisms. Furthermore, a complete explanation may not always possible, because some biophysical phenomena will co-occur with hydrogen bonds. For example, a protein that lacks one side of a salt bridge differs from one with a complete salt bridge because it might lack a hydrogen bond donor or because it might lack a charged amino acid. We see HBcompare as one tool in an *Analytic Ensemble* that would eventually be complemented by other methods—both holistic and atomistic—that focus on other mechanisms, such as electrostatic isopotentials [16]. Together, these tools might assemble explanations for mechanisms that achieve specific binding.

HBcompare classifies patterns of hydrogen bonds using graph convolutional networks (GCNs), which make use of the symmetrically normalized graph Laplacian to compute vertex embeddings and to evaluate vertex similarity [19]. Recent works [20,21] have shown that GCNs are useful for automating feature learning from graph-structured data compared to traditional methods, such as convolutional neural networks (CNN). HBcompare adapts existing GCN approaches by constructing a molecular graph for each protein to aggregate neighborhood information. As a result, HBcompare performs accurate graph classification and avoids sensitivity to the input order of graph vertices, which can be a challenge for existing methods.

In this paper, we evaluated the effectiveness of HBcompare for classifying protein binding preferences on several protein superfamilies. Each superfamily was selected because it contained well defined subfamilies with different binding preferences, where differences in specificity hinge on differences in hydrogen bonding patterns. These superfamilies include groups of subfamilies from the tRNA-synthetases, the alpha-amylases, and the serine proteases. Our computational results explore how accurately HBcompare performed classifications consistent with experimentally established binding preferences. We also examine how HBcompare would perform in a more holistic setting, integrated with atomic coordinates, and compare its performance to existing methods on the same kinds of features. These results point to the importance of considering the distinct applications of both holistic and atomistic techniques.

2. Methods

2.1. Constructing Molecular Graphs with HBondFinder

HBcompare represents hydrogen bond topologies using molecular graphs. We define a molecular graph as an undirected graph $G = (V, E, \mathbf{A})$. The nodes or vertices $V = \{v_i\}_{i=1}^{N}$ are atoms that are hydrogen bond donors and/or acceptors. The edges E are hydrogen bonds, identified between one donor and one acceptor atom. Since donors and acceptors may be positioned to participate in one of several possible hydrogen bonds, the resulting graph may be more than a collection of disconnected donor-acceptor pairs. Finally, a weighted adjacency matrix $\mathbf{A}$ describes the weights $\mathbf{A}_{ij}$ of edges between nodes i and j.

To generate molecular graphs from protein structures, we developed HBondFinder, which uses geometric criteria to determine the set of all possible hydrogen bonds. Beginning with a standard chain from the Protein Data Bank [22], we prepare the data by first removing all ligands, ions, hydrogens and water molecules. Hydrogens specifically are removed because their positions are not always solved in an experimental crystal structure, leaving some amino acids with incomplete protonation. Thus, for uniformity, we model the positions of all hydrogens using the reduce tool from MolProbity [23], assuming biological pH. We then use the element of each atom, its position within an amino acid and residue names, which define the type of amino acid, to identify all atoms that are hydrogen bond donors, donor hydrogens, hydrogen bond acceptors, and acceptor antecedants. These four atoms appear in pairs on each end of the hydrogen bond. The nodes of the molecular graph are defined by any atom that is a donor, acceptor, or both.

HBondFinder defines the edges of the graph by finding all donor-acceptor pairs that satisfy our hydrogen bond criteria, which are inspired by the HBPlus program [24]. This process is accelerated with a lattice-based geometric data structure [25] that allows us to rapidly search for all atoms of a specific identity that are within a radius of a given point. This search allows us to find all combinations of the four critical atoms of a hydrogen bond: "D", the hydrogen bond donor, "H", the donor hydrogen, "A", the acceptor, and "AA", the acceptor antecedent. From these combinations, we enforce our criteria: First, the D-A distance must be within 3.9 Å, and the H-A distance must be within 2.5 Å. In addition, the angles D-H-A, H-A-AA, D-A-AA, where the middle member is the node of the angle, must all exceed 90 degrees. If these four atoms satisfy the constraints, then a hydrogen bond could exist and we add an edge to the graph, and a weight of 1.0 to the adjacency matrix, between donor and acceptor. All weights on the adjacency matrix are otherwise zero. We refer to graphs with these binary weights as *coordinate-free molecular graphs*.

To compare the predictive value of coordinate-free molecular graphs to a maximally similar representation that incorporates atomic coordinates, we also created a second kind of molecular graph called a *coordinate-based molecular graph*. These graphs are identical except that the edges recorded in the adjacency matrix, between donors and acceptors that can form a hydrogen bond, are weighted by the Euclidean distance in angstroms.

2.2. HBcompare

Overview. We hypothesize that molecular graphs with similar topology and class labels will describe proteins with similar binding preferences. These proteins are expected to exhibit different numbers of atoms, different amino acids, different numbers of hydrogen bond donors and acceptors, and also some variation in edge topology. The classification task performed by HBcompare begins with a set of molecular graphs $\{G_1, \cdots, G_M\}$, each assigned a subfamily class label $\{y_i\}_{i=1}^{M}$. HBcompare performs whole-graph analysis on an input graph G_i to learn an embedding $\mathbf{e}_{G_i}$ and predict its subfamily label y_i (Figure 1).

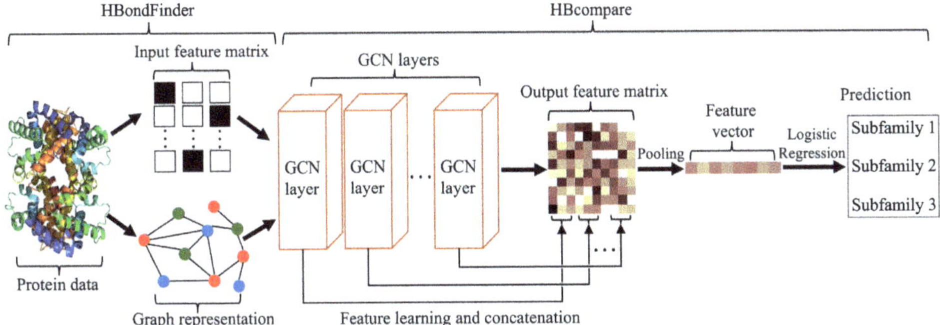

Figure 1. The HBcompare model. As input, HBondFinder takes protein structures and constructs the feature matrix and graph representation. Next, these data are analyzed using GCN layers and their results are concatenated to generate the output feature matrix, which is vectorized via graph pooling and fed to a logistic regression (LR) classifier.

Consider the general multi-layer GCN model with the following propagation rule for graph-structured data [19]:

$$\mathbf{X}^{(l)} = \sigma(\hat{\mathbf{A}}\mathbf{X}^{(l-1)}\mathbf{W}^{(l)}), \tag{1}$$

where $\hat{\mathbf{A}} \in \mathbb{R}^{N\times N}$ is the normalized adjacency matrix of the graph G with added self-connections, i.e., $\hat{\mathbf{A}} = \mathbf{D}^{-\frac{1}{2}}(\mathbf{A}+\mathbf{I}_N)\mathbf{D}^{-\frac{1}{2}}$, $\mathbf{D}$ is the degree matrix, $\mathbf{W}^{(l)} \in \mathbb{R}^{D^{(l-1)}\times D^{(l)}}$ is the layer-specific weight matrix with trainable parameters, and $\sigma(\cdot)$ is a nonlinear activation function. $\mathbf{X}^{(l-1)} \in \mathbb{R}^{N\times D^{(l-1)}}$ is the input of the l-th layer, and $\mathbf{X}^{(l)} \in \mathbb{R}^{N\times D^{(l)}}$ is the output of the l-th layer. Naturally, $\mathbf{X}^{(0)}$ is the initial node feature matrix.

In the following, we show how the propagation rule of GCN in Equation (1) can be extended to multiplex models, thereby enabling HBcompare to learn graph representations across multiple graphs with different orders and sizes of nodes.

Node feature construction. Unfortunately, the initial node features are not available. To solve this issue, we notice that every node of a molecular graph is labelled a hydrogen bond donor, acceptor, or both, we adopted the one-hot encoding strategy on node labels [26] to construct the input node feature matrix $\mathbf{X}^{(0)} \in \mathbb{R}^{N\times 3}$.

Multi-GCN model. After the initial node representations are obtained, each molecular graph can be represented by $G = (V, E, \mathbf{A}, \mathbf{X}^{(0)})$. To explain how the multi-GCN model works, we first analyze the propagation Equation (1) and factorize it into feature aggregation (FA) and feature transformation (FT) following [27].

Feature aggregation. To learn the node representation $\mathbf{X}^{(l)}$ of the l-th layer, in the first step GCN follows the neighborhood aggregation strategy to smooth nodes' representations over a graph by

$$\hat{\mathbf{X}}^{(l)} = \hat{\mathbf{A}}\mathbf{X}^{(l-1)}, \tag{2}$$

This means that the role of $\hat{\mathbf{A}}$ in GCN is to aggregate the neighborhood information of a node for updating its embedding. This design of GCN is suitable for hydrogen bond data analysis. First, the learning process and the ultimate classification of graphs with similar topologies is performed independent of the order in which the nodes are described. Second, the GCN approach is unaffected by graphs with sparse edges, where classification is more difficult. Finally, noise in hydrogen positions, which may affect whether a hydrogen bond is considered to exist near its length and angle limits, is also unlikely to affect classification.

Feature transformation. After FA, in the second step GCN conducts FT in the l-th layer, which consists of linear and nonlinear transformations:

$$\mathbf{X}^{(l)} = \sigma(\hat{\mathbf{X}}^{(l)}\mathbf{W}^{(l)}) \tag{3}$$

The weight matrix $\mathbf{W}^{(l)}$ can adjust the output features, which is equivalent to feature selection and combination. Intuitively, if the same weight matrix is used for different graphs, then we can project them into the common feature space with the same dimension to perform group analysis.

Based on the above analysis, we generalize the propagation rule in Equation (1) to the following form for multi-graph embedding.

$$\mathbf{X}_i^{(l)} = \sigma(\hat{\mathbf{A}}_i \mathbf{X}_i^{(l-1)} \mathbf{W}^{(l)}), \quad \forall i \in \{1, 2, \cdots, M\} \tag{4}$$

where $\hat{\mathbf{A}}_i$ is the normalized adjacency matrix of the i-th graph G_i, $\mathbf{X}_i^{(l-1)}$ and $\mathbf{X}_i^{(l)}$ are its corresponding input and output embeddings of nodes in the l-th layer, and $\mathbf{W}^{(l)}$ is the trainable weight matrix shared by all graphs.

To obtain the vector representation $\mathbf{e}_{G_i}$ of the entire graph G_i, a general and straightforward practice [28,29] is to aggregate the embedded node features of the last GCN layer. However, the extracted information from each layer could also be useful to supplement the graph structure—especially for the molecular graphs that are sparse and the initial information of nodes is not rich. Thus, we adopt the concatenation strategy [30] to exploit features from all layers at multiple scales to contribute to the characterization of the graph, and let the classifier decide which of the features are useful. More specifically, we concatenate the node features $\mathbf{X}_i^{(l)}$ from all layers to get the final node representation matrix

$$\mathbf{X}_i^{all} = [\mathbf{X}_i^{(1)}, \mathbf{X}_i^{(2)}, \cdots, \mathbf{X}_i^{(L)}], \tag{5}$$

where $\mathbf{X}_i^{all} \in \mathbb{R}^{N_i \times \sum_{l=1}^{L} D^{(l)}}$, with each row corresponding to a node and each column corresponding to a feature, and $N_i = |V_i|$ is the number of nodes for the i-th graph G_i.

Whole-graph training. Based on the node representations, we are able to design different task-specific loss functions to train the overall multi-GCN model in the same way as of training GCN. Since for the protein family identification problem, we have access to all nodes from the entire datasets and the node labels are available, we can adopt the learning method in [20,31] to make full use of the node-level information and also capture the substructures within each graph to improve classification accuracies. Specifically, given the node label set $\mathbf{Y}$ for all nodes, the training process for multi-GCN is then formulated as:

$$\min_{\mathbf{W}^{(1)}, \cdots, \mathbf{W}^{(L)}, \mathbf{\Theta}} \quad Loss(\{\mathbf{X}_1^{all}, \cdots, \mathbf{X}_M^{all}\}, \mathbf{\Theta}, \mathbf{Y}), \tag{6}$$

where $\mathbf{\Theta} \in \mathbb{R}^{\sum_{l=1}^{L} D^{(l)} \times C}$ is the linear classification matrix, C is the number of classes in the classification problem, and $Loss(\cdot)$ is the cross-entropy loss function for multi-class classification.

Graph embedding and classification. There are several ways to get the graph-level outputs using node features, such as concatenation, mean pooling, and max pooling operators [29]. In our task, graphs are not aligned across different subjects and each graph may have an arbitrary number of nodes. Thus, the average pooling technique is used [30] here to obtain the embedding $\mathbf{e}_{G_i}$ of the entire graph G_i, which allows us to eliminate the dependence on the node order and size. Mathematically, for each graph G_i, we can formalize the mean pooling of node features as

$$\mathbf{e}_{G_i} = \frac{1}{N_i} \sum_{v \in V_i} [\mathbf{x}_{i_v}^{(1)}, \mathbf{x}_{i_v}^{(2)}, \cdots, \mathbf{x}_{i_v}^{(L)}], \quad \forall i \in \{1, 2, \cdots, M\} \tag{7}$$

Finally, we apply the logistic regression (LR) classifier based on the above whole-graph embedding vectors $\{\mathbf{e}_{G_i}\}_{i=1}^{M}$ and associated protein subfamily class labels $\{y_i\}_{i=1}^{M}$ as input for prediction.

2.3. Datasets Used in This Study

To evaluate HBcompare as a classifier, we constructed datasets based on protein superfamilies with three criteria. First, we selected superfamilies that contained subfamilies with distinct ligand binding preferences. Second, we selected only superfamilies and subfamilies where differences in binding preferences are experimentally established to rely on variations in hydrogen bonding patterns. Finally, proteins in each superfamily were selected with the same overall fold.

These criteria enable our datasets to test the overall hypothesis. The first two criteria are required for evaluating HBcompare as a classifier of hydrogen bonding topologies. The third ensures that the classification task is not trivial, because subfamilies with different folds have very different hydrogen bond topologies that can be easily distinguished. The general properties of the constructed protein datasets are summarized in Table 1 and details are described as below.

Table 1. Primary and Auxiliary data sets used in this study.

Dataset	Superfamily E.C. Class	Pivot Structure	Subfamily	E.C. Class	Number of Structures
Primary-1 (P1)	Glycosidases 3.2.1.*	1aqm	Alpha Amylase	3.2.1.1	30
			Beta Amylase	3.2.1.2	30
Primary-2 (P2)	Serine Proteases 3.4.21.*	1ghz	Chymotrypsin	3.4.21.1	40
			Trypsin	3.4.21.4	40
			Elastase	3.4.21.36	40
			Thrombin	3.4.21.5	37
			Coagulation factor Xa	3.4.21.6	39
Primary-3 (P3)	Aminoacyl-tRNA Synthetases 6.1.1.*	6rlt	Ser-tRNA Synthetase	6.1.1.11	24
			Thr-tRNA Synthetase	6.1.1.3	23
Auxiliary-1 (A1)	Serine Proteases (subset)	1ggd	Chymotrypsin	3.4.21.1	40
			Trypsin	3.4.21.4	40
Auxiliary-2 (A2)	Glycosidases, Serine Proteases (A1 + P1)	2xfy	Alpha Amylase	3.2.1.1	40
			Beta Amylase	3.2.1.2	40
			Chymotrypsin	3.4.21.1	30
			Trypsin	3.4.21.4	30

Primary protein datasets. Our criteria identified the glycosidases, the serine proteases, the aminoacyl-tRNA synthetases, and several subfamilies of each (Table 1). We used the Enzyme Commission Classification index [32] of each subfamily to identify the protein data bank (PDB) [33] structure of every constituent protein. To avoid the overrepresentation of well studied proteins with many available structures, we removed one member of any pair of proteins with greater than 95% sequence identity. We also removed any structures labeled as mutants to avoid misclassifying proteins with deactivating mutations (Table 2). After this filtration, molecular graphs were generated on the remaining structures using the method in Section 2.1.

There are 303 structures across all primary datasets. 298 structures were derived from X-ray crystallography, and five were produced by nuclear magnetic resonance spectroscopy. Xray structure resolutions ranged from 0.81 Å to 3.5 Å, with an average of 2.05 Å, a median of 2.0 Å, and a standard deviation 0.443 Å. 291 out of 303 structures have resolution less than or equal to 3.0 A, and 261 out of 303 structures have resolution less than or equal to 2.5 A. The number of proteins observed in each subfamily of each dataset was generally similar, requiring no additional treatment to to balance the datasets.

In Primary-1 (P1), the glycosidase superfamily proteins conserve an alpha/beta barrel fold where they hydrolyze the glycosidic bonds of polysaccharide chains. The alpha and beta amylase subfamilies hydrolyze the intermediate and the terminal bonds, respectively, of these chains, and recognize them in part through differences in hydrogen bonding [34,35].

In Primary-2 (P2), the PA clan of the serine protease superfamily exhibit a chymotrypsin-like fold and catalyze the cleavage of peptide bonds. They share a catalytic triad at the center of an extensive hydrogen bonding network that also plays a crucial role in stabilizing substrate backbones for efficient substrate hydrolysis [36].

In Primary-3 (P3), the aminoacyl-tRNA synthetases catalyze the attachment of a transfer RNA and an amino acid in preparation for protein translation. The seryl- and threonyl-tRNA Synthetase share an anti-parallel beta-sheet fold [37] but coordinate their amino acid substrates through different patterns in hydrogen bonding [38,39].

Auxiliary datasets. We also developed two variations on our original datasets to evaluate the performance of HBcompare. Noting that the serine protease dataset has five subfamilies, we developed a two-subfamily variation, using only the chymotrypsin and trypsin subfamilies. This variation allowed us to evaluate how HBcompare performed on a classification problems with different numbers of categories. We created a second dataset to evaluate the scenario where some subfamilies have different folds, and thus radically different hydrogen bond topologies. We combined the glycosidases and the serine proteases into a single artificial superfamily. Using two subfamilies of each of the joined superfamilies, we assess if the substantial differences between the superfamilies obscure the subtler differences between subfamilies.

Table 2. Average properties of proteins in all datasets.

Dataset	# Proteins	# Subfamilies	Avg. # Nodes	Avg. # Edges
P1	60	2	826	578
P2	196	5	402	241
P3	47	2	901	573
A1	80	2	372	201
A2	140	4	568	363

2.4. Comparison with Existing Methods

Directly comparing HBcompare against existing methods is difficult, because HBcompare uses only the topology of hydrogen bonds while existing methods for comparing protein structures generally require atomic coordinates and other data. For this reason, we performed two separate comparisons. First, to demonstrate the fitness of HBcompare as a tool for coordinate-free graph classification, we compare the performance of HBcompare against several modern graph classification techniques that also use only graph topology. Second, to understand how classification by hydrogen bond topology performs relative to classification by atomic coordinates, we modified all methods, including HBcompare, to incorporate coordinate-based molecular graphs (see Section 2.1).

Our first comparison study includes a convolutional neural network (CNN), a graph kernel-based comparison method (GK), and principal component analysis based methods (PCA, 2DPCA, and PCA-NF). These methods use hydrogen bond topology alone via an analysis of node adjacency matrices, but they have never been applied for the coordinate-free comparison of hydrogen bond topologies. As such, they require modifications for direct comparison. The need for small modifications demonstrates, qualitatively, a degree of unsuitability for the problem of topological comparison relative to HBcompare, which does not require such modification.

First, CNN, PCA and 2DPCA are sensitive to variations in input order, while GCNs are not. To minimize this sensitivity, dataset proteins were structurally aligned to an arbitrarily selected pivot structure to produce a 1-to-1 mapping between most amino acids, ensuring that all proteins could be indexed in the same order. Structural alignments were performed with ska [1], which is designed for identifying distantly related proteins with subtle similarities in their folds. In this application, where we are considering datasets of closely related proteins with nearly identical folds, ska easily generated 1-to-1 mappings appropriate for our comparison.

Second, CNN, PCA and 2DPCA also require input data to have the same number of nodes, because the features they consider cannot have varying dimensionality. To resolve this issue, we trimmed all molecular graphs to contain exactly 600 nodes, a quantity chosen because the largest connected component of all graphs in our dataset would not be altered. This trimming was possible without disrupting the topology of the graph because all structures contain a large number of donors and acceptors that are uninvolved in a hydrogen bond. In the molecular graph they are singleton nodes, and they contribute no distinguishing information to the topological character of the graph overall. By removing some of these nodes as necessary, we were able to trim larger graphs to exactly 600 nodes. Graphs that had fewer than 600 nodes, such as those in P2, had singleton nodes added to arrive at exactly 600 nodes.

Our second comparison study adds the protein structure comparison algorithm Ska and the sequence comparison algorithm Clustalw [40]. These classic methods benchmark the performance of HBcompare against existing comparison techniques in structural bioinformatics. GK, CNN, PCA and 2DPCA remain, but they are provided coordinate-based rather than coordinate-free molecular graphs as input.

The CNN model [41] utilizes shared weights for common feature extraction, and also local reception fields to take advantage of the local structure of input data. In our case, we trained an end-to-end CNN model with fully connected network (FCN) classifier that takes adjacency matrices **A** as input and outputs the corresponding graph classes.

The GK method [42] applies the Weisfeiler Lehman (WL) kernel to calculate similarities between graphs [43,44]. Each vertex is labelled with its original vertex label and the label of its neighbors, resulting in a representation of graphical neighborhoods of each vertex. The WL kernel goes through n iterations until WL kernels are unchanged for successive iterations. This kernel is then fed into a support vector machine (SVM) to measure the graph classification performance.

The PCA method [45] for comparing graphs learns a common projection matrix via singular value decomposition (SVD) by vectorizing the submatrices to perform feature extraction. Similar to our HBCompare model, the extracted graph feature vectors are passed to the LR classifier. Furthermore, to investigate the effectiveness of using one-hot encoding labels as the node feature input for GCN, we also concatenate the features extracted by PCA and the GCN node features. This variation, PCA-NF, adds the donor/acceptor status of each graph node to the topology being classified.

The 2DPCA method [46] avoids vectorization of input submatrices by learning pairwise projection matrices for feature extraction and dimensionality reduction. The extracted feature matrices are then vectorized and fed to the LR classifier for prediction.

The ska [1] algorithm finds corresponding secondary structure elements between two proteins to build detailed correspondences between backbone atom coordinates, which are required. The atomic correspondences are used to compute least root mean square difference (RMSD) between backbone atoms. As a measure of geometric similarity, RMSD is lower between proteins that are more similar. Using ska, we generated an all-vs-all matrix of RMSD distances between all proteins of each dataset. Viewed as a set of column vectors, this matrix is decomposed into training and test sets and the training sets are used to train an LR classifier via five fold validation, similar to [47]. Finally, the test set is passed to the classifier to form predictions.

Clustalw [40] is the classic sequence-based comparison algorithm that measures similarity between the sequences of amino acids that define two proteins. It applies dynamic programming to build correspondences between amino acid sequences and then measure the percentage of sequence identity. Higher percentages are generated by protein pairs with similar sequences of amino acids, and lower percentages indicate proteins that are more different. These percentages are subtracted from 100 so that smaller values indicate more similar proteins, and then used to populate an all-vs-all matrix that is treated in the same way as the RMSDs are for ska.

2.5. Implementation Details

All models were implemented in Python 3.6 with Tensorflow 1.15 for the deep learning backend. The validation of our method was performed by randomly and uniformly splitting each dataset and each subclass by a 4:1 ratio. The split results in a larger training set (80% of the data) and a smaller test set (20% of the data). Since the subclasses were split uniformly, the approximate balance of the subclasses in each dataset was preserved in each split. The performance of all classifiers reported in Tables 3 and 4 is an average and a standard deviation of accuracy, f1-score, and AUC-ROC computed from 10 such random splits. We evaluate predictions as correct if the prediction agrees with the class label and incorrect if the prediction does not agree with the label. We report accuracy (acc) as the ratio of correct predictions to total predictions, $\frac{Correct}{Correct+Incorrect}$.

We performed parameter tuning on all methods using 5-fold validation on the training set. Since this training set is held separate from the testing set, no data leakage influences the classifier performance reported. Training was performed for 50 epochs per fold, and parameters associated with the highest accuracy fold were used for evaluation on the corresponding test set. We used the Adam optimizer [48] and selected the learning rate l_r from $\{5e^{-4}, 1e^{-4}, 5e^{-3}, 1e^{-3}\}$.

For the design of HBcompare we considered between 1 and 6 GCN layers, and batch sizes in the range $\{1, 2, 4, 8, 16\}$. To build the CNN model, we varied the number of filters in the set $\{6, 12, 18, 24, 30\}$, and the number of strides in the set $\{1, 2, 4, 8, 16\}$. The total number of parameters in the network was 384. The number of layers, epochs, the batch size, and learning rate are selected for the CNN model in the same manner as HBcompare. For the other compared methods, we also carefully tuned their parameters and use the same data splits and the same 5-fold cross-validation scheme. All experiments were performed on a 8-core machine with 32 GB RAM.

3. Results

During training, we observed converging improvements in accuracy relative to training time and number of epochs. These observations are illustrated for all datasets in Appendix A, Figures A1 and A2. By dividing the data sets into non-overlapping training and testing sets, we found that classification accuracy of HBcompare for training and testing quickly converged towards a stable accuracy performance and remains at this performance level regardless of added epochs past the saturation point. This is shown in Appendix A, Figure A3. Collectively, these observations suggest that overtraining is not a major concern for the accuracy of HBcompare on our datasets.

Overall, using only hydrogen bond topology, HBcompare had a high degree of classification accuracy. The classification accuracy of HBcompare averaged from 85.0% to 92.3% on all folds of all primary datasets (Table 3, right column, top three rows). The standard deviation in accuracy across all folds ranged from 4.8% to 7.7%. The F1 score averaged between 84.8% and 92.2%, and the area under the ROC curve (AUC-ROC) averaged between 90.6% and 92.3%.

In comparison to existing coordinate-free methods, HBcompare was 11.38% more accurate, had 12.17% greater F1 score, and had 9.92% higher AUC, on average, than the second best method, PCA-NF, across all data sets. Standard deviations in HBcompare accuracy, F1 score and AUC were also generally the same or lower than existing methods. Overall, HBcompare had the best classification performance of all methods on all primary datasets (Table 3, top three rows).

Table 3. Average classification accuracy and F1 score (avg ± std) of compared methods using only hydrogen bond topology, across all cross-validation folds. The *set*(#) column indicates the dataset and the number of subfamilies it contains. The *stat* column indicates rows with either classifier accuracy or F1 score. The highest value in each row is bolded.

set(#)	stat	GK	PCA	PCA-NF	2DPCA	CNN	HBCompare
	Acc	68.8 ± 1.1	76.7 ± 15.3	81.7 ± 12.2	73.3 ± 8.2	85.0 ± 15.3	**92.3** ± **7.0**
P1(2)	F1	69.3 ± 1.1	74.7 ± 17.6	79.8 ± 15.6	73.1 ± 8.1	83.1 ± 18.5	**92.2** ± **6.8**
	AUC-ROC	68.8 ± 1.1	76.7 ± 15.3	81.7 ± 12.2	73.3 ± 8.2	85.0 ± 15.3	**92.3** ± **6.7**
	Acc	38.1 ± 1.5	63.3 ± 4.6	67.8 ± 4.0	65.3 ± 6.9	68.6 ± 3.1	**85.0** ± **4.8**
P2(5)	F1	41.8 ± 1.4	62.0 ± 5.0	67.2 ± 3.9	65.4 ± 7.0	68.6 ± 3.4	**84.8** ± **5.0**
	AUC-ROC	27.0 ± 1.1	76.8 ± 2.8	79.7 ± 2.4	69.1 ± 4.3	80.2 ± 2.0	**90.6** ± **3.1**
	Acc	58.7 ± 3.4	61.6 ± 10.7	68.0 ± 9.6	59.1±12.0	68.0 ± 9.6	**91.3** ± **7.7**
P3(2)	F1	60.2 ± 3.7	60.9 ± 10.8	66.3 ± 10.7	58.4 ± 12.7	67.3 ± 9.2	**91.2** ± **8.5**
	AUC-ROC	58.1 ± 3.4	62.0 ± 10.2	67.0 ± 9.9	60.0 ± 11.3	68.0 ± 8.6	**91.5** ± **8.2**
	Acc	76.6 ± 1.2	90.0 ± 5.0	**93.8** ± **4.0**	90.0 ± 3.1	88.1 ± 4.4	91.8 ± 5.5
A1(2)	F1	77.8 ± 1.5	89.9 ± 5.0	**93.7** ± **4.0**	89.9±3.1	88.0 ± 4.4	91.7 ± 5.6
	AUC-ROC	76.6 ± 1.2	90.0 ± 5.0	**93.8** ± **4.0**	90.0 ± 3.1	88.1 ± 4.4	91.8 ± 5.5
	Acc	52.2 ± 0.9	75.7 ± 4.7	80.0 ± 2.9	72.9 ± 5.3	73.6 ± 3.6	**86.8** ± **5.4**
A2(4)	F1	50.2 ± 2.0	74.3 ± 5.2	79.3 ± 3.1	70.1 ± 5.1	73.3 ± 3.6	**86.3** ± **6.7**
	AUC-ROC	54.6 ± 1.0	83.4 ± 3.3	86.3 ± 2.1	81.3 ± 3.0	82.4 ± 1.7	**90.9** ± **4.3**

Auxiliary-1 simplified the multi-class classification problem by removing three of the five subfamilies in Primary-2. As a result, on Auxiliary-1, all comparison methods were significantly more accurate, with PCA-NF outperforming HBcompare slightly (93.8% vs. 91.8%). The fact that HBcompare significantly outperforms other methods on the five categories of Primary-2 suggest that it is more robust to the multi-class classification problem.

On Auxiliary-2, which combined two subfamilies from each of Primary-1 and Auxiliary-1, HBcompare outperformed other methods by at least 6.8%. In this case, where some subfamilies are far more similar than others, HBcompare did not lose discriminating power, performing only slightly worse than it did on Primary-1 and on Auxiliary-1 despite two additional categories.

Since HBcompare operates with only hydrogen bond topology, we also asked how HBcompare and other graph-based methods would perform if atomic coordinates were included (Table 4). Again, on all primary datasets, HBcompare outperformed existing methods, with accuracy averaging from 2.1% to 14.9% above existing methods. Unsurprisingly, since these comparisons used representations of both hydrogen bond topology and also atomic coordinates, GK, PCA, 2DPCA, CNN, and HBcompare all performed the same or better than their coordinate-free counterparts. Classifications using only sequence identity or structure similarity underperformed.

On Auxiliary-1, the addition of atomic coordinates into the graph representation resulted in slightly superior classification accuracy for HBcompare (93.8%) relative to PCA-NF (91.3%). As in the coordinate-free scenario, GK, PCA, 2DPCA and CNN all performed similarly. On Auxiliary-2, HBcompare was again more accurate (88.4%).

Table 4. Average classification accuracy and F1 score (avg ± std) using both hydrogen bond topology and coordinate information, across all folds. The *set* column indicates the dataset. The *stat* column indicates rows with either classifier accuracy or F1 score. The highest value in each row is bolded.

Set	stat	Clustalw	Ska	GK	PCA	PCA-NF	2DPCA	CNN	HBcompare
P1	Acc	75.0 ± 17.7	63.3 ± 4.6	68.3 ± 1.2	83.3 ± 12.9	86.7 ± 11.3	83.3 ± 9.1	88.3 ± 8.5	**92.3 ± 7.2**
	F1	79.5 ± 11.7	66.9 ± 5.3	68.8 ± 1.2	81.5 ± 16.3	85.8 ± 12.8	82.4 ± 10.6	87.8 ± 9.6	**92.8 ± 7.1**
P2	Acc	80.0 ± 8.1	68.8 ± 7.7	37.3 ± 1.1	70.4 ± 3.8	72.4 ± 3.9	74.5 ± 5.9	70.4 ± 3.0	**83.6 ± 6.3**
	F1	80.8 ± 7.1	70.1 ± 8.1	41.7 ± 1.8	69.5 ± 5.5	72.0 ± 4.7	74.8 ± 5.6	70.8 ± 3.1	**83.9 ± 6.5**
P3	Acc	60.8 ± 3.4	63.3 ± 2.6	58.7 ± 2.4	74.7 ± 12.8	76.7 ± 12.6	74.7 ± 12.8	76.8 ± 8.9	**90.6 ± 6.8**
	F1	69.2 ± 2.3	73.4 ± 7.7	60.3 ± 2.8	73.8 ± 13.1	76.2 ± 12.7	73.8 ± 13.1	76.4 ± 8.7	**90.5 ± 6.9**
A1	Acc	60.7 ± 2.5	50.0 ± 8.1	76.3 ± 1.3	91.3 ± 6.4	91.3 ± 6.4	92.5 ± 6.1	91.9 ± 5.6	**93.8 ± 4.4**
	F1	67.6 ± 3.8	53.9 ± 6.8	77.3 ± 1.5	91.2 ± 6.4	91.2 ± 6.4	92.5 ± 6.1	91.8 ± 5.6	**93.9 ± 4.4**
A2	Acc	86.6 ± 14.5	81.1 ± 8.0	52.1 ± 0.5	75.0 ± 6.4	77.9 ± 5.2	80.7 ± 5.3	80.4 ± 6.4	**88.4 ± 6.4**
	F1	87.6 ± 13.3	82.3 ± 6.8	49.5 ± 1.3	73.7 ± 6.5	76.7 ± 5.4	80.0 ± 5.4	80.1 ± 6.7	**88.2 ± 6.5**

3.1. Hyperparameter Analysis

In training HBcompare, we considered a range of batch sizes and GCN layers, both of which can influence classifier performance. Adding more GCN layers expands the graph neighbourhood within which the node features are averaged [49]. These findings are plotted in Figure 2. We observed that accuracy was maximized with batch size 4 and with 3 GCN layers, using these parameters in HBcompare.

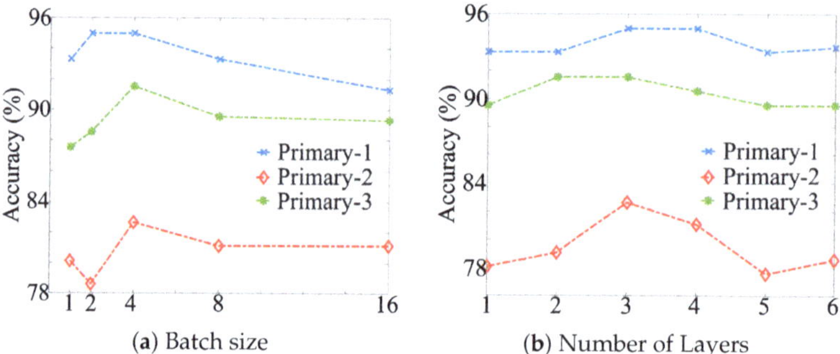

Figure 2. Influence of the number of layers (**a**), and of the batch size (**b**) on the classification accuracy of HBcompare. Accuracy is shown on all three primary datasets (blue, red and green lines), and was highest for batch size 4 and for 3 GCN layers.

3.2. Feature Concatenation

In our HBcompare model, we concatenate the output of all GCN layers to obtain the final feature representation (Figure 1). To evaluate the effectiveness of this concatenation strategy, we compare the implementation of HBcompare with and without feature concatenation in Table 5 using only hydrogen bond topology. We observed that HBcompare can benefit from the concatenation strategy, which helps to aggregate more information when the input node feature size is small.

Table 5. Average classification accuracy of HBcompare model with and without concatenation strategy using only hydrogen bond topology across all folds. The more accurate method is bolded.

Method (Acc)	Primary-1	Primary-2	Primary-3	Auxillary-1	Auxillary-2
Hbcompare with concatenation	**91.3**	**80.6**	**89.6**	90.8	**87.1**
Hbcompare without concatenation	88.6	78.2	89.1	**91.8**	86.4

4. Discussion

We have presented HBcompare, a GCN-based algorithm for classifying protein structures based exclusively on hydrogen bonding topology. Once trained on a group of closely related subfamilies that perform the same function on different preferred ligands, HBcompare addresses the problem where a novel protein structure or model is to be classified into one of the subfamilies. HBcompare should be retrained to make classifications into different subfamilies.

Since it only examines hydrogen bond topology, accurate classifications implicate the importance of hydrogen bonds in achieving the binding preferences of the predicted subfamily. This novel capability contrasts from holistic representations, which do not implicate specific mechanisms.

To evaluate HBcompare, we performed specificity classification experiments on protein superfamilies that achieve distinct binding preferences based on differences in hydrogen bonding. On nonredundant subsets of the glucosidases, serine proteases, and tRNA synthetase superfamilies, the average accuracy of HBcompare was 92.3%, 85.0% and 91.3%. As a tool for classifying hydrogen bond topologies, HBcompare is a capable classifier. When we adapted several modern techniques to the topology-only classification problem, we observed that HBcompare was more accurate in all but one case, where PCA with node features outperformed HBcompare 93.8% versus 91.8%. This classification performance was well within the variations observed in different training folds, indicating comparable performance between PCA-NF and HBcompare, rather than a superior performance of one over the other. Furthermore, it is important to note that CNN, GK, PCA, PCA-NF and 2DPCA all require a structural alignment to produce a 1-to-1 mapping between most amino acids, ensuring that all proteins could be indexed in the same order. CNN, PCA, PCA-NF and 2DPCA also require input graphs to have the same number of nodes. Our comparison included a preprocessing step that maximizes their comparability in this study, but in truly experimental settings, accurate preprocessing could not be guaranteed, further limiting the applicability of these alternative methods. The same challenges do not apply to HBcompare, which is unaffected by input order or graph size, making it more applicable in experimental settings and often more accurate than existing methods.

We also compared HBcompare to conventional coordinate-based approaches. In comparison to ska, a coordinate-based method that does not use hydrogen bonding topology (Table 4), coordinate-free HBcompare (Table 3) was an average of 20.6% more accurate on all datasets. These findings demonstrate that hydrogen bond topology contributes information that is complementary to conventional structural approaches.

Finally, we modified HBcompare to consider both atomic coordinates and also hydrogen bond topology. In a comparison to the same methods above, each modified to incorporate both data types, HBcompare was 2.1% to 14.9% more accurate on average (Table 4). This result demonstrates that combining hydrogen bond topology and atomic coordinates enhances subfamily classification at the cost of being able to implicate hydrogen bonds as a mechanism.

As a first step in the atomistic analysis of hydrogen bond topology, HBcompare has considerable potential for novel applications. Where specificity mechanisms are unknown, HBcompare can detect when hydrogen bonding distinguishes between isoforms with different binding preferences without influences from other structural properties. This capability can focus experimental scrutiny on hydrogen bonding when it correlates with specificity. Combined with structural models, HBcompare could be applied to identify mutations that change bond topology to resemble proteins with different binding preferences. Together with other sources of information, HBcompare could thus support efforts in protein engineering and in annotating binding specificity mechanisms.

Author Contributions: Conceptualization, L.H. and B.Y.C.; methodology, Z.K. and L.H.; software, Z.K. and L.H; validation, J.Z.T., Z.K., O.A., L.H. and B.Y.C.; formal analysis, Z.K. and L.H.; resources, B.Y.C.; data curation, O.A. and B.Y.C.; writing—original draft preparation, Z.K., J.Z.T., L.H. and B.Y.C.; writing—review and editing, J.Z.T., Z.K., L.H. and B.Y.C.; visualization, J.Z.T. and Z.K.; supervision, L.H. and B.Y.C.; project administration, L.H. and B.Y.C.; funding acquisition, B.Y.C. All authors have read and agreed to the published version of the manuscript.

Funding: This work was funded by NIH grant R01GM123131 to Brian Y. Chen.

Institutional Review Board Statement: Not applicable.

Informed Consent Statement: Not applicable.

Data Availability Statement: HBcompare and HbondFinder have been made open source. The five datasets used in this study are available here. These links were accessed on 28 Oct 2022.

Acknowledgments: The authors would like to acknowledge the timely and insightful suggestions of Houliang Zhou relating to the training of classifiers in Tensorflow.

Conflicts of Interest: The authors declare no conflict of interest. The funders had no role in the design of the study; in the collection, analyses, or interpretation of data; in the writing of the manuscript, or in the decision to publish the results.

Appendix A. Protein Datasets Used in This Work

The following tables catalog the protein structures used in this work, based on Protein Data Bank codes. The two datasets Auxiliary-1 and Auxiliary-2 are generated from these sets (see Table 1), so they are not listed again here.

Table A1. Dataset: Primary-1.

Subfamily A: Alpha-Amylase (3.2.1.1)									
1amy	1aqm	1b2y	1bpl	1bvn	1clv	1e3x	1eh9	1g94	1hny
1ht6	1hvx	1hx0	1jae	1ji1	1jxk	1kxq	1l0p	1mfu	1mwo
1ose	1p6w	1pif	1rpk	1smd	1tmq	1u2y	1ua3	1ud2	1uh3
Subfamily B: Beta-Amylase (3.2.1.2)									
1b90	1b9z	1btc	1bya	1byb	1byc	1byd	1cqy	1fa2	1j0y
1j0z	1j10	1j11	1j12	1j18	1q6c	1vem	1wdp	2laa	2lab
2xff	2xfr	2xfy	2xg9	2xgb	2xgi	3voc	5bca	5wqs	5wqu

Table A2. Dataset: Primary-2.

Subfamily A: Chymotrypsin (3.4.21.1)									
1ab9	1acb	1afq	1cbw	1cgi	1chg	1cho	1dlk	1eq9	1ggd
1gha	1ghb	1gl0	1gl1	1gmc	1gmh	1hja	1mtn	1oxg	1p2m
1yph	2cga	2jet	3bg4	3cp7	3gch	3ru4	3t62	3wy8	4cha
4gch	4q2k	4vgc	5cha	5gch	5j4q	6di8	6gch	7gch	8gch
Subfamily B: Trypsin (3.4.21.4)									
1a0j	1aks	1ane	1bit	1bju	1bra	1brc	1btp	1bzx	1c1n
1co7	1d6r	1eja	1ezx	1f5r	1fmg	1fn8	1fni	1fxy	1fy4
1fy8	1g36	1gdn	1ghz	1h4w	1h9i	1hj8	1jir	1mbq	1mts
1os8	1ox1	1ppc	1pq5	1qb1	1trm	1trn	1utj	1xvm	2f3c
Subfamily C: Elastase (3.4.21.36)									
1b0e	1bma	1btu	1c1m	1e34	1eai	1eas	1eat	1eau	1ela
1elb	1elc	1esa	1est	1fle	1fzz	1gvk	1gwa	1h9l	1hax
1hay	1hb0	1hv7	1inc	1jim	1l0z	1l1g	1lka	1lkb	1lvy
1mmj	1nes	1okx	1qgf	1qix	1qnj	1qr3	1uo6	1uvo	1uvp
Subfamily D: Thrombin (3.4.21.5)									
1a2c	1aoh	1avg	1awf	1bbr	1bcu	1hah	1hrt	1id5	1ihs
1mkw	1mu6	1nrp	1sr5	1ta2	1tb6	1tbq	1tbr	1tmb	1ucy
1uma	1vit	1vr1	1ycp	1ypm	2a1d	2c8z	2jh5	2ocv	2pf1
2pf2	2pgb	2pux	2r2m	2thf	2v3h	3edx			
Subfamily E: Coagulation factor Xa (3.4.21.6)									
1apo	1c5m	1ccf	1ezq	1f0s	1fjs	1g2l	1hcg	1iod	1ioe
1iqf	1kig	1ksn	1lpg	1mq5	1xkb	2bmg	2boh	2bok	2bq6
2cji	2ei7	2fzz	2g00	2h9e	2j94	2p16	3sw2	3tk5	4a7i
4bti	4btt	4bxw	4y6d	5jqy	5jtc	5k0h	5voe	5vof	

Table A3. Dataset: Primary-3.

Subfamily A: Ser-tRNA Synthetase (6.1.1.11)									
1ses	1set	1sry	2dq0	2zr2	2zr3	3lsq	3lss	3qne	3qo5
3qo8	3vbb	6gir	6h9x	6hdz	6he1	6he3	6hhy	6hhz	6r1m
6r1n	6r1o	6rlt	6rlv						
Subfamily B: Thr-tRNA Synthetase (6.1.1.3)									
1tje	1tke	1tkg	1tky	1wwt	1y2q	2hl0	2hl1	3pd2	3pd3
3pd4	3pd5	3ugq	3uh0	4eo4	4hwo	4hwp	4hwr	4hws	4hwt
4p3n	4ttv	4yye							

Appendix B. Additional Results

The following figures show additional data on HBcompare performance over all data sets.

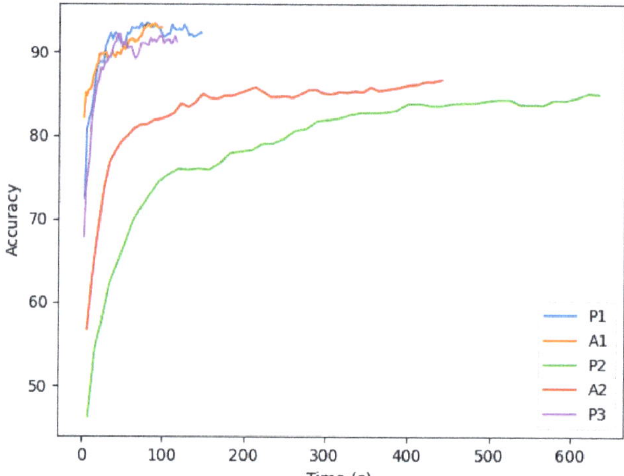

Figure A1. Accuracy over time.

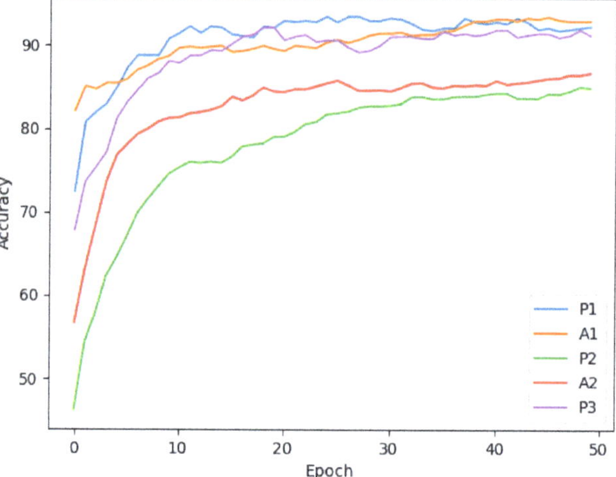

Figure A2. Accuracy over number of epochs.

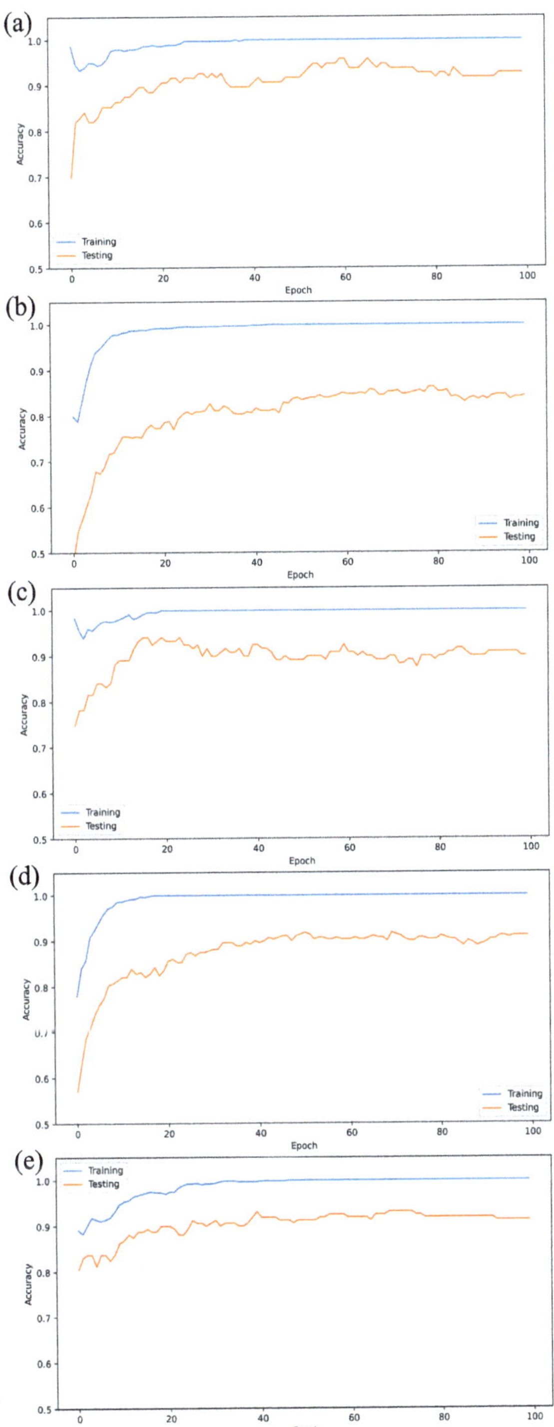

Figure A3. Learning Curve: Training and testing accuracy for P1 (**a**), P2 (**b**), P3 (**c**), A1 (**d**) and A2 (**e**).

References

1. Yang, A.S.; Honig, B. An integrated approach to the analysis and modeling of protein sequences and structures. I. Protein structural alignment and a quantitative measure for protein structural distance. *J. Mol. Biol.* **2000**, *301*, 665–678. [CrossRef] [PubMed]
2. Madej, T.; Gibrat, J.F.; Bryant, S.H. Threading a database of protein cores. *Proteins Struct. Funct. Bioinform.* **1995**, *23*, 356–369. [CrossRef] [PubMed]
3. Shindyalov, I.N.; Bourne, P.E. An alternative view of protein fold space. *Proteins Struct. Funct. Bioinform.* **2000**, *38*, 247–260. [CrossRef]
4. Bryant, D.H.; Moll, M.; Finn, P.W.; Kavraki, L.E. Combinatorial clustering of residue position subsets predicts inhibitor affinity across the human kinome. *PLoS Comput. Biol.* **2013**, *9*, e1003087. [CrossRef]
5. Kolodny, R.; Petrey, D.; Honig, B. Protein structure comparison: Implications for the nature of 'fold space', and structure and function prediction. *Curr. Opin. Struct. Biol.* **2006**, *16*, 393–398. [CrossRef] [PubMed]
6. Krishna, S.S.; Grishin, N.V. Structural drift: A possible path to protein fold change. *Bioinformatics* **2005**, *21*, 1308–1310. [CrossRef] [PubMed]
7. Chen, B.Y.; Fofanov, V.Y.; Bryant, D.H.; Dodson, B.D.; Kristensen, D.M.; Lisewski, A.M.; Kimmel, M.; Lichtarge, O.; Kavraki, L.E. The MASH pipeline for protein function prediction and an algorithm for the geometric refinement of 3D motifs. *J. Comput. Biol.* **2007**, *14*, 791–816. [CrossRef]
8. Sael, L.; La, D.; Li, B.; Rustamov, R.; Kihara, D. Rapid comparison of properties on protein surface. *Proteins Struct. Funct. Bioinform.* **2008**, *73*, 1–10. [CrossRef]
9. Rosen, M.; Lin, S.L.; Wolfson, H.; Nussinov, R. Molecular shape comparisons in searches for active sites and functional similarity. *Protein Eng.* **1998**, *11*, 263–277. [CrossRef]
10. Kinoshita, K.; Nakamura, H. Identification of the ligand binding sites on the molecular surface of proteins. *Protein Sci.* **2005**, *14*, 711–718. [CrossRef]
11. Liang, J.; Woodward, C.; Edelsbrunner, H. Anatomy of protein pockets and cavities: Measurement of binding site geometry and implications for ligand design. *Protein Sci.* **1998**, *7*, 1884–1897. [CrossRef] [PubMed]
12. Fischer, D.; Wolfson, H.; Lin, S.L.; Nussinov, R. Three-dimensional, sequence order-independent structural comparison of a serine protease against the crystallographic database reveals active site similarities: Potential implications to evolution and to protein folding. *Protein Sci.* **1994**, *3*, 769–778. [CrossRef] [PubMed]
13. Chen, B.Y.; Honig, B. VASP: A volumetric analysis of surface properties yields insights into protein-ligand binding specificity. *PLoS Comput. Biol.* **2010**, *6*, e1000881. [CrossRef]
14. Chen, B.Y. Vasp-e: Specificity annotation with a volumetric analysis of electrostatic isopotentials. *PLoS Comput. Biol.* **2014**, *10*, e1003792. [CrossRef] [PubMed]
15. Zhou, Y.; Li, X.P.; Chen, B.Y.; Tumer, N.E. Ricin uses arginine 235 as an anchor residue to bind to P-proteins of the ribosomal stalk. *Sci. Rep.* **2017**, *7*, 42912. [CrossRef]
16. Quintana, F.M.; Kong, Z.; He, L.; Chen, B.Y. DeepVASP-E: A Flexible Analysis of Electrostatic Isopotentials for Finding and Explaining Mechanisms that Control Binding Specificity. In *Pacific Symposium on Biocomputing 2022*; World Scientific: Singapore, 2022; pp. 56–67.
17. Artymiuk, P.J.; Poirrette, A.R.; Grindley, H.M.; Rice, D.W.; Willett, P. A graph-theoretic approach to the identification of three-dimensional patterns of amino acid side-chains in protein structures. *J. Mol. Biol.* **1994**, *243*, 327–344. [CrossRef]
18. Jumper, J.; Evans, R.; Pritzel, A.; Green, T.; Figurnov, M.; Ronneberger, O.; Tunyasuvunakool, K.; Bates, R.; Žídek, A.; Potapenko, A.; et al. Highly accurate protein structure prediction with AlphaFold. *Nature* **2021**, *596*, 583–589. [CrossRef]
19. Kipf, T.N.; Welling, M. Semi-supervised classification with graph convolutional networks. *arXiv* **2016**, arXiv:1609.02907.
20. Nguyen, D.; Nguyen, T.; Phung, D. Universal Self-Attention Network for Graph Classification. *arXiv* **2019**, arXiv:1909.11855.
21. Cai, R.; Chen, X.; Fang, Y.; Wu, M.; Hao, Y. Dual-Dropout Graph Convolutional Network for Predicting Synthetic Lethality in Human Cancers. *Bioinformatics* **2020**, *36*, 4458–4465. [CrossRef]
22. Berman, H.M.; Westbrook, J.; Feng, Z.; Gilliland, G.; Bhat, T.N.; Weissig, H.; Shindyalov, I.N.; Bourne, P.E. The protein data bank. *Nucleic Acids Res.* **2000**, *28*, 235–242. [CrossRef] [PubMed]
23. Chen, V.B.; Arendall, W.B.; Headd, J.J.; Keedy, D.A.; Immormino, R.M.; Kapral, G.J.; Murray, L.W.; Richardson, J.S.; Richardson, D.C. MolProbity: All-atom structure validation for macromolecular crystallography. *Acta Crystallogr. Sect. D Biol. Crystallogr.* **2010**, *66*, 12–21. [CrossRef] [PubMed]
24. McDonald, I.K.; Thornton, J.M. Satisfying hydrogen bonding potential in proteins. *J. Mol. Biol.* **1994**, *238*, 777–793. [CrossRef]
25. Georgiev, G.D.; Dodd, K.F.; Chen, B.Y. Precise parallel volumetric comparison of molecular surfaces and electrostatic isopotentials. *Algorithms Mol. Biol.* **2020**, *15*, 11. [CrossRef]
26. Hamilton, W.L.; Ying, R.; Leskovec, J. Representation learning on graphs: Methods and applications. *arXiv* **2017**, arXiv:1709.05584.
27. You, Y.; Chen, T.; Wang, Z.; Shen, Y. L2-GCN: Layer-Wise and Learned Efficient Training of Graph Convolutional Networks. In Proceedings of the IEEE Conference on Computer Vision and Pattern Recognition (CVPR), Seattle, WA, USA, 13–19 June 2020; pp. 2124–2132. [CrossRef]
28. Ying, R.; You, J.; Morris, C.; Ren, X.; Hamilton, W.L.; Leskovec, J. Hierarchical graph representation learning with differentiable pooling. *arXiv* **2018**, arXiv:1806.08804.

29. Zhang, M.; Cui, Z.; Neumann, M.; Chen, Y. An end-to-end deep learning architecture for graph classification. In Proceedings of the AAAI Conference on Artificial Intelligence, New Orleans, LA, USA, 2–7 February 2018; Volume 32.
30. Xu, K.; Hu, W.; Leskovec, J.; Jegelka, S. How powerful are graph neural networks? *arXiv* **2018**, arXiv:1810.00826.
31. Narayanan, A.; Chandramohan, M.; Venkatesan, R.; Chen, L.; Liu, Y.; Jaiswal, S. graph2vec: Learning distributed representations of graphs. *arXiv* **2017**, arXiv:1707.05005.
32. Jeske, L.; Placzek, S.; Schomburg, I.; Chang, A.; Schomburg, D. BRENDA in 2019: A European ELIXIR core data resource. *Nucleic Acids Res.* **2019**, *47*, D542–D549. [CrossRef]
33. Rose, P.W.; Prlić, A.; Altunkaya, A.; Bi, C.; Bradley, A.R.; Christie, C.H.; Costanzo, L.D.; Duarte, J.M.; Dutta, S.; Feng, Z.; et al. The RCSB protein data bank: Integrative view of protein, gene and 3D structural information. *Nucleic Acids Res.* **2016**, *45*, D271–D281.
34. MacGregor, E.A.; Janeček, Š.; Svensson, B. Relationship of sequence and structure to specificity in the α-amylase family of enzymes. *Biochim. Biophys. Acta (BBA)-Protein Struct. Mol. Enzymol.* **2001**, *1546*, 1–20. [CrossRef]
35. Monroe, J.D.; Storm, A.R. The Arabidopsis β-amylase (BAM) gene family: Diversity of form and function. *Plant Sci.* **2018**, *276*, 163–170. [CrossRef] [PubMed]
36. Hedstrom, L. Serine protease mechanism and specificity. *Chem. Rev.* **2002**, *102*, 4501–4524. [CrossRef] [PubMed]
37. Perona, J.J.; Rould, M.A.; Steitz, T.A. Structural basis for transfer RNA aminoacylation by Escherichia coli glutaminyl-tRNA synthetase. *Biochemistry* **1993**, *32*, 8758–8771. [CrossRef] [PubMed]
38. Belrhali, H.; Yaremchuk, A.; Tukalo, M.; Berthet-Colominas, C.; Rasmussen, B.; Bösecke, P.; Diat, O.; Cusack, S. The structural basis for seryl-adenylate and Ap4A synthesis by seryl-tRNA synthetase. *Structure* **1995**, *3*, 341–352. [CrossRef]
39. Arnez, J.G.; Moras, D. Structural and functional considerations of the aminoacylation reaction. *Trends Biochem. Sci.* **1997**, *22*, 211–216. [CrossRef]
40. Larkin, M.A.; Blackshields, G.; Brown, N.P.; Chenna, R.; McGettigan, P.A.; McWilliam, H.; Valentin, F.; Wallace, I.M.; Wilm, A.; Lopez, R.; et al. Clustal W and Clustal X version 2.0. *bioinformatics* **2007**, *23*, 2947–2948. [CrossRef]
41. Lawrence, S.; Giles, C.L.; Tsoi, A.C.; Back, A.D. Face recognition: A convolutional neural-network approach. *IEEE Trans. Neural Netw.* **1997**, *8*, 98–113. [CrossRef]
42. Shervashidze, N.; Vishwanathan, S.; Petri, T.; Mehlhorn, K.; Borgwardt, K. Efficient graphlet kernels for large graph comparison. *Artif. Intell. Stat.* **2009**, *5*, 488–495.
43. Hofmann, T.; Schölkopf, B.; Smola, A.J. Kernel methods in machine learning. *Ann. Stat.* **2008**, *36*, 1171–1220. [CrossRef]
44. Shervashidze, N.; Schweitzer, P.; Van Leeuwen, E.J.; Mehlhorn, K.; Borgwardt, K.M. Weisfeiler-lehman graph kernels. *J. Mach. Learn. Res.* **2011**, *12*, 2539–2561.
45. Turk, M.; Pentland, A. Eigenfaces for recognition. *J. Cogn. Neurosci.* **1991**, *3*, 71–86. [CrossRef] [PubMed]
46. Lu, H.; Plataniotis, K.N.; Venetsanopoulos, A.N. MPCA: Multilinear principal component analysis of tensor objects. *IEEE Trans. Neural Netw.* **2008**, *19*, 18–39. [PubMed]
47. Yuan, B.; Heiser, W.; de Rooij, M. The δ-machine: Classification based on distances towards prototypes. *J. Classif.* **2019**, *36*, 442–470. [CrossRef]
48. Kingma, D.P.; Ba, J. Adam: A method for stochastic optimization. *arXiv* **2014**, arXiv:1412.6980.
49. Rahimi, A.; Cohn, T.; Baldwin, T. Semi-supervised user geolocation via graph convolutional networks. *arXiv* **2018**, arXiv:1804.08049.

Article

RetroComposer: Composing Templates for Template-Based Retrosynthesis Prediction

Chaochao Yan [1], Peilin Zhao [2], Chan Lu [2], Yang Yu [2] and Junzhou Huang [1,*]

[1] Computer Science and Engineering, University of Texas at Arlington, Arlington, TX 76019, USA
[2] Tencent AI Lab, Shenzhen 518054, China
* Correspondence: jzhuang@uta.edu

Abstract: The main target of retrosynthesis is to recursively decompose desired molecules into available building blocks. Existing template-based retrosynthesis methods follow a template selection stereotype and suffer from limited training templates, which prevents them from discovering novel reactions. To overcome this limitation, we propose an innovative retrosynthesis prediction framework that can compose novel templates beyond training templates. As far as we know, this is the first method that uses machine learning to compose reaction templates for retrosynthesis prediction. Besides, we propose an effective reactant candidate scoring model that can capture atom-level transformations, which helps our method outperform previous methods on the USPTO-50K dataset. Experimental results show that our method can produce novel templates for 15 USPTO-50K test reactions that are not covered by training templates. We have released our source implementation.

Keywords: drug discovery; retrosynthesis; reaction template; machine learning; recurrent neural network; and graph neural network

Citation: Yan, C.; Zhao, P.; Lu, C.; Yu, Y.; Huang, J. RetroComposer: Composing Templates for Template-Based Retrosynthesis Prediction. *Biomolecules* **2022**, *12*, 1325. https://doi.org/10.3390/biom12091325

Academic Editors: Cameron Mura and Lei Xie

Received: 24 June 2022
Accepted: 15 September 2022
Published: 19 September 2022

1. Introduction

Retrosynthesis plays a significant role in organic synthesis planning, in which target molecules are recursively decomposed into available commercial building blocks. This analysis mode was firstly formulated in the pioneering work [1,2] and now is one of the fundamental paradigms in modern chemical society. Since then numerous retrosynthesis prediction algorithms have been proposed to aid or even automate the retrosynthesis analysis. However, the performance of existing methods is still not satisfactory. The massive search space is one of the major challenges of retrosynthesis considering that on the order of 10^7 compounds and reactions [3] have been reported in synthetic–organic knowledge. The other challenge is that there are often multiple viable retrosynthesis pathways and it is challenging to decide the most appropriate route since the feasibility of a route is often compounded by several factors, such as reaction conditions, reaction yield, potential toxic byproducts, and the availability of potential reactants [4].

Most of existing machine-learning-powered retrosynthesis methods focus on the single-step version. These methods are broadly grouped into template-based and template-free major categories. Templates-free methods [4–9] usually rely on deep learning models to directly generate reactants. One effective strategy is to formulate the retrosynthesis prediction as a sequence translation task and generate SMILES [10] sequences directly using sequence-to-sequence models such as Seq2Seq [5], SCROP [6], and AT [11]. SCROP [6] proposes to use a second transformer to correct the initial wrong predictions. Translation-based methods are simple and effective, but lack interpretability behind the prediction. Another representative paradigm is to first find a reaction center and split the target accordingly to obtain hypothetical units called synthons, and then generate reactants incrementally from these synthons, such as in RetroXpert [4], G2Gs [7], RetroPrime [12], and GraphRetro [13].

On the other hand, template-based methods are receiving less attention than the rapid surge of template-free methods. Template-based methods conduct retrosynthesis based on either hand-encoded rules [14] or automatically extracted retrosynthesis templates [15]. Templates encode the minimal reaction transformation patterns, and are straightforwardly interpretable. The key procedure is to select applicable templates to apply to targets [15–18]. Template-based methods have been criticized for the limitation that they can only infer reactions covered by training templates and cannot discover novel reactions [4,19].

In this work, we propose a novel template-based single-step retrosynthesis framework to overcome the mentioned limitation. Unlike previous methods only selecting from training templates, we propose to compose templates with basic template building blocks (molecule subgraphs) extracted from training templates. Specifically, our method composes templates by first selecting appropriate product and reactant molecule subgraphs iteratively, and then annotates atom transformations between the selected subgraphs. This strategy enables our method to discover novel templates from training subgraphs, since the reaction space of our method is the exponential combination of these extracted template subgraphs. What is more, we design an effective reactant scoring model that can capture atom-level transformation information. Thanks to the scoring model, our method achieves state-of-the-art (SOTA) Top-1 accuracy of 54.5% and 65.9% on the USPTO-50K dataset without and with reaction types, respectively. Our contributions are summarized as: (1) we propose a first-ever template-based retrosynthesis framework to compose templates, which can discover novel reactions beyond the training data; (2) we design an effective reactant scoring model that can capture atom-level transformations, which contributes significantly to the superiority of our method; (3) the proposed method achieves 54.5% and 65.9% Top-1 accuracy on the benchmark dataset USPTO-50K without and with reaction types, respectively, which establishes the new SOTA performance. The code is available at https://github.com/uta-smile/RetroComposer (accessed on 10 September 2022).

2. Related Work

Recently there has been an increasing amount of work using machine learning methods to solve the retrosynthesis problem. These methods can be categorized into template-based [15–18,20] and template-free approaches [4,5,7,13,21]. Template-based methods extract templates from training data and build models to learn the corresponding relationship between products and templates. RetroSim [15] selects the templates based on the fingerprint similarity between products and reactions. NeuralSym [16] uses a neural classification model to select corresponding templates. However, this method does not scale well with an increasing number of templates. To mitigate the problem, [20] adopts a multi-scale classification model to select templates according to a manually defined template hierarchy. GLN [18] proposes a graph logic network to model the decomposed template hierarchy by first selecting reaction centers within the targets, and then only consider templates that contain the selected reaction centers. The decomposition strategy can reduce the search space significantly. GLN models the relationship between reactants and templates jointly by applying selected templates to obtain reactants, which are also used to optimize the model simultaneously.

Template-free methods do not rely on retrosynthesis templates. Instead, they construct models to predict reactants from products directly. Translation-based methods [6,11,22,23] use SMILES to represent molecules and treat the problem as a sequence-to-sequence task. MEGAN [8] treats the retrosynthesis problem as a graph transformation task, and trains the model to predict a sequence of graph edits that can transform the product into the reactants. To imitate a chemist's approach to the retrosynthesis, two-step methods [4,7,12,13] first perform reaction center recognition to obtain synthons by disconnecting targets according to the reaction center, and then generate reactants from the synthons. G2Gs [7] treats the reactant generation process as a series of graph editing operations and utilizes a variational graph generation model to implement the generation process. RetroXpert [4] converts the synthon into SMILES to generate reactants as a translation task. GraphRetro [13] also adopts a similar framework and generates the reactants by attaching leaving groups to synthons. Dual model [9] proposes a general energy-based model framework that integrates both sequence- and graph-based models, and performs consistent training over forward and backward prediction directions.

3. Preliminary Knowledge

3.1. Retrosynthesis and Template

Single-step retrosynthesis predicts a set of reactant molecules given a target product, as shown in Figure 1a. Note that the product and reactant molecules are atom-mapped, which ensures that every product atom is uniquely mapped to a reactant atom. Templates are reaction rules extracted from chemical reactions. They are composed by reaction centers and encode the atom and bond transformations during the reaction process. The illustrated template in Figure 1b consists of a product subgraph (upper) and reactant subgraphs (lower). The subgraph patterns are highlighted in pink within the corresponding molecule graphs.

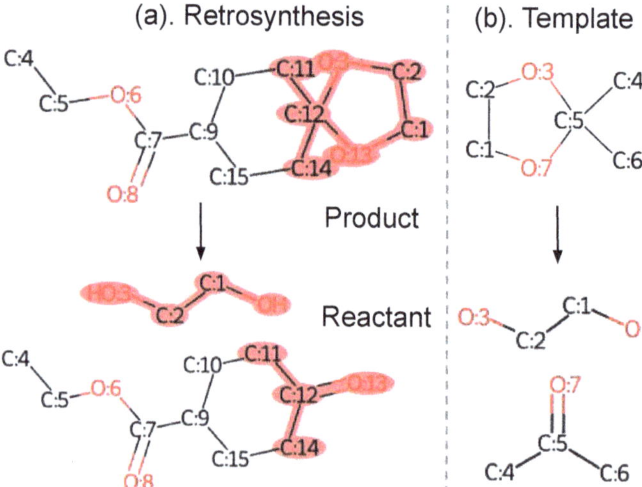

Figure 1. A retrosynthesis example from USPTO-50K dataset and its template extracted using an open-source toolkit. Note that the product and reactant are atom-mapped. The product and reactant subgraphs in (**b**) are highlighted in pink within the product and reactant molecule graphs in (**a**), respectively.

3.2. Molecule Graph Representation

The graph representation of a molecule or subgraph pattern is denoted as $G(\mathcal{V}, \mathcal{E})$, where $\mathcal{V}$ and $\mathcal{E}$ are the set of graph nodes (atoms) and edges (bonds), respectively. Following previous work [4,18], each bond is represented as two directed edges. Initial node and edge features can be easily collected for the learning purpose.

3.3. Graph Attention Networks

Graph neural networks [24] are especially good at learning node- and graph-level embeddings of molecule data. In this work, we adapt graph attention networks (GATs) [25] to incorporate bond features. The GAT layer updates a node embedding by aggregating its neighbor's information. The modified GAT concatenates edge embeddings with the associated incoming node embeddings before each graph message passing. The input of the GAT layer is node embeddings $\{v_i | \forall i \in \mathcal{V}\}$ and edge features $\{e_{i,j} | (i,j) \in \mathcal{E}\}$, and the output updated node embeddings $\{v_i' | \forall i \in \mathcal{V}\}$. Each node embedding is updated with a shared parametric function t_θ:

$$v_i' = t_\theta(v_i, \text{AGGREGATE}(\{[v_j||e_{i,j}]|\forall j \in \mathcal{N}(i)\})), \tag{1}$$

where $\mathcal{N}(i)$ are neighbor nodes of v_i and $||$ is the concatenation operation. The AGGREGATE of GAT adopts an attention-based mechanism to adaptively weight the neighbor information. A scoring function $c(i,j)$ computes the importance of the neighbor node j to node i:

$$c(i,j) = \text{LeakyReLU}(w^T[\mathbf{W}_1 v_i||\mathbf{W}_1 v_j||\mathbf{W}_2 e_{i,j}]), \tag{2}$$

where w is a learnable vector parameter and each $\mathbf{W}$ is a learnable matrix parameter. These importance scores are normalized using the Softmax function across the neighbor nodes $\mathcal{N}(i)$ of the node i to obtain attention weights:

$$\alpha(i,j) = \text{Softmax}_j(c(i,j)) = \frac{\exp(c(i,j))}{\sum_{j' \in \mathcal{N}(i)} \exp(c(i,j'))}. \tag{3}$$

The modified GAT instances t_θ and updates the node embedding as the non-linear function σ activated weighted-sum of the transformed embeddings of its neighbor nodes:

$$v_i' = \sigma(\sum_{j \in \mathcal{N}(i)} \alpha(i,j) * \mathbf{W}_3[\mathbf{W}_1 v_j||\mathbf{W}_2 e_{i,j}]). \tag{4}$$

GAT is usually stacked by multiple layers and enhanced with multi-head attention [26]. Please refer to [25] for more details.

3.4. Graph-Level Embedding

After obtaining the output node embeddings from the GAT, a graph READOUT operation can be used to obtain the graph-level embedding. Inspired by [27], we aggregate and concatenate the output node embeddings from all GAT layers to learn structure-aware node representations from different neighborhood ranges:

$$\text{emb}_G = \text{READOUT}(\{v_{i,1}||v_{i,2}||...||v_{i,L}|\forall i \in \mathcal{V}\}). \tag{5}$$

where $v_{i,l}$ is the output embedding of node i after the lth GAT layer. The READOUT can be any permutation-invariant operation (e.g., mean, sum, max). We adopt the global soft attention layer from [28] as the READOUT function for molecule graphs due to its excellent performance.

4. Methods

We propose to compose retrosynthesis templates from a predefined set of template building blocks; then, these composed templates are applied to target products to obtain the associated reactants. Unlike previous template-based methods [15–18] only selecting from training templates, our method can discover novel templates that are beyond the training templates. To further improve the retrosynthesis prediction performance, we design a scoring model to evaluate the suitability of product and candidate reactant pairs. The scoring procedure acts as a verification step and it plays a significant role in our method.

The overall pipeline of our method is shown in Figure 2. Our method tackles retrosynthesis in two stages. The first stage is to compose retrosynthesis templates with a TCM, which composes retrosynthesis templates by selecting template building blocks and then assembling them. In the second stage, the obtained templates are applied to the target product to generate the associated reactants. After that, we utilize a powerful RSM to evaluate the generated reactants for each product. During evaluation, the probability scores of both stages are linearly combined to rank Top-K reactant predictions. In following sections, we will detail each stage of our method.

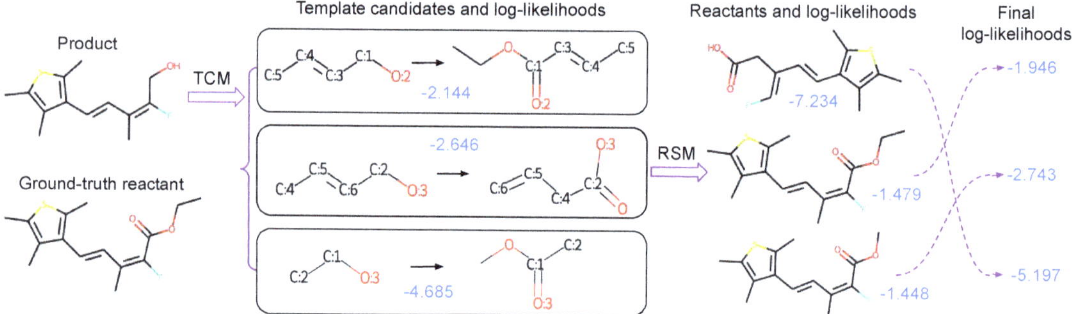

Figure 2. The overall pipeline of our proposed method. Given the desired product as shown at the top left, single-step retrosynthesis finds the ground-truth reactant as shown at the bottom left. Numbers indicated in blue are the corresponding log-likelihoods of our models, and the log-likelihoods of the template composer model (TCM) and the reactant scoring model (RSM) are combined to obtain the final ranking of the reactants. In this example, combining log-likelihoods of TCM and RSM helps to find the correct Top-1 reactant.

4.1. Compose Retrosynthesis Templates

Template-based retrosynthesis methods are criticized for their limitation of not generalizing to unseen reactions, since all existing template-based methods follow a similar procedure to select applicable templates from the extracted training templates. To overcome the above limitation, we propose a different pipeline to find template candidates. As illustrated in Figure 3, our method first selects product and reactant subgraphs sequentially from the corresponding subgraph vocabularies, which is detailed in Section 4.1.1. Then, these selected subgraphs are assembled into templates with properly assigned atom mappings, as detailed in Section 4.1.4. As far as we know, this is the first attempt to compose retrosynthesis templates instead of simple template selection. During evaluation, a beam search algorithm [29] is utilized to find Top-K predicted templates. Reactants can be obtained by applying templates to the target molecule.

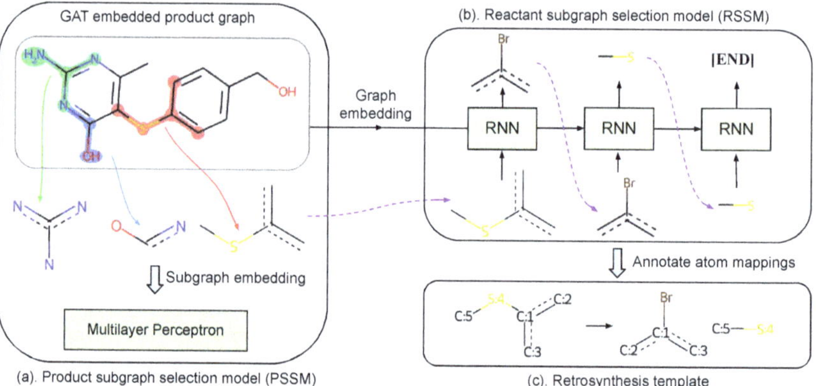

Figure 3. The workflow of our template composer model: (**a**) selecting a proper product subgraph from product subgraph candidates with PSSM, (**b**) selecting reactant subgraphs sequentially from reactant subgraph vocabulary with RSSM, and (**c**) annotating atom mappings between the product and reactant subgraphs to obtain a template.

4.1.1. Subgraph Selection

We denote a subgraph pattern as f, the product and reactant subgraphs for a template as f_p and f_r, respectively, and the product and reactant subgraph vocabulary for the dataset as $\mathcal{F}_P$ and $\mathcal{F}_R$, respectively. To build the product subgraph vocabulary $\mathcal{F}_P$ and reactant subgraph vocabulary $\mathcal{F}_R$, retrosynthesis templates extracted from training data are split into separate subgraphs to collect unique subgraph patterns. We build separate vocabularies for the product and reactant subgraphs due to their essential difference. Product subgraphs represent reaction centers and are more generalizable, while reactant subgraphs may contain extra leaving groups, which are more specific to the reaction type and the desired target. We find this strategy works well in practice.

4.1.2. Product Subgraph Selection

To compose retrosynthesis templates for a desired target, the first step is to choose proper f_p from the vocabulary $\mathcal{F}_P$. In this work, we focus on the single-product reactions; therefore, there is only a single product subgraph pattern. Note that there may be multiple viable retrosynthesis templates for each reaction, so each target may have several applicable product subgraphs. The set of applicable product subgraphs are denoted as $\mathcal{F}_a$. Starting with any applicable product subgraph in $\mathcal{F}_a$ may obtain a applicable retrosynthesis template for the target. Here, $\mathcal{F}_a \subseteq \mathcal{F}_P$ because all applicable product subgraphs must be in the vocabulary $\mathcal{F}_P$.

Each product molecule graph G_p contains only a limited set of candidate subgraphs $\mathcal{F}_c$ predefined in the vocabulary $\mathcal{F}_P$. Three candidate subgraphs are illustrated in Figure 3a. The candidate subgraphs for each target can be obtained offline by checking the existence of every product subgraph from $\mathcal{F}_P$ in the product graph G_p. Therefore, we only need to consider the candidate subgraphs $\mathcal{F}_c$ to guide the selection process [18] when selecting a product subgraph. Here, $\mathcal{F}_a \subseteq \mathcal{F}_c \subseteq \mathcal{F}_P$ since the candidate subgraphs $\mathcal{F}_c$ must contain all applicable subgraphs.

In this situation, the product subgraph selection can be regarded as a multi-label classification problem and starting with any applicable product subgraph in $\mathcal{F}_a$ can obtain a viable retrosynthesis template. However, it is not optimal to train the product subgraph selection model with binary cross-entropy loss (BCE) as in the multi-label classification setting, since it predicts the applicability score independently for each $f \in \mathcal{F}_c$ without considering their interrelationship. Note that the absolute applicability scores of subgraphs in $\mathcal{F}_c$ do not matter here; what really matters is the ranking of these applicability scores, since the beam search is adopted to find a series of template candidates during model inference. While a Softmax classifier can consider the relationship of all subgraphs in $\mathcal{F}_c$, it cannot be directly applied to PSSM, since it is not suitable for the multi-label case. Inspired by Softmax, we propose a novel negative log-likelihood loss for the PSSM:

$$L_{\text{PSSM}} = \log \frac{\arg\min_{f \in \mathcal{F}_a} o_f}{\arg\min_{f \in \mathcal{F}_a} o_f + \sum_{f \in \mathcal{F}_c \setminus \mathcal{F}_a} o_f}, \tag{6}$$

where o_f is the exponential of PSSM output logits for subgraphs in $\mathcal{F}$, $|\mathcal{F}|$ is the size of $\mathcal{F}$, and $\setminus$ is set subtraction. In the above loss function, the numerator is the minimal exponential output for all applicable subgraphs in $\mathcal{F}_a$, which is considered as the ground-truth class proxy in the Softmax classifier. The extra item in denominator is the summation of exponential output of all inapplicable subgraphs in $\mathcal{F}_c$. The intuition is that we always optimize the PSSM to increase the prediction probability for the least probable applicable subgraph, so the model is driven to generate large scores for all applicable subgraphs $\mathcal{F}_c$ while considering interrelationships of candidate subgraphs. The novel loss outperforms BCE loss in our experiments. Detailed experimental comparison results between the proposed loss function Equation (6) and BCE loss can be found in the experiment section.

PSSM scores candidate subgraphs $\mathcal{F}_c$ based on their subgraph embeddings. As shown in Figure 3a, to obtain subgraph embeddings, the nodes of product molecule graph G_p

are first encoded with the modified GAT that is detailed in Section 3.3. The embedding emb_f of the subgraph f is gathered as the average embedding of subgraph $f-$associated nodes in G_p, and then these embeddings are fed into a multilayer perceptron (MLP) for subgraph selection. Here, for a subgraph f, the READOUT function is implemented as the arithmetic average for its simplicity and efficiency. Note that this is different from GLN [18], in which product graphs and candidate subgraphs are considered as separate graphs and embedded independently. Our strategy to reuse node embeddings is more efficient and can learn more informative subgraph embedding since the neighboring structure of a subgraph is also incorporated during the message passing procedure of GAT. Besides, our method can naturally handle multiple equivalent subgraph situations in which the same subgraph appears multiple times within the product graph.

4.1.3. Reactant Subgraph Selection

The second step of the subgraph selection is to choose reactant subgraphs f_r from the vocabulary $\mathcal{F}_R$, which is ordered according to the subgraph frequency in training data, so that f_r is also determinedly ordered. With minor notation abuse, f_r also denotes an ordered sequence of reactant subgraphs in the following content.

Since the number of reactant subgraphs is undetermined, we build the reactant subgraph selection model based on the recurrent neural network (RNN), as illustrated in Figure 3b, and formulate reactant subgraph selection as the sequence generation. The hidden state of RNN is initialized from the product graph embedding emb_{G_p} as defined in Equation (5) to explicitly consider the target product, and the start token is the product subgraph f_p selected in the previous procedure (Section 4.1.2). Furthermore, an extra end token [END] is appended to reactant subgraph sequence f_r. At each time step, the RNN output is fed into a MLP for the token classification. For the start token f_p, we reuse product subgraph embeddings obtained previously (Section 4.1.2) since we find it provides better performance than embedding the token in the traditional one-hot embedding manner.

4.1.4. Annotate Atom Mappings

Given f_p and f_r, the final step is to annotate the atom mappings between f_p and f_r to obtain the retrosynthesis template, as shown in Figure 3c. A subgraph pattern f can also be represented in the SMARTS string, and we use open source toolkit Indigo's (https://github.com/epam/Indigo (accessed on 20 March 2022)) automap() function to build atom mappings. We empirically find about 70% of USPTO-50K training templates can be successfully annotated with correct atom mappings. To remedy this deficiency, we keep a memo of training templates and associated f_p and f_r. During evaluation, the predicted f_p and f_r are processed with automap() if not found in the memo.

4.2. Score Predicted Reactants

After a retrosynthesis template is composed, reactants can be easily obtained by applying the template to the target using RunReactants from RDKit [30] or the run_reaction() function from RDChiral [31]. To achieve superior retrosynthesis prediction performance, it is important to verify that the predicted reactants can generate the target successfully. The verification is achieved by scoring the reactants and target pair, which is formulated as a multi-class classification task where the true reactant set is the ground-truth class.

To serve the verification purpose, we build a reactant scoring model based on the modified GAT. Product molecule graph G_p and reactant molecule graph G_r are first input into a GAT to learn atom embeddings. Since the target and generated reactants are atom-mapped as in Figure 1a, for each atom in G_p, we can easily find its associated atom in G_r. Inspired by WLDN [32], we define a fusion function $\mathrm{F}(n_a^p, n_{a'}^r)$ to combine embeddings of a product atom a and its associated reactant atom a':

$$\mathrm{F}(n_a^p, n_{a'}^r) = \mathbf{W}_4(n_a^p - n_{a'}^r)||\mathbf{W}_5(n_a^p + n_{a'}^r), \tag{7}$$

where || indicates the concatenation operation and W is a matrix that halves the node embedding dimension so that the concatenated embedding restores the original dimension.

The fused atom embeddings are regarded as new atom features of G_p, which are input into another GAT to learn the graph-level embedding emb_G. In this way, the critical difference between the product and reactant can be better captured since our RSM can incorporate higher order interactions between fused atom embeddings through the message passing process of GAT. Previous retrosynthesis methods score reactants by modeling the compatibility of reactant and product at the molecule level without considering the atom-level embedding.

The graph-level embedding emb_G is then fed into a simple MLP composed of two fully-connected layers to output a compatibility score. The final probability score is obtained by applying a Softmax function to the compatibility scores of all candidate reactants associated to the target.

Our scoring model is advantageous since it operates on atom-level embeddings and is sensitive to local transformations between the product and reactants, while the existing method GLN [18] takes only molecule-level representations as the input. Therefore, GLN cannot capture atom-level transformations and has a weaker distinguishing ability.

The log-likelihoods of our TCM and RSM model predictions are denoted as $l_{TCM} = \log(\mathcal{P}(\mathcal{T}|P))$ and $l_{RSM} = \log(\mathcal{P}(R|P))$, respectively. The predicted reactants are finally ranked according to the linear combination value of $\lambda * l_{TCM} + (1-\lambda) * l_{RSM}$, $0 \leq \lambda \leq 1$. The formulation can be understood as:

$$\begin{aligned} &\lambda * \log(\mathcal{P}(\mathcal{T}|P)) + (1-\lambda) * \log(\mathcal{P}(R|P)) \\ &= \log(\mathcal{P}(\mathcal{T}|P)^{\lambda} * \mathcal{P}(R|P)^{1-\lambda}), \end{aligned} \tag{8}$$

where $\mathcal{P}(\mathcal{T}|P)$ is the probability of that the template $\mathcal{T}$ is applicable to the given product P and $\mathcal{P}(R|P)$ is the probability of the reactant set R for the given product P. When combined together, $\mathcal{P}(\mathcal{T}|P) * \mathcal{P}(R|P)$ approximates the joint probability distribution $\mathcal{P}(\mathcal{T}, R|P)$. Hyper-parameter λ regulates the relative importance of $\mathcal{P}(\mathcal{T}|P)$ and $\mathcal{P}(R|P)$. The optimal λ can be determined by the validation.

5. Experiment and Results

5.1. Dataset and Preprocessing

Our method is evaluated on the standard benchmark dataset USPTO-50K [33] under two settings (with or without reaction types) to demonstrate its effectiveness. USPTO-50K is derived from USPTO granted patents [34] and is composed of 50,000 reactions annotated with 10 reaction types. More detailed dataset information can be found in the Appendix A.1. We split reaction data into training/validation/test sets at an 8:1:1 ratio, in the same way as previous work [15,18]. Since the original annotated mapping numbers in the USPTO dataset may result in unexpected information leakage (https://github.com/uta-smile/RetroXpert (accessed on 20 March 2022)), we first preprocess the USPTO reactions to re-assign product mapping numbers according to the canonical atom order, as suggested by RetroXpert [4]. The atom and bond features are similar to the previous work [4] and reaction types are converted into one-hot vectors concatenated with the original atom features.

Following RetroXpert [4], we extract templates from training reactions using RDChiral [31]. We can obtain 10386 unique templates in total for the USPTO-50K training data and 94.08% of test reactions are covered by these training templates. The gathered templates are split into product and reactant subgraphs, from which mapping numbers are further removed to obtain the subgraph vocabularies $\mathcal{F}_P$ of size 7766 and $\mathcal{F}_R$ of size 4391.

For each target molecule, we find its candidate subgraphs $\mathcal{F}_c$ using graph matching algorithms and applicable templates by checking if the ground-truth reactant can be obtained when each training template is applied to the target. The applicable subgraphs $\mathcal{F}_a$ then can be obtained easily from the acquired applicable templates. Since the exact graph matching process might be time-consuming, we extract the fingerprint for each molecule/sub-molecule to filter those impossible subgraphs. For the subgraph screening purpose, we adopt the PatternFingerprint from RDKit and use a fingerprint size of 1024.

5.2. Evaluation

Following previous methods [4,18], we use beam search [29] to find Top-50 template predictions during evaluation, which are applied to targets to collect candidate reactants. The collected reactants and targets are the experimental data for RSM. The predicted reactants are finally ranked according to the combined log-likelihood of TCM and RSM. The evaluation metric for retrosynthesis prediction is the Top-K exact match accuracy, which is the percentage of reactions where the ground truth reactant set is within the top K predictions.

5.3. Implementation

Our model is implemented using PyTorch [35] and PyTorch Geometric [36]. The adapted GAT model is built based on the source implementation of Pretrain-GNN [37]. The TCM model is composed of a modified GAT and a simple RNN model. The embedding dimension is set as 300 for all embeddings for simplicity. The number of GAT layers is six. We adopt GRU [38] as the RNN implementation in TCM; the number of GRU layers is two and both its embedding and hidden size are 300. We add a self-loop to each graph node following [4,18]. We use the parametric rectified linear unit (PReLU) [39] comprehensively as the activation function in our model. We replace the original batch normalization [40] layer with a layer normalization [41] layer after each GAT layer, since we find layer normalization provides better performance in our experiments. We adopt Equation (5) as the graph READOUT operation. A simple MLP is applied to product subgraph embeddings to select the proper product subgraph. The MLP is composed of two linear layers, and the PReLU activation function is placed between the two linear layers. We also use a dropout [42] layer with a dropout rate of 0.3 in the MLP.

The RSM model is composed of two GATs and a MLP head, and the GAT uses the same settings as in the TCM except that each GAT is composed of three layers. Product and reactant graphs are embedded with the first GAT model. Note that for reactions with multiple reactants, we regard the disconnected molecule graphs as a single large graph. Once the fused atom embeddings are obtained, the new product molecule graphs with fused atom embeddings are input into the second GAT. The composition of the MLP head is similar to that in TCM. The RSM model is also trained in multi-process mode for acceleration.

Both TCM and RSM are optimized with the Adam [43] optimizer with default settings, and the initial learning rates are 0.0003 and 0.00005 for TCM and RSM, respectively. The learning rate is adjusted with the CosineAnnealingLR scheduler during training. The models are trained in multi-process mode on a single GTX 1080 Ti GPU for acceleration. TCM is trained with batch size 32; it only takes about two hours to train TCM for 80 epochs. RSM training takes about 6 hours for 20 epochs. The final model parameters are saved and loaded later for inference. We repeat all experiments three times and report the mean performance as default. We find our model is quite robust to the hyper-parameters, and most of the model settings are copied from [37] as they are given. We slightly tune the model hyper-parameters, such as learning rate and batch size, manually on validation data to achieve the best results.

5.4. Main Results

We decide the optimal value of λ according to validation performance. Specifically, we set λ as 0.4 for both experimental settings (with/without reaction types). We use these optimal settings in all experiments unless explicitly stated. Detailed ablation study about λ are included in Section 5.4.3.

5.4.1. Retrosynthesis Prediction Performance

We compare our RetroComposer with existing methods on the standard benchmark dataset USPTO-50K, and report comparison results in Table 1. The results of RetroXpert have been updated by the authors (https://github.com/uta-smile/RetroXpert (accessed on 20 March 2022)). For both evaluation settings (with or without reaction types), our method outperforms previous methods by a significant margin in seven out of eight compared Top-K metrics.

Table 1. Retrosynthesis evaluation results (%) on USPTO-50K. Existing methods are grouped into two categories. Our method RetroComposer belongs to the template-based methods. The best results in each column are highlighted in bold. RetroXpert* results have been updated by the authors in their GitHub repository (https://github.com/uta-smile/RetroXpert (accessed on 20 March 2022)).

Methods	Without Reaction Types				With Reaction Types			
	Top-1	**Top-3**	**Top-5**	**Top-10**	**Top-1**	**Top-3**	**Top-5**	**Top-10**
Template-free methods								
SCROP [6]	43.7	60.0	65.2	68.7	59.0	74.8	78.1	81.1
G2Gs [7]	48.9	67.6	72.5	75.5	61.0	81.3	86.0	88.7
MEGAN [8]	48.1	70.7	78.4	86.1	60.7	82.0	87.5	**91.6**
RetroXpert* [4]	50.4	61.1	62.3	63.4	62.1	75.8	78.5	80.9
RetroPrime [12]	51.4	70.8	74.0	76.1	64.8	81.6	85.0	86.9
AT [11]	53.5	-	81.0	85.7	-	-	-	-
GraphRetro [13]	53.7	68.3	72.2	75.5	63.9	81.5	85.2	88.1
Dual model [9]	53.6	70.7	74.6	77.0	65.7	81.9	84.7	85.9
Template-based methods								
RetroSim [15]	37.3	54.7	63.3	74.1	52.9	73.8	81.2	88.1
NeuralSym [16]	44.4	65.3	72.4	78.9	55.3	76.0	81.4	85.1
GLN [18]	52.5	69.0	75.6	83.7	64.2	79.1	85.2	90.0
Ours	**54.5**	**77.2**	**83.2**	**87.7**	**65.9**	**85.8**	**89.5**	91.5
TCM only	49.6	71.7	80.8	86.4	60.9	82.3	87.5	90.9
RSM only	51.8	75.7	82.4	87.3	64.3	84.8	88.9	91.4

Specially, our RetroComposer achieves 54.5% Top-1 accuracy without reaction types, which improves on the previous best template-based method GLN [18] significantly by 2.0% and also outperforms existing SOTA template-free methods Dual model and GraphRetro. Besides, our method achieves 77.2% Top-3 accuracy, which improves on the Top-3 accuracy 70.8% of RetroPrime [12] by 6.4%, and 87.7% Top-10 accuracy, which improves on the Top-10 accuracy 85.7% of AT [11] by 2.0%.

When reaction types are given, our method also obtains the best Top-1 accuracy, 65.9%, among all methods and outperforms GLN by 1.7%. Compared with template-free methods GraphRetro and Dual model, our method outperforms the SOTA Dual model (65.7%) by 0.2% and GraphRetro significantly by 2.0% in Top-1 accuracy. As for the Top-10 accuracy, our method achieves 91.5%, which is slightly lower than 91.6% of MEGAN [8].

As the ablation study, we report results with only TCM or RSM. With only either TCM or RSM, the model performance is largely degraded. Without reaction types, TCM only achieves 49.6% Top-1 accuracy while RSM achieves only 51.8%. With reaction types, TCM only achieves 60.9% Top-1 accuracy while RSM achieves only 64.3%. Since TCM and RSM score retrosynthesis from different perspectives and are complementary, their results can be combined to achieve the best performance. Particularly, our method achieves 54.5% and 65.9% Top-1 accuracy when combining TCM and RSM according to Equation (8).

The superior performance demonstrates the effectiveness of our method. Particularly, the superiority of our method is more significant in real world applications where reaction types are unknown. What is more, our Top-10 accuracy is already quite high. This indicates that our method can usually find the best reactant set for the target in a few candidates.

This is especially important for multi-step retrosynthesis scenarios, in which the number of predicted reaction paths may grow exponentially with the retrosynthesis path length.

5.4.2. Ablation Study of PSSM Loss

We experimentally show that our proposed loss function Equation (6) for PSSM outperforms the BCE loss. For all ablation experiments, we find the optimal value of hyper-parameter λ independently and report the best results for a fair comparison. The comprehensive experimental results are reported in Table 2.

Without given reaction types, our method with Equation (6) as PSSM loss achieves the best Top-1 and Top-3 accuracy results, outperforming the BCE loss in Top-1 and Top-3 accuracy by 1.4% and 1.5%, respectively. With known reaction types, our method with Equation (6) as PSSM loss outperforms BCE loss by 0.6% in Top-1 accuracy. While BCE loss can achieve better Top-5 and Top-10 results in both settings, our proposed loss function Equation (6) can achieve better Top-1 accuracy. The retrosynthesis prediction emphasizes more Top-1 accuracy, therefore, we adopt Equation (6) as the PSSM loss in our method.

For all experiments, combining the TCM and RSM scores can always achieve the best performance, which proves the effectiveness of our strategy.

Table 2. Ablation study results (%) of two different PSSM loss functions: our proposed Equation (6) and BCE. The bold indicates the best results.

Types	L_{PSSM}	Methods	Top-1	Top-3	Top-5	Top-10
Without	Equation (6)	Ours	**54.5**	**77.2**	83.2	87.7
		TCM only	49.6	71.7	80.8	86.4
		RSM only	51.8	75.7	82.4	87.3
	BCE	Ours	53.1	77.1	**83.8**	**89.2**
		TCM only	46.5	69.9	78.5	86.9
		RSM only	51.2	75.7	82.9	88.6
With	Equation (6)	Ours	**65.9**	85.8	89.5	91.5
		TCM only	60.9	82.3	87.5	90.9
		RSM only	64.3	84.8	88.9	91.4
	BCE	Ours	65.3	**85.9**	**90.3**	**92.6**
		TCM only	58.5	81.8	87.6	91.5
		RSM only	64.2	85.4	89.6	92.4

5.4.3. Ablation Study of Hyper-Parameter λ

We conduct the ablation study of λ and report results in Table 3; when $\lambda = 0.4$, the best Top-1 accuracy is achieved for both settings. Note that with only RSM ($\lambda = 0$), the Top-1 accuracy 64.3% already outperforms the previous best template-based method GLN of 63.2% [18] with given reaction types. This demonstrates the effectiveness of our RSM. With only TCM ($\lambda = 1.0$), the performance has an appreciable gap with the existing methods. In our method, each generated set of subgraphs may have multiple associated templates due to the uncertainty of product subgraphs and atom transformations. Therefore, there may be multiple top-tier predictions that cannot be distinguished with only TCM. With a little help from RSM ($\lambda = 0.9$), these top-tier predictions can be differentiated and the Top-1 accuracy significantly boosted.

The l_{RSM} indicates the likelihood of retrosynthesis templates, while l_{TCM} scores each reaction by looking at the detailed atom transformations. These two terms are complementary and combined together to achieve the best performance.

Table 3. Top-1 accuracy (%) with different λ values. The bold indicates the best results.

λ	0	0.1	0.2	0.3	0.4	0.5	0.6	0.7	0.8	0.9	1.0
Without types	51.8	53.3	53.9	**54.5**	**54.5**	54.4	54.1	53.6	53.0	52.3	49.6
With types	64.3	65.2	65.6	65.7	**65.9**	**65.9**	65.6	65.1	64.7	64.4	60.9

5.4.4. Novel Templates

Different from existing methods, our method can find novels templates that are not in training data. Our model predicts different templates based on different possible reaction centers for a given target. For example, an amide formation template and alkylation template may both be applied in the same target molecule, and our model can predict suitable templates very well and give reasonable corresponding reactants for such cases. For the 5.92% of test reactions that are not covered by training templates, our algorithm can predict relevant templates very well for most reaction types, although it fails in some heterocyclic formation reactions. This is because there are very few reaction data on such reactions in USPTO-50K. Particularly, our method successfully discovers chemically valid templates for 15 uncovered test reactions, which confirms that our method can find novel reactions. Two such examples are illustrated in Figure 4.

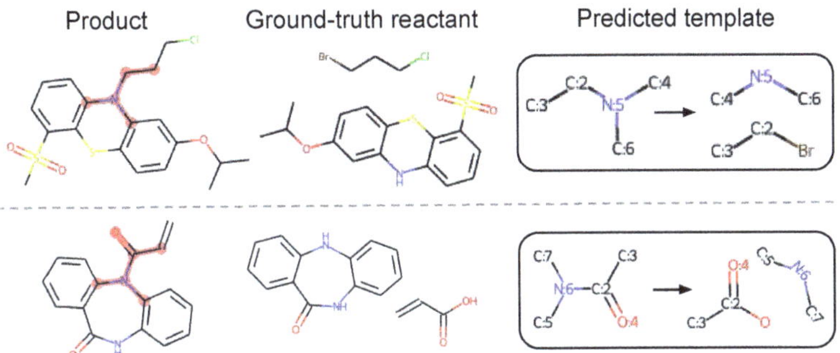

Figure 4. Our method successfully finds valid templates for two test reactions that are not covered by training data. The matched product subgraphs are highlighted in pink for better visualization.

6. Discussion and Conclusions

In this work, we propose a novel template-based retrosynthesis prediction framework that composes templates by selecting and assembling molecule subgraphs. Besides, experimental results confirm that the proposed strategy can discover novel reactions. Although currently our method can find only a few novel templates, we believe our method can inspire the community to explore further in this direction to improve models' ability to find more novel reactions. To further improve the ranking accuracy, we present a novel reactant scoring model to rank candidate reactants by taking into account atom-level transformations. Our method significantly outperforms previous methods and sets new SOTA performance on the USPTO-50K, which proves the effectiveness of our method.

We tried to adapt our method to run on the USPTO-full dataset [34], but find it needs non-trivial effort to manually handle edge cases due to noisy reactions (such as wrong mapping numbers) from USPTO-full, since our methods rely on correct mapping numbers to extract templates as well as build the reactant scoring model. We have released our source implementation and encourage the community to help adapt our method to the USPTO-full dataset.

Author Contributions: Conceptualization, C.Y. and P.Z.; methodology, C.Y. and P.Z.; formal analysis, Y.Y.; investigation, C.L.; writing—original draft preparation, C.Y. and C.L.; writing—review and editing, C.Y. and P.Z.; supervision, J.H.; project administration, J.H.; funding acquisition, J.H. All authors have read and agreed to the published version of the manuscript.

Funding: This work was partially supported by US National Science Foundation IIS-1553687 and Cancer Prevention and Research Institute of Texas (CPRIT) award (RP190107).

Institutional Review Board Statement: Not applicable.

Informed Consent Statement: Not applicable.

Data Availability Statement: The experimental dataset USPTO-50K can be downloaded at http://pubs.acs.org/doi/suppl/10.1021/acs.jcim.6b00564/suppl_file/ci6b00564_si_002.zip (accessed on 20 March 2022).

Conflicts of Interest: The authors declare no conflict of interest. The funders had no role in the design of the study; in the collection, analyses, or interpretation of data; in the writing of the manuscript, or in the decision to publish the results.

Appendix A

Appendix A.1. USPTO-50K Dataset Information

The USPTO-50K consists of 50,000 reactions that are annotated with 10 reaction types; the detailed distribution of reaction types is displayed in the below Table A1. The imbalanced reaction type distribution makes the retrosynthesis prediction more challenging.

Table A1. Distribution of 10 recognized reaction types.

Type	Reaction Type Name	Number of Reactions
1	Heteroatom alkylation and arylation	15,204
2	Acylation and related processes	11,972
3	C-C bond formation	5667
4	Heterocycle formation	909
5	Protections	672
6	Deprotections	8405
7	Reductions	4642
8	Oxidations	822
9	Functional group interconversion	1858
10	Functional group addition (FGA)	231

We can extract 10,386 unique templates from the training data, and 94.08% of test reactions are covered by these templates. For each product molecule, there are an average of 35.19 candidate subgraphs, which are denoted as $\mathcal{F}_c$ in Section 4.1.2. Among these subgraphs, there are an average of 2.02 applicable subgraphs denoted as $\mathcal{F}_a$ for each target.

Table A2. Statistical results of templates and reactions. # is the short for "number".

# total templates	10,386
# unique product subgraphs	7766
# unique reactant subgraphs	4391
Test reactions coverage by training templates	94.08%
Average # contained product subgraphs per mol	35.19
Average # applicable product subgraphs per mol	2.02
Average # templates per reaction	2.23
Average # reactants per reaction	1.71

Appendix A.2. Atom and Bond Features

Following [4], we use similar bond and atom features to build molecule graphs as listed in Tables A3 and A4. These features can be easily extracted using the chemistry toolkit RDKit.

Table A3. Bond features used in our method. These features are one-hot encoding.

Feature	Description	Size
Bond type	Single, double, triple, or aromatic.	4
Conjugation	Whether the bond is conjugated.	1
In ring	Whether the bond is part of a ring.	1
Stereo	None, any, E/Z or cis/trans.	6

Table A4. Atom features used in our method. All features are one-hot encoding, except the atomic mass is a real number scaled to be on the same order of magnitude. The reaction type is applicable for type conditional setting.

Feature	Description	Size
Atom type	Type of atom (ex. C, N, O), by atomic number.	17
# Bonds	Number of bonds the atom is involved in.	6
Formal charge	Integer electronic charge assigned to atom.	5
Chirality	Unspecified, tetrahedral CW/CCW, or other.	4
# Hs	Number of bonded Hydrogen atom.	5
Hybridization	sp, sp2, sp3, sp3d, or sp3d2.	5
Aromaticity	Whether this atom is part of an aromatic system.	1
Atomic mass	Mass of the atom, divided by 100.	1
Reaction type	The specified reaction type if it exists.	10

References

1. Corey, E.J.; Wipke, W.T. Computer-assisted design of complex organic syntheses. *Science* **1969**, *166*, 178–192. [CrossRef] [PubMed]
2. Corey, E.J. The logic of chemical synthesis: Multistep synthesis of complex carbogenic molecules (Nobel Lecture). *Angew. Chem. Int. Ed. Engl.* **1991**, *30*, 455–465. [CrossRef]
3. Gothard, C.M.; Soh, S.; Gothard, N.A.; Kowalczyk, B.; Wei, Y.; Baytekin, B.; Grzybowski, B.A. Rewiring chemistry: Algorithmic discovery and experimental validation of one-pot reactions in the network of organic chemistry. *Angew. Chem. Int. Ed.* **2012**, *51*, 7922–7927. [CrossRef] [PubMed]
4. Yan, C.; Ding, Q.; Zhao, P.; Zheng, S.; Yang, J.; Yu, Y.; Huang, J. RetroXpert: Decompose Retrosynthesis Prediction Like A Chemist. *Adv. Neural Inf. Process. Syst.* **2020**, *33*, 11248–11258.
5. Liu, B.; Ramsundar, B.; Kawthekar, P.; Shi, J.; Gomes, J.; Luu Nguyen, Q.; Ho, S.; Sloane, J.; Wender, P.; Pande, V. Retrosynthetic reaction prediction using neural sequence-to-sequence models. *ACS Cent. Sci.* **2017**, *3*, 1103–1113. [CrossRef]
6. Zheng, S.; Rao, J.; Zhang, Z.; Xu, J.; Yang, Y. Predicting Retrosynthetic Reactions using Self-Corrected Transformer Neural Networks. *J. Chem. Inf. Model.* **2020**, *60*, 47–55. [CrossRef]
7. Shi, C.; Xu, M.; Guo, H.; Zhang, M.; Tang, J. A Graph to Graphs Framework for Retrosynthesis Prediction. *arXiv* **2020**, arXiv:2003.12725.
8. Sacha, M.; Błaż, M.; Byrski, P.; Włodarczyk-Pruszyński, P.; Jastrzebski, S. Molecule Edit Graph Attention Network: Modeling Chemical Reactions as Sequences of Graph Edits. *J. Chem. Inf. Model.* **2021**, *61*, 3273–3284. [CrossRef]
9. Sun, R.; Dai, H.; Li, L.; Kearnes, S.; Dai, B. Towards understanding retrosynthesis by energy-based models. *Adv. Neural Inf. Process. Syst.* **2021**, *34*, 10186–10194.
10. Weininger, D. SMILES, a chemical language and information system. 1. Introduction to methodology and encoding rules. *J. Chem. Inf. Comput. Sci.* **1988**, *28*, 31–36. [CrossRef]
11. Tetko, I.V.; Karpov, P.; Van Deursen, R.; Godin, G. State-of-the-art augmented NLP transformer models for direct and single-step retrosynthesis. *Nat. Commun.* **2020**, *11*, 5575. [CrossRef] [PubMed]
12. Wang, X.; Li, Y.; Qiu, J.; Chen, G.; Liu, H.; Liao, B.; Hsieh, C.; Yao, X. RetroPrime: A Diverse, plausible and Transformer-based method for Single-Step retrosynthesis predictions. *Chem. Eng. J.* **2021**, *420*, 129845. [CrossRef]
13. Somnath, V.R.; Bunne, C.; Coley, C.; Krause, A.; Barzilay, R. Learning graph models for retrosynthesis prediction. *Adv. Neural Inf. Process. Syst.* **2021**, *34*, 9405–9415.
14. Szymkuć, S.; Gajewska, E.P.; Klucznik, T.; Molga, K.; Dittwald, P.; Startek, M.; Bajczyk, M.; Grzybowski, B.A. Computer-Assisted Synthetic Planning: The End of the Beginning. *Angew. Chem. Int. Ed.* **2016**, *55*, 5904–5937. [CrossRef]
15. Coley, C.W.; Rogers, L.; Green, W.H.; Jensen, K.F. Computer-assisted retrosynthesis based on molecular similarity. *ACS Cent. Sci.* **2017**, *3*, 1237–1245. [CrossRef]
16. Segler, M.H.; Waller, M.P. Neural-symbolic machine learning for retrosynthesis and reaction prediction. *Chem.-Eur. J.* **2017**, *23*, 5966–5971. [CrossRef]
17. Segler, M.H.; Preuss, M.; Waller, M.P. Planning chemical syntheses with deep neural networks and symbolic AI. *Nature* **2018**, *555*, 604–610. [CrossRef]

18. Dai, H.; Li, C.; Coley, C.; Dai, B.; Song, L. Retrosynthesis Prediction with Conditional Graph Logic Network. In Proceedings of the Advances in Neural Information Processing Systems, Vancouver, BC, Canada, 8–14 December 2019; pp. 8870–8880.
19. Segler, M.H.; Waller, M.P. Modelling chemical reasoning to predict and invent reactions. *Chem.-Eur. J.* **2017**, *23*, 6118–6128. [CrossRef]
20. Baylon, J.L.; Cilfone, N.A.; Gulcher, J.R.; Chittenden, T.W. Enhancing retrosynthetic reaction prediction with deep learning using multiscale reaction classification. *J. Chem. Inf. Model.* **2019**, *59*, 673–688. [CrossRef]
21. Tu, Z.; Coley, C.W. Permutation invariant graph-to-sequence model for template-free retrosynthesis and reaction prediction. *arXiv* **2021**, arXiv:2110.09681.
22. Irwin, R.; Dimitriadis, S.; He, J.; Bjerrum, E.J. Chemformer: A Pre-Trained Transformer for Computational Chemistry. *Mach. Learn. Sci. Technol.* **2021**, *3*, 015022. [CrossRef]
23. Mao, K.; Xiao, X.; Xu, T.; Rong, Y.; Huang, J.; Zhao, P. Molecular graph enhanced transformer for retrosynthesis prediction. *Neurocomputing* **2021**, *457*, 193–202. [CrossRef]
24. Gilmer, J.; Schoenholz, S.S.; Riley, P.F.; Vinyals, O.; Dahl, G.E. Neural message passing for quantum chemistry. In Proceedings of the 34th International Conference on Machine Learning-Volume 70, Sydney, Australia, 6–11 August 2017; pp. 1263–1272.
25. Veličković, P.; Cucurull, G.; Casanova, A.; Romero, A.; Liò, P.; Bengio, Y. Graph Attention Networks. In Proceedings of the 6th International Conference on Learning Representations, Vancouver, BC, Canada, 30 April–3 May 2018.
26. Vaswani, A.; Shazeer, N.; Parmar, N.; Uszkoreit, J.; Jones, L.; Gomez, A.N.; Kaiser, Ł.; Polosukhin, I. Attention is all you need. In Proceedings of the Advances in Neural Information Processing Systems, Long Beach, CA, USA, 4–9 December 2017; pp. 5998–6008.
27. Xu, K.; Li, C.; Tian, Y.; Sonobe, T.; Kawarabayashi, K.i.; Jegelka, S. Representation learning on graphs with jumping knowledge networks. In Proceedings of the International Conference on Machine Learning, Stockholm, Sweden, 10–15 July 2018; pp. 5453–5462.
28. Li, Y.; Tarlow, D.; Brockschmidt, M.; Zemel, R. Gated graph sequence neural networks. *arXiv* **2015**, arXiv:1511.05493.
29. Tillmann, C.; Ney, H. Word reordering and a dynamic programming beam search algorithm for statistical machine translation. *Comput. Linguist.* **2003**, *29*, 97–133. [CrossRef]
30. Landrum, G. RDKit: Open-Source Cheminformatics. 2021. Available online: https://github.com/rdkit/rdkit/tree/Release_2021_03_1 (accessed on 14 September 2022).
31. Coley, C.W.; Green, W.H.; Jensen, K.F. RDChiral: An RDKit wrapper for handling stereochemistry in retrosynthetic template extraction and application. *J. Chem. Inf. Model.* **2019**, *59*, 2529–2537. [CrossRef]
32. Jin, W.; Coley, C.; Barzilay, R.; Jaakkola, T. Predicting organic reaction outcomes with weisfeiler-lehman network. In Proceedings of the Advances in Neural Information Processing Systems, Long Beach, CA, USA, 4–9 December 2017.
33. Schneider, N.; Stiefl, N.; Landrum, G.A. What's what: The (nearly) definitive guide to reaction role assignment. *J. Chem. Inf. Model.* **2016**, *56*, 2336–2346. [CrossRef]
34. Lowe, D.M. Extraction of Chemical Structures and Reactions from the Literature. Ph.D. Thesis, University of Cambridge, Cambridge, UK, 2012.
35. Paszke, A.; Gross, S.; Massa, F.; Lerer, A.; Bradbury, J.; Chanan, G.; Killeen, T.; Lin, Z.; Gimelshein, N.; Antiga, L.; et al. Pytorch: An imperative style, high-performance deep learning library. *Adv. Neural Inf. Process. Syst.* **2019**, *32*, 8026–8037.
36. Fey, M.; Lenssen, J.E. Fast graph representation learning with PyTorch Geometric. *arXiv* **2019**, arXiv:1903.02428.
37. Hu, W.; Liu, B.; Gomes, J.; Zitnik, M.; Liang, P.; Pande, V.; Leskovec, J. Strategies for pre-training graph neural networks. *arXiv* **2019**, arXiv:1905.12265.
38. Cho, K.; Van Merriënboer, B.; Bahdanau, D.; Bengio, Y. On the properties of neural machine translation: Encoder-decoder approaches. *arXiv* **2014**, arXiv:1409.1259.
39. He, K.; Zhang, X.; Ren, S.; Sun, J. Delving deep into rectifiers: Surpassing human-level performance on imagenet classification. In Proceedings of the IEEE International Conference on Computer Vision, Santiago, Chile, 7–13 December 2015; pp. 1026–1034.
40. Ioffe, S.; Szegedy, C. Batch normalization: Accelerating deep network training by reducing internal covariate shift. In Proceedings of the International Conference on Machine Learning, Lille, France, 7–9 July 2015; pp. 448–456.
41. Ba, J.L.; Kiros, J.R.; Hinton, G.E. Layer normalization. *arXiv* **2016**, arXiv:1607.06450.
42. Srivastava, N.; Hinton, G.; Krizhevsky, A.; Sutskever, I.; Salakhutdinov, R. Dropout: A simple way to prevent neural networks from overfitting. *J. Mach. Learn. Res.* **2014**, *15*, 1929–1958.
43. Kingma, D.P.; Ba, J. Adam: A method for stochastic optimization. *arXiv* **2014**, arXiv:1412.6980.

Article

The Pharmacorank Search Tool for the Retrieval of Prioritized Protein Drug Targets and Drug Repositioning Candidates According to Selected Diseases

Sergey Gnilopyat, Paul J. DePietro, Thomas K. Parry and William A. McLaughlin *

Department of Medical Education, Geisinger Commonwealth School of Medicine, 525 Pine Street, Scranton, PA 18509, USA
* Correspondence: wmclaughlin@som.geisinger.edu; Tel.: +570-504-9633; Fax: +570-504-9636

Abstract: We present the Pharmacorank search tool as an objective means to obtain prioritized protein drug targets and their associated medications according to user-selected diseases. This tool could be used to obtain prioritized protein targets for the creation of novel medications or to predict novel indications for medications that already exist. To prioritize the proteins associated with each disease, a gene similarity profiling method based on protein functions is implemented. The priority scores of the proteins are found to correlate well with the likelihoods that the associated medications are clinically relevant in the disease's treatment. When the protein priority scores are plotted against the percentage of protein targets that are known to bind medications currently indicated to treat the disease, which we termed the pertinency score, a strong correlation was observed. The correlation coefficient was found to be 0.9978 when using a weighted second-order polynomial fit. As the highly predictive fit was made using a broad range of diseases, we were able to identify a general threshold for the pertinency score as a starting point for considering drug repositioning candidates. Several repositioning candidates are described for proteins that have high predicated pertinency scores, and these provide illustrative examples of the applications of the tool. We also describe focused reviews of repositioning candidates for Alzheimer's disease. Via the tool's URL, https://protein.som.geisinger.edu/Pharmacorank/, an open online interface is provided for interactive use; and there is a site for programmatic access.

Keywords: protein database; search tool; prioritization algorithm; drug repositioning

Citation: Gnilopyat, S.; DePietro, P.J.; Parry, T.K.; McLaughlin, W.A. The Pharmacorank Search Tool for the Retrieval of Prioritized Protein Drug Targets and Drug Repositioning Candidates According to Selected Diseases. *Biomolecules* **2022**, *12*, 1559. https://doi.org/10.3390/biom12111559

Academic Editors: Cameron Mura and Lei Xie

Received: 27 September 2022
Accepted: 22 October 2022
Published: 26 October 2022

1. Introduction

Proteins are currently being mapped to diseases in a comprehensive manner within open, online databases [1–4]. There is also an ongoing expansion of resources that document the associations of proteins with medications [2,5–12]. With the availability of these two types of information, proteins can serve as connection points between diseases and medications. Such connections have the potential to predict new indications for existing medications via the process of drug repurposing, also called drug repositioning [13].

To aid with the process of drug repurposing, estimates of the likelihoods that drug repurposing candidates could be effective new treatments have been made using various approaches. These approaches include network-based approaches, text-based approaches, and semantics-based approaches, as reviewed by Xue et al. [14]. A central component of some of these approaches is a gene prioritization algorithm that uses similarity profiling, and such algorithms offer demonstrated applicability to the prioritization of proteins associated with diseases [15]. The proteins' priority or rank scores, in turn, mirror the likelihoods that they may be useful as medication targets for the treatment of a selected disease [16]. Since the priority scores of the protein targets can be linked to their associated medications, the result can be the uncovering of novel predictions regarding which medications may be most effectively repurposed [17].

Methods that prioritize proteins involved in diseases may use the presence of co-occurring words or database terms as part of the prioritization algorithm. Consider, for example, the use of co-occurring database terms in the PolySearch tool [18]. We previously implemented a tool called KB-Rank that considers the co-occurrences of a diverse set of functional annotations to prioritize proteins associated with selected diseases [19]. More example methods that prioritize protein–disease datasets using protein functions include ToppGene [20], TargetMine [21], and network methods [22,23]. Examples of the integration of gene prioritization methods into an online search tool with the goal of aiding drug repositioning efforts have been implemented in RepurposeDB [24] and Project Rephetio [25].

To further enable drug repurposing efforts, we present the Pharmacorank search tool. To prioritize the candidates, an unsupervised gene prioritization algorithm is implemented, which may be classified as a similarity profiling or a data fusion method [26]. This method utilizes the diverse set of protein functions and annotations that are available via the UniProt database [3].

To validate and benchmark the search tool's accuracy, we calculated the percentages of medication–protein target pairs that were already known to be involved in the disease's treatment for each priority range, and we termed these values the pertinency scores. An optimal correlation was sought between the priority scores and the pertinency scores. The equation of best fit then served as a predictive model that takes the priority score of each of the protein targets and predicts the pertinency score. We interpret the pertinency scores as being estimates of the likelihoods that the medications associated with the protein targets will be clinically relevant in a selected disease's treatment.

With the goal of helping to spur drug repositioning efforts within the greater scientific community, the predictive model between the priority score and pertinency score was applied to all diseases described in Disease Ontology (DO) [27], and drug repositioning candidates were made available for each of the diseases described in DO that map to an ICD-9-CM or ICD-10-CM code. To highlight the results, a few of the drug repositioning candidates with the highest predicted pertinency scores across all the diseases are further described. We also provide focused manual reviews of the drug repositioning candidates that are identified as possible future treatments for Alzheimer's disease (AD).

2. Materials and Methods

2.1. Overview

The overall steps of the method to obtain prioritized proteins and medications to inform drug repurposing studies may be summarized as: (1) retrieve the proteins associated with each identified disease; (2) rank the proteins with priority scores calculated using protein functions; (3) obtain the correspondences between medications and proteins along with correspondences between medications and diseases from public databases to determine whether each protein interacts with a medication already known to treat the disease; (4) perform validation studies regarding the contribution that each type of protein function has in generating a priority score that discerns whether a protein interacts with a medication that is already used to treat the disease; (5) derive a predictive, quantitative relationship between the priority score and pertinency score, where the pertinency score is the percentage of protein targets known to interact with medications that are already known to treat the disease; (6) apply the resulting predictive relationship between the priority score and pertinency score to all proteins and medications associated with the disease; (7) identify a recommended pertinency score threshold for the end-user; and (8) review the protein targets and medications that have high predicted pertinency scores for consideration in future drug repositioning studies.

2.2. Assemble Protein–Disease Datasets (Overall Step 1)

A set of proteins associated with each disease was obtained based on information from multiple sources. Known correspondences were retrieved based on the Online Mendelian

Inheritance in Man (OMIM) phenotypic descriptions [1] for a total of 4646 correspondences. Further, an additional 1798 correspondences between diseases and proteins were obtained using the Kyoto Encyclopedia of Genes and Genomes (KEGG) disease name assignments [2]. The integrated protein–disease datasets from DisGeNET [28] added an additional 54,226 correspondences. The total number of protein–disease correspondences was therefore 60,670. With nonredundant protein–disease datasets derived collectively from these multiple sources, the result was a comprehensive dataset for each DO entry.

2.2.1. Implement the Prioritization Algorithm Using Protein Functions (Overall Step 2, First Part)

The functions used in the prioritization algorithm were retrieved from UniProt files and from coordinating resources that describe protein function. The types of functions include UniProt keywords, Gene Ontology terms, Enzyme Commission (EC) numbers, InterPro assignments, SUPERFAMILY assignments, small molecule interaction assignments from Chemical Entities of Biological Interest (ChEBI) [11], and UniProt residue features.

There were three types of annotations used from Gene Ontology (GO): molecular function, cellular component, and biological process. GO terms were obtained through UniProt GOA, which had granular GO annotations and excluded those higher up within the GO hierarchy when identified by the same technique [29,30]. All ChEBI entries except those that mapped to ChEMBL entries were utilized as function assignments. Broadly, the functions used here are functional characteristics that are shared among two or more proteins. Functional characteristics that could only possibly be attributed to one protein were excluded. For example, point mutations were excluded as functions since a point mutation would not be shared with other proteins.

2.2.2. Calculate the Priority Scores (Overall Step 2, Second Part)

After the identification of the proteins of each protein–disease dataset, the priority scores of the proteins and their associated medications were calculated. An outline of the prioritization algorithm is shown in Figure 1. Raw priority scores were first calculated separately according to each type of function. For this purpose, the total number of proteins in the protein–disease dataset with a specific function was found, and this number was assigned to the specific function. A protein's raw priority score according to the type of function was then the sum of the numbers assigned for each specific function for which the protein was known to have. Mathematically, the raw priority score of a protein using a type of function is equivalent to the dot product of two equal length 1D arrays, rho and mu. The array mu, μ, has an entry for each of the specific functions found among the proteins in the protein–disease dataset with the value that is equal to the number of UniProt accession codes (proteins) in the protein–disease dataset that had that specific function. Each protein in the protein–disease dataset also had its own associated binary array rho, ρ, which consisted of one if the specific function was attributed to that protein or zero otherwise. The formula for the raw priority score for a protein regarding a type of function was calculated as the dot product of ρ and μ:

$$Pscore_{raw}^{FuncType} = \rho \cdot \mu$$

To calculate the raw priority score, a Python dictionary was created, where the keys were the unique identifiers of the specific functions in the protein–disease dataset and the values were the corresponding total numbers of proteins that had each specific function. The raw priority score for a protein regarding a type of function was then the sum of the values associated with each of the specific functions that were attributed to that protein. This is numerically equivalent to the dot product but avoids what we found to be a more computationally intense and more error-prone task of creating and multiplying the 1D arrays ρ and μ

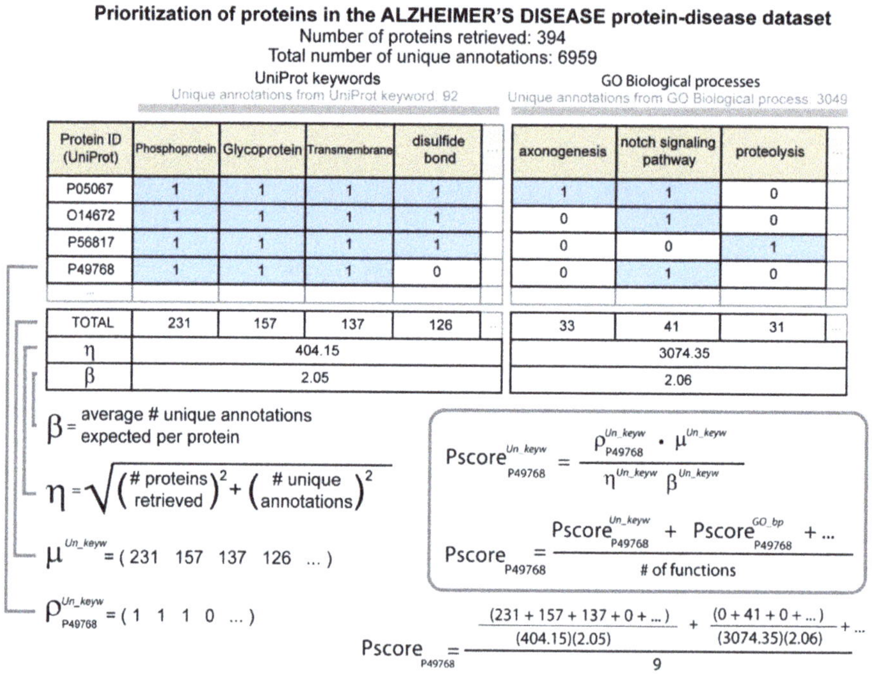

Prioritization of proteins in the ALZHEIMER'S DISEASE protein-disease dataset

Number of proteins retrieved: 394

Total number of unique annotations: 6959

	UniProt keywords (Unique annotations from UniProt keyword: 92)					GO Biological processes (Unique annotations from GO Biological process: 3049)			
Protein ID (UniProt)	Phosphoprotein	Glycoprotein	Transmembrane	disulfide bond	...	axonogenesis	notch signaling pathway	proteolysis	...
P05067	1	1	1	1		1	1	0	
014672	1	1	1	1		0	1	0	
P56817	1	1	1	1		0	0	1	
P49768	1	1	1	0		0	1	0	
...									
TOTAL	231	157	137	126	...	33	41	31	...
η	404.15					3074.35			
β	2.05					2.06			

β = average # unique annotations expected per protein

$$\eta = \sqrt{\left(\begin{matrix}\text{\# proteins}\\ \text{retrieved}\end{matrix}\right)^2 + \left(\begin{matrix}\text{\# unique}\\ \text{annotations}\end{matrix}\right)^2}$$

$$\mu^{Un\_keyw} = (231 \quad 157 \quad 137 \quad 126 \quad \ldots)$$

$$\rho^{Un\_keyw}_{P49768} = (1 \quad 1 \quad 1 \quad 0 \quad \ldots)$$

$$\text{Pscore}^{Un\_keyw}_{P49768} = \frac{\rho^{Un\_keyw}_{P49768} \cdot \mu^{Un\_keyw}}{\eta^{Un\_keyw}\,\beta^{Un\_keyw}}$$

$$\text{Pscore}_{P49768} = \frac{\text{Pscore}^{Un\_keyw}_{P49768} + \text{Pscore}^{GO\_bp}_{P49768} + \ldots}{\text{\# of functions}}$$

$$\text{Pscore}_{P49768} = \frac{\frac{(231+157+137+0+\ldots)}{(404.15)(2.05)} + \frac{(0+41+0+\ldots)}{(3074.35)(2.06)} + \ldots}{9}$$

Figure 1. Diagram describing how priority scores are calculated. The protein–disease dataset for AD is used in the example. The proteins are identified by their corresponding accession codes in UniProt. A unit matrix is then created where the rows are the UniProt entries, and the columns are specific functions. The 1D array rho, ρ, is a binary array that represents the presence or absence of each function for a given UniProt entry. The 1D array mu, μ, holds the total number of UniProt entries for each function among the proteins of the protein–disease dataset. The dot product of rho and mu produces a raw priority score. The factor eta, η, is one of the variables that is used to normalize the raw priority score. Eta is calculated by summing in quadrature the total number of UniProt entries and the total number of functions associated with the disease for a given type of function. A second normalization factor beta, β, is the average number of unique functions per protein of a given type of function. The priority score of a protein is the mean of the normalized priority scores that were calculated separately using each of the different types of functions.

The raw priority score of each protein was normalized using the total number of proteins in the protein–disease dataset and the total number of distinct specific functions of the type of function under consideration that were represented in the protein–disease dataset. The normalization factor eta, η, was obtained with the following formula:

$$\eta^{FuncType} = \sqrt{p^2 + a^2}$$

where the variables p and a, respectively, are the number of proteins retrieved in the protein–disease dataset and the number of distinct specific functions represented in the protein–disease dataset with reference to the type of function.

We inferred that η was proportional to the combined error associated with the measurements of the number of proteins in the protein–disease dataset and the numbers of unique specific functions represented for proteins in the protein–disease dataset. Based on this inference, the normalization then follows a standard procedure of dividing by the total error [31], where the total error was obtained by adding the contributing errors in quadrature. Since the numbers of proteins and functions can vary greatly across diseases, the normalization factor η helped to ensure that the values of the priority scores were on a comparable scale for the different diseases.

The rate at which functional features were assigned to proteins varied greatly according to the type of function. We therefore introduced the second normalization factor beta, β, which considered the average number of functions per protein for each type of function separately. Upon application of both normalization factors, the resulting priority score of a protein was calculated with the following formula:

$$Pscore_{norm}^{FuncType} \frac{Pscore_{raw}^{FuncType}}{\eta^{FuncType}.\beta^{FuncType}}$$

The final normalized priority score of a protein for a given disease was then the average of the normalized priority scores calculated when using each type of function separately. In the formula, n is in the number of different types of functions considered, which was 9. As described above, the types of functions were UniProt keywords, the three types of Gene Ontology terms (molecular function, cellular component, and biological process), Enzyme Commission (EC) numbers, InterPro assignments, SUPERFAMILY assignments, small molecule interactions from ChEBI, and UniProt residue features.

$$Pscore_{norm}^{Average} \frac{\sum Pscore_{raw}^{FuncType}}{n}$$

2.3. Select Diseases Used for the Validation Studies (Overall Step 3)

To select the diseases from Disease Ontology that were to be used in the validation studies, the following procedure was implemented. The correspondences between medications and indications were retrieved from three sources: Medication-Indication (MEDI-C) resource [32], DrugCentral [12], and ChEMBL [7]. All medications were classified as being on the market with a status of being in phase 4, and each medication did not have a flag indicating it had been withdrawn from the market as per the annotations in DrugCentral. For each indication, the diagnostic code(s) from the International Classification of Diseases, Clinically Modified, was obtained from the 9th and 10th editions, which are referred to, respectively, as the ICD-9-CM and ICD-10-CM codes. The DO term description that corresponded to each diagnostic code was subsequently obtained using the correspondences that are available in the DO OBO file.

Each of the diseases analyzed were required to have at least one medication currently on the market for the disease's treatment, which interacted with a protein in the corresponding protein–disease dataset. To ascertain whether a retrieved protein interacted with a known medication for a selected disease, the list of medications for that disease was compared to the list of medications associated with the protein obtained from ChEMBL [7]. Diseases that were nonspecific such as "cancer" or "skin disease" were excluded; these are listed as the vague diseases in the Supplementary Material.

2.4. Evaluate the Contributions of the Types of Functions to the Priority Score Accuracy (Overall Step 4)

To evaluate the utility of the priority score for discerning medication–protein target pairs that are clinically useful for the treatment of a selected disease, receiver operator characteristic (ROC) curves were generated. The positives were defined as those proteins in the protein–disease dataset that interacted with one or more medications currently known to treat the disease. The rest of the proteins in the protein–disease dataset were the negatives.

Fixed values of the priority scores that corresponded to each protein were used as thresholds. The sensitivity and specificity values at each threshold for each of the different protein–disease datasets were obtained. The area under the curve (AUC) for each ROC for each protein–disease dataset was calculated, and an average AUC across the protein–disease datasets was obtained. The AUC calculations were performed using a Python script developed for this purpose. For the analyses, each protein–disease dataset was required to have 30 or more proteins to ensure there was ample data to estimate the sensitivity and specificity measurements at each threshold. Each protein–disease dataset used for validation was also required to have at least one protein that was known to be the target of a medication used to treat the corresponding disease.

To identify medications that may be repositioned to treat a selected disease, we first identified those medications already known to treat the disease. A match was sought between the ChEMBL medication identifier and the medication identifier in MEDI-C or DrugCentral. Matching was carried out by one of the following ways: match the identification codes between RxNorm Ingredient ID [6] of MEDI-C to ChEMBL ID with the normalized names for clinical drugs (RxNorm) database, table 'RXNCONS'O; or match the generic medication name with the text match. Since the MEDI-C resource contained medication–indication pairs for which the medication was a combination of two or more drugs, each drug of each combination was connected separately.

2.5. Relationship between the Priority Score and Pertinency Score (Overall Step 5 and 6)

The relationship between the priority score and percentage of proteins that interact with medications currently known to treat a selected disease, referred to as the pertinency score, was assessed. To generate a plot of priority scores versus pertinency scores, six equal intervals were considered along the full range of priority scores. Six cross-fold validation sets, each containing 16.6% of the diseases of the full validation set, were generated for each of the intervals. For each fold, the percentage of proteins targeted by medications known to treat the selected disease was calculated. A scatter plot of the average of these percentages, called the pertinency scores, versus the average priority scores was generated. A fit of the scatter plot was constructed using a weighted, second-degree polynomial using the lm package R [33]. The average inverse of the variances of the pertinency scores was used as a weighting factor in the scatter plot [34]. These variances were based on the six values that were obtained using the six folds of the validation set.

2.6. Identification of a Threshold for the Pertinency Score (Overall Step 7)

To provide the end-user with a threshold for the pertinency score, tests were conducted using the drug to disease correspondences from MEDI-1, which was created in 2013, and the drug to disease correspondences from MEDI-2. The goal was to empirically identify the pertinency score range(s) of the drug to protein to disease tuples that were identified using MEDI-2 data but were absent when using MEDI-1 data. The study protocol was run using only the MEDI-1 data and only the MEDI-2 data separately.

We then examined the number of new entries for different pertinency score ranges to observe empirically where the new entries fell. To normalize the numbers for each range, the number of new drug/protein/disease tuples for each pertinency score range were divided by the total number of protein targets that fell within each corresponding pertinency score range. We then plotted a bar chart of these ratios to empirically observe the cut-off point, where there was a large increase in the estimated likelihood of success.

2.7. The Retrieval of Results on the Pharmacorank Site (Overall Step 8)

After completing the validation and prediction studies, we expanded the diseases considered and further applied the resultant prediction mode based on the second-degree polynomial fit of the priority score versus pertinency score. For the presentation of the predicted pertinency scores on the Pharmacorank website, analyses of all diseases described in Disease Ontology that mapped to either an ICD-9-CM or ICD-10-CM code were performed. The pertinency score was calculated for each protein in each protein–disease dataset regarding all corresponding proteins within Swiss-Prot, which does not include TrEMBL.

3. Results

3.1. Validation Studies of the Priority Score

We first tested the accuracy of the priority score regarding its ability to discern, among all the proteins of a protein–disease dataset, those that were targets of medications already known to treat the disease. ROC curves were obtained for each disease, where the positives are the known medication targets and the negatives are the rest of the proteins within the protein–disease dataset. All protein–disease datasets with thirty or more proteins that had one or more proteins that were identified as interacting with a currently used medication for the disease were included. There were 513 diseases that met these criteria. The list of these 513 diseases in the validation set is provided in the Supplementary Material. The results presented here are based on the 3 August 2022 timestamp of the UniProt/Swiss-Prot data.

The effect on the AUC values of using each type of function separately for the calculation of the priority scores is shown in Table 1. We observe that when using the SUPERFAMILY or ChEBI assignments only, their resultant AUC values were not significantly higher than an AUC of 0.5, which is the value that corresponds to no discrimination. These two types of functions were therefore removed from the formulation of the priority score. The average AUC value obtained when all types of functions except SUPERFAMILY and ChEBI were retained was 0.68661.

Table 1. Area under the curve (AUC) values for the retrieval of known targets of the queried diseases. The positives are proteins known to be targeted by medications currently used to treat the disease. The negative comparison set is represented by proteins of the protein–disease dataset that are not known to be targeted by a medication used to treat the disease. The average AUC value per analysis was determined by obtaining the mean AUC when using the 513 diseases of the validation set.

Type of Function(S) Used in The Priority Score Formulation	Average AUC and Standard Deviation	*p*-Value of *t*-Test Relative to 0.5
All	0.68936 ± 0.25888	1.08310×10^{-49}
All but ChEBI and SUPERFAMILY	0.68661 ± 0.26522	9.66598×10^{-47}
UniProt keywords only	0.65680 ± 0.25919	1.22968×10^{-36}
GO molecular function only	0.65522 ± 0.24997	3.12006×10^{-38}
GO biological process only	0.65329 ± 0.26250	1.42324×10^{-34}
UniProt residue features	0.62528 ± 0.267669	7.01153×10^{-24}
GO cellular component	0.61153 ± 0.27507	1.02508×10^{-18}
InterPro	0.58814 ± 0.00050	3.54596×10^{-14}
Enzyme commission (EC) number	0.53205 ± 0.25601	0.00570
SUPERFAMILY identifier	0.47983 ± 0.25644	0.07545
ChEBI	0.42085 ± 0.25987	1.55528×10^{-11}

3.2. Predictive Relationship between the Priority Score and Pertinency Score

We modeled the relationship between the priority score of a protein in a protein–disease dataset and its pertinency score. As described in the methods, six equal intervals along the priority score range were identified, and the average priority score for each of the intervals was calculated. Then, for each priority score interval, the fraction of proteins in the interval that were targets of medications currently used to treat the disease, referred to as the pertinency score, was calculated. To obtain the average and standard error estimates for the priority scores and pertinency scores, a six-fold cross-validation was implemented as described in the methods.

A weighted least-squares fit with a second-degree polynomial was found to have the following equation: $y = 1.399x^2 - 0.110x + 0.015$. The correlation coefficient for the fit was 0.9978, which indicates a highly predictive relationship, as shown in Figure 2. The polynomial equation for the fit was subsequently applied to estimate the pertinency score of a protein for a selected disease given its priority score for that disease.

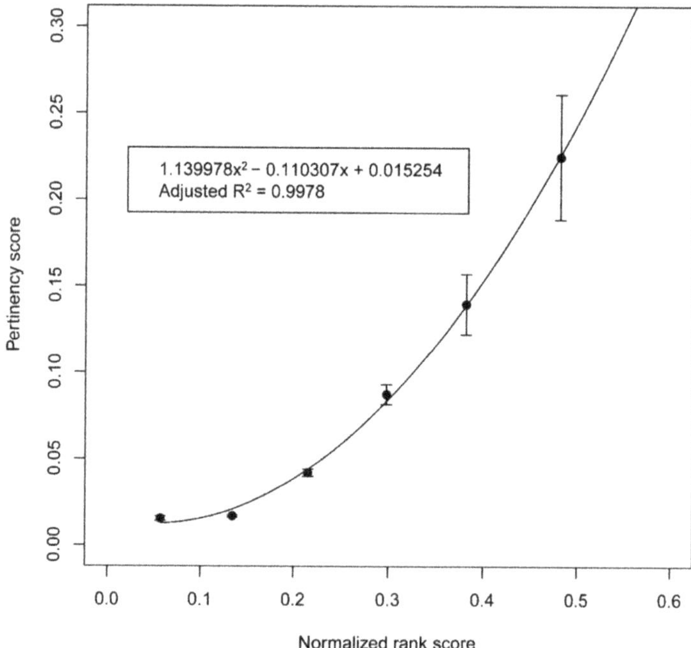

Figure 2. Plot of the pertinency scores versus priority scores. The priority scores of proteins in the protein–disease datasets were collected into six equally spaced intervals along the priority score range. The points on the ordinate of the plot are the means of the fractions of proteins that interact with medications currently known to treat the corresponding disease, termed the pertinency score, for the six folds of the validation set for each of the priority score intervals. The error bars are the standard errors of the means across the six folds. The curved black line indicates the fit using a weighted least squares regression with a second-order polynomial. The resulting equation is $y = 1.399x^2 - 0.110x + 0.015$, and the correlation coefficient is 0.9978.

The pertinency score is interpreted as an estimate of the probability that the medication would be relevant in the treatment of the corresponding disease. To be clinically useful in the disease's treatment, the medication would need to oppose the aberrant function of the protein, which contributes to the disease. This determination is left to the user. The tool estimates the strength of the association, but the user must then determine if the drug can theoretically be indicated or contraindicated. For example, if a protein's function increases in the disease state and the medication inhibits the corresponding protein function, the medication would likely be clinically useful in the disease's treatment. If the medication further increases the aberrant protein function or the medication exacerbates a loss of function that is associated with the disease, the medication would be relevant as a possible contraindication. The direction of the effect of the medication and the direction of the aberrant protein function in the disease mechanism therefore need to be ascertained to know whether the drug's effect would be useful in the control the disease [35].

3.3. Estimation of an Empirical Threshold for the Pertinency Score

As described in the methods, the analyses were conducted separately using only known drug to disease correspondences from MEDI-1 and then using only the drug to disease correspondence from MEDI-2. The pertinency scores of the drug/protein/disease tuples found from the MEDI-2 analyses but not from the MEDI-1 analyses were used to estimate the pertinency score range(s) where the tuples with a high likelihood of successful development would lie in the future.

Figure 3 shows a bar chart with the ratios of the pertinency scores of new drug/protein/disease tuples to the number of protein targets for each of the corresponding pertinency score range bins. For the range of 0–0.1, there were 4790 new tuples and 100,369 protein targets, which gave a ratio of 0.048. For the range of 0.1–0.2, there were 1231 new tuples and 7019 protein targets, giving a ratio of 0.175. For 0.2–0.3, there were 96 tuples and 625 protein targets, giving a ratio of 0.154. For 0.3–0.4, our analysis detected no new tuples. Within 0.4–0.5, there were 2 tuples and 75 protein targets, giving a ratio of 0.027. Based on the large jump when moving to the 0.1–0.2 bin, and a review of the rest of the chart, we infer that a threshold of 0.1 or above for the pertinency score captures new drugs effectively while eliminating most of the protein targets from consideration. We make the inference that new drug repositioning candidates would also likely have pertinency scores above the 0.1 threshold and have the highest likelihoods of ultimately becoming useful drugs.

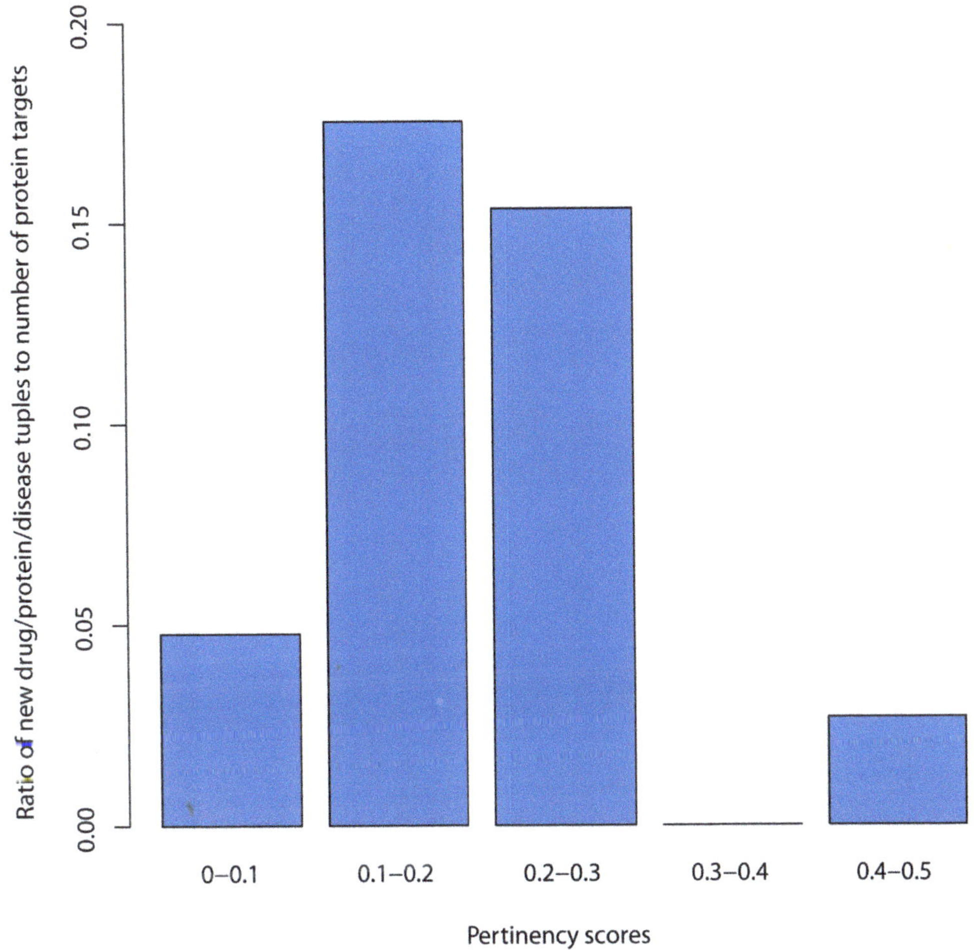

Figure 3. Bar chart of the ratio of the number of drug/protein/disease tuples to the number of proteins targets for each pertinency score bin.

3.4. Three Illustrative Examples of Repositioning Candidates

We reviewed the examples of the top repositioning candidates with the highest priority scores and therefore the highest predicted pertinency scores across all the diseases analyzed. The total number of unique diseases was 4041. The idea is that by considering those candidates with the highest predicted pertinency scores across all the diseases, a focus would be placed on those most likely to be clinically relevant. We selected three medications among the top predicted repositioning candidates to gain insight as to whether these candidates would likely be clinically useful. Additionally, these examples help to illustrate what one would expect to encounter when carrying out such reviews.

The main goal of each individual candidate review was to identify cases where the drug has an opposing effect to that of the aberrant function of the protein. This boils down to finding cases where there is a gain of function of the disease protein and the drug inhibits the corresponding function. Alternatively, it corresponds to cases where the protein has a loss of function variation, and the drug increases the corresponding function of the protein. Cases where the drug enhances a gain of function of a disease or where the drug inhibits the function of a protein that had a corresponding loss of function would likely be contraindicated.

Three proteins and repositioning candidates selected from across all the diseases are listed in Table 2. A top repositioning candidate is sotorasib for the possible treatment of linear sebaceous syndrome. Sotorasib inhibits GTPase KRas. Consider that while linear sebaceous syndrome is usually associated with a benign skin lesion, more severe phenotypes such as malignant tumors may also manifest [36]. Linear sebaceous syndrome can also be associated with cerebral, ocular, or skeletal defects, which together are referred to as Schimmelpenning syndrome. The mutant GTPase KRas of the disease state has a higher proportion of HRAS-GTP activity than that found in wild-type cells [36]. The inhibitor sotorasib may therefore offer a treatment to reduce the relatively high activity seen of GTPase KRas in the disease state [37].

Table 2. Illustrative candidates for drug repositioning across all diseases. For each medication, the protein target, the current use, and the putative indication are listed. The estimated pertinency score for the possible indication is also shown.

Medication	Target	UniProt ID	Current Use(s)	Possible Indication(s)	Pertinency Score
Sotorasib	GTPase KRas	P01116	Non-small cell lung cancer with KRAS G12C mutation	Linear nevus sebaceous syndrome	0.5861
Pyrimethamine	Beta-hexosamididase subunit beta	P07686	Toxoplasmosis	GM2 gangliosides	0.4671
Tolcapone	Genome polyprotein	P29990	Parkinson's disease	Dengue hemorraghagic fever	0.1942

For the review of the second candidate, consider GM2 gangliosidosis, which is progressive lysosomal storage disease marked by the accumulation of GM2 gangliosides in neuronal cells. This condition is caused by loss of function variations in beta-hexosaminidase subunit beta protein, and the phenotype of GM2 gangliosidosis is indistinguishable from that of Tay–Sachs disease. A possible repositioning candidate is pyrimethamine, which is a pharmacological chaperone (PC) that can stabilize the conformation of the mutant protein [38]. This allows the protein to pass quality control, avoid degradation, and continue to function. Although pyrimethamine can cross the blood–brain barrier and increase the beta-hexosaminidase activity, clinical trials have described limited impact on the manifestations of the disease in the central nervous system [39,40].

A third repositioning candidate is tolcapone for the possible treatment of dengue hemorrhagic fever. Tolcapone inhibits the serine protease function of NS3. The dengue virus type 2 NS3 protein is one of the cleavage products of a large 3391-amino-acid glycoprotein from the dengue virus. The combination of several proteins into one large genome glycoprotein that is subsequently cleaved into functional smaller proteins has apparently enabled all the proteins of this genome glycoprotein to have high priority scores since all the functional annotations of all the proteins (cleavage products) that make up the glycoprotein would have been used in the prioritization algorithm. Further, given that there are multiple strains of the dengue virus, which means repeats of similar glycoproteins of the dengue virus are described in UniProt, these glycoproteins have high calculated priority scores. Nonetheless, the NS3 protein is described as being a target of tolcapone, and tolcapone is reported as a hit from a high-throughput screening with a Ki value range of 0.61–1.25 μM [41,42]. These results point to possible treatment of the viral infection through the possible derivation of tolcapone as a hit compound and the subsequent steps required for drug development.

3.5. Examples of Repositioning Candidates for Alzheimer's Disease

A list of three protein targets and their associated medications to consider for repositioning studies for AD is provided in Table 3. One repurposing candidate is insulin, which binds to the insulin receptor. Insulin is used to treat type 1 diabetes mellitus and type 2 diabetes mellitus [43]. The insulin receptor has a high predicted pertinency score, and insulin has demonstrated disease-modifying activity that opposes the disease mechanism. A recent study by Keller et al. found that intranasal insulin has demonstrated clinical benefit based on a phase 2 clinical trial [44]. Relative to the control group, the insulin-treated group showed beneficial changes in CSF immune/inflammatory/vascular markers. Beneficial changes in cognition, brain volume, and both amyloid and tau concentrations were also observed. The authors conclude that intranasal insulin may promote a compensatory immune response that is associated with the therapeutic benefit.

Table 3. Protein targets or biochemical pathways together with drug repositioning candidates for the possible treatment of Alzheimer's disease.

Medication	Target/Pathway	Uniport ID	Current Use(s)	Pertinency Score
Insulin	Isoform short of insulin receptor	P06213-2	Types 1 and 2 diabetes mellitus	0.1666
Riluzole	Alpha-synuclein	P37840	Amyotrophic lateral sclerosis	0.1637
Diminazene aceturate (DIZE) to increase ACE2 activity	Neuroinflammation pathway of angiotensin-converting enzyme (ACE1)	P12821	Trypanosomiasis	0.1442

A second repurposing candidate is riluzole, which has the ability to reduce the alpha-synuclein protein aggregation seeds [45]. The current indication for riluzole is amyotrophic lateral sclerosis [43]. There is evidence that the pathology of AD is linked to alpha-synuclein via multiple mechanisms that include asymptomatic accumulation of Aβ plaques and tau hyperphosphorylation [46]. A clinical trial demonstrated a strong correlation between riluzole treatment, cognitive measures, and brain metabolism in those with AD. The changes in brain metabolism included a slower rate of cerebral glucose metabolism decline [47,48]. Further, in a mouse model, riluzole impacted some immune-related pathways that are implicated in AD [49].

We found that angiotensin-converting enzyme 1 (ACE) has a relatively high pertinency score. Some evidence suggests that ACE2 inhibitors are associated with a slower rate of cognitive decline [50], but this evidence appears to be inconclusive. Alternatively, evidence also points to a protective effect of angiotensin-converting enzyme 1 against AD. Specifically, the function of ACE within the cerebrum is needed for a protective effect in AD, and the associated function of ACE is possibly independent of its contributions to the control of blood pressure [51]. In addition to the inhibitors retrieved here via the ChEMBL mappings to ACE, we observe that there are known activators of angiotensin-converting enzyme 2 (ACE2) [52,53]. Additionally, we observe that the neuroinflammation pathways involved in neurodegeneration can have associated decreases in ACE2 activity and increases in ACE1 activity [54]. Further, in a mouse model of AD, the ACE2 activator diminazene aceturate (DIZE) reduced the levels of Aβ1-42, hyperphosphorylated tau, and pro-inflammatory cytokines in the brain [55]. DIZE is a veterinary drug used to treat blood-transmitted protozoan parasites such as trypanosoma, and it has also been used to treat human trypanosomiasis without major toxicity [56]. This example highlights the need to review each target that has a high pertinency score and see which medications, regardless of whether they are currently represented in the ChEMBL mappings, may oppose the overall disease pathway.

4. Discussion

4.1. Pharmacorank's Possible Role in Enabling Drug Repositioning

Drug repositioning can improve treatment outcomes and reduce the cost of drug development [57]. Marketed drugs have already been through clinical trials, so the number of trials that would be required during the drug repositioning process would be reduced. Such a reduction can save approximately 2 years of time and 40% of the overall cost of drug development [58].

Computational approaches for drug repositioning prioritize their identified candidates based on their estimated likelihoods of success [17,59]. The estimates of the success rates can be made using the known, clinically used medication–indication pairs. These estimates can further aid in the selection of the drug repositioning candidates that are to be moved forward through drug development. This information has demonstrated importance for modeling purposes [60] and validation purposes [61].

As described through our manual reviews, the results of searches across different diseases with Pharmacorank can be collated, and drug repositioning candidates with the highest predicted pertinency scores can be identified for review. These candidates may constitute the lowest hanging fruits, where a focus on further drug discovery and development may be placed [62]. Further, when there is identification of protein targets with a determined three-dimensional structure or accurate structural models, computational approaches that use structural information, such as for rational drug design [63], docking, and/or virtual screening, may be readily applied [64].

We anticipate that the Pharmacorank search tool will complement other open technologies that are available to aid in the identification of new possible treatments for orphan diseases. There is a need for software applications that identify drug repositioning candidates for orphan diseases [65,66]. Our findings regarding the repositioning candidates for the possible treatments of linear nevus sebaceous syndrome, dengue hemorrhagic fever, and GM2 gangliosides highlights examples for orphan diseases. The need to fully bring to light easily searchable and viable drug repositioning candidates for orphan diseases is apparent. The long-term goal would be to improve clinical outcomes for these conditions. The identification of new uses for old drugs may also add value by enabling a drug to enter into a new market for the treatment of an orphan disease, which may extend the patent life of the drug [67,68].

4.2. Relation to Other Tools for Drug Repositioning

Drug repositioning methods span a variety of different experimental and computational approaches [67,69,70], which may be grouped according to whether they predict new interactions between medications and proteins or just prioritize the proteins involved in the disease. Those that predict new physical interactions between medications and protein targets may validate their results by testing the predicted interactions using ligand-binding assays [71]. In contrast, methods that prioritize known medication–protein target pairs may validate predictions using cross-validation studies or by evaluating their effects on disease phenotypes using animal models [72]. In the absence of performing clinical trials, both types of methods may be validated using evidence from the literature that describes the possible clinical usefulness of medications for a selected disease.

The application of Pharmacorank for drug repositioning may be classified into the latter group of methods, since the priority scores of medication–protein target pairs are described, and no new physical interactions between protein targets and medications are predicted. Comparable methods include ToppGene [20], TargetMine [21], a network method by Emig et al. [22], and DrugNet [73]. ToppGene discusses drug repositioning candidates in the context of evidence from literature references that support their plausibility. DrugNet takes validation a step further by performing cross-validation of the predicted medication–indication pairs using data from clinical trials. The use of DrugNet reports an AUC value of 0.836 when the positives were medication–indication pairs found to be in clinical trials and the negatives were randomly selected drugs.

Regarding the network method reported by Emig et al., the positives were those proteins known to be targeted by a medication in a clinical trial for the treatment of the disease and the negatives were randomly selected protein targets [22]. The Emig study reports AUC values that range between 0.63 and 0.93 for different diseases. In a study by Kissa et al. [74], algorithmic approaches for unsupervised prioritization of drug repositioning candidates were also described. The validation sets include positives that were approved medication–protein target pairs for each disease in question [74]. The negatives were random pairs of drugs with targets. They report an overall AUC value of 0.84 for the discovery of medication–indication pairs using the Pointwise Mutual Information algorithm.

For the Pharmacorank search tool, the following validation approach was undertaken: the priority score was used to discern proteins targeted by medications used to treat the disease from all the other proteins associated with the disease. The Pharmacorank search tool similarly falls into the category of methods that utilize AUC values for validation, and it therefore falls into the category of methods that use both sensitivity- and specificity-based validation (SSV) as described by Brown and Patel [75]. Here, the priority score is used to discern proteins targeted by medications used to treat the disease from all the other proteins associated with the disease. This was carried out based on the coordinated mapping of the medication–indication pairs in MEDI-C and DrugCentral with the medication–protein target pairs in ChEMBL. For the described previous studies, we observe that the values for AUCs are higher than those described in the current study. One difference in the approaches is that random proteins were part of the comparison sets of the previous described studies and not all these random proteins were deemed to be directly involved in the disease.

For example, in the Emig study, all the proteins of the comparison set have a score that relates each protein to the disease but not all the proteins were deemed to be directly involved in the disease. In our studies, we note that our negatives do not include many random proteins not directly associated with the condition. Specifically, our negatives are proteins directly involved in the disease but are not known to interact with a known drug to treat the disease. Since we are contrasting the functions of proteins directly involved in the disease that interact with medications known to treat the disease versus the functions of protein directly involved in the disease that do not interact with medications known to treat the disease, we infer that the method enabled a more precise identification of those functions most relevant to make the protein a viable target whose corresponding function could be modified as part of the disease's treatment.

We interpret the AUC values obtained here as being more relevant and applicable to drug repositioning. This is because discerning the most clinically useful targets among the proteins known to be directly involved in the disease is likely what will be most useful in practice. Drug discovery and development projects focus on the proteins known to be involved in the disease and then try to figure out which of those will be the most effective drug targets.

4.3. Significance of the Relationship between the Priority Score and Pertinency Score

One of the take home messages from the analyses is that there is a predictive phenomenological relationship between the priority score and pertinency score. Knowledge of the set of proteins involved in a disease along with their normal biological functions can be used to quantitatively predict the likelihood that a given protein and its associated medications would be relevant in the treatment of the selected disease.

We note that in the relationship between the priority score and pertinency score, as the priority score reaches relatively high values, the standard deviations of the pertinency scores increase. One reason for this observation was noted earlier, which is that some proteins are part of polyproteins, such as what is typically found in viruses. These proteins will receive high priority scores since all the protein functions would be attributed to a single UniProt entry that represents the entire polyprotein prior to cleavage. Furthermore, since there are multiple viral strains represented in UniProt, protein functions would be repeated multiple times in the prioritization algorithm, thus artificially increasing the corresponding priority scores. Of note is that the predictive relationship between the average priority score versus the average pertinency score holds.

The observation that this strong relationship between the priority score and pertinency score exists is significant since the data assembled in UniProt and the data from the clinical trials of medications are of disparate origins. The information in UniProt regarding the characterization of the normal functions of proteins is independent of the testing and validation of medications brought to market via the required clinical trials.

When considering the selection of drug repositioning candidates to consider for a selected disease, we recommend starting with those with the highest predicted pertinency scores that are likely to oppose the disease pathway. In many cases, the pertinency score may not reach above 10–20%, but those with the highest predicted pertinency scores are recommended to be the ones to reviewed first. There are many factors that contribute to this low rate such as, simply, not all the drugs have been successfully developed yet so the percentage of proteins that are targeted by clinically useful medications has not achieved its maximum value.

As described in the results section, we recommend that the end-user consider a threshold of 0.1 for the pertinency score. We further note that the current success rates for drug development from the downstream points of the beginning of the clinical trial to the point of receiving marketing approval has remained about 10–20% [76]. Although the pertinency score is not on the same scale, we surmise that a drug repositioning candidate with a pertinency score above 10% should garner attention as these are predicted to have likely paths to success that is possibly better than many of the drugs that have reached the stage of entering clinical trials.

Now, the Pharmacorank site reveals that there are hundreds of drug repositioning candidates across a wide range of conditions that meet the 0.1 threshold. These candidates may be particularly ripe for the clinical trials that are required in their development for the selected conditions. As an additional repurposing example candidate that meets the 0.1 threshold, we observe that probucol has a pertinency score of 0.102 through its interaction with vascular cell adhesion protein 1 in the condition multiple sclerosis. We note that probucol has been shown to reduce neural cell apoptosis after cellular injury [77]. This example further highlights the potential of identifying viable drug repositioning candidates for future drug development studies as guided by the threshold value of the pertinency score.

4.4. Access

The web interface for the Pharmacorank search engine is available at the URL http://protein.som.geisinger.edu/Pharmacorank/. Automated monthly updates of the searchable content and the predictive model are enabled based on the routine updates of the information in UniProt. All analyses are pre-run prior to making them available on the website to make the queries fast since it is then just a matter of looking up the pre-computed results. For interactive queries, an autocomplete tool identifies the corresponding Disease Ontology term as a disease name is being typed. For each protein and drug retrieved, links are provided to the corresponding entries in UniProt and ChEMBL for further information.

Programmatic access is enabled to retrieve the results of precomputed searches via the URL http://protein.som.geisinger.edu/Pharmacorank/Downloads/.

5. Conclusions

The Pharmacorank search tool provides a means to retrieve protein medication targets and their associated medications that are either known or predicted to be relevant in the treatment of a selected disease. The results of searches are prioritized using an objective algorithm that considers each protein target's complement of functions. The functions are derived from the broad collection of function descriptions in UniProt. Different types of functions are collectively used in the formulation of the priority scores. We find a quantitative, predictive relationship between the resulting priority score of a medication–protein target pair and its probability of being clinically relevant in the treatment of the selected disease.

To facilitate drug repositioning efforts across a wide range of diseases, the disease terms and phrases described in Disease Ontology were analyzed. The medications associated with the retrieved proteins were considered as drug repositioning candidates if they were not yet used to treat the queried disease and were likely to oppose the disease mechanism. We anticipate that the drug repositioning candidates described here and those found subsequently through the updated search tool will ultimately be clinically relevant for their predicted indications, which may lead to cost savings and a reduction in disease burden. An emergent feature of the search tool is that the repositioning candidates most likely to be clinically relevant across a wide range of diseases could be readily identified since the priority scores and pertinency scores are normalized across all the diseases analyzed.

Supplementary Materials: The following supporting information can be downloaded at: https://www.mdpi.com/article/10.3390/biom12111559/s1, Pharmcorank_Supplementary_Data: Validation Set, Spreadsheet 1; Vague Diseases, Spreadsheet 2.

Author Contributions: Conceptualization, W.A.M. and P.J.D.; methodology, W.A.M., P.J.D. and S.G.; software, P.J.D., S.G. and T.K.P.; validation, W.A.M., P.J.D., S.G. and T.K.P.; data curation, W.A.M., P.J.D. and S.G.; writing—original draft preparation, W.A.M. and P.J.D.; writing—review and editing, W.A.M., P.J.D., S.G. and T.K.P.; project administration, W.A.M.; funding acquisition, W.A.M. All authors have read and agreed to the published version of the manuscript.

Funding: This research was funded in part by the National Institute of General Medical Sciences, grant number 5U01GM093324-02, and in part by the Marquardt Foundation for Alzheimer's Research.

Institutional Review Board Statement: Not applicable.

Informed Consent Statement: Not applicable.

Data Availability Statement: The results for the study with the timestamp at time of publication are available at https://protein.som.geisinger.edu/Pharmacorank/Downloads/publication_data/. Updates to the results upon further application of the described algorithms are available at the URL https://protein.som.geisinger.edu/Pharmacorank.

Acknowledgments: We thank James Basting for advice and expertise in the implementation of the Pharmacorank website.

Conflicts of Interest: The authors declare no conflict of interest.

References

1. McKusick, V.A. Mendelian Inheritance in Man and its online version, OMIM. *Am. J. Hum. Genet.* **2007**, *80*, 588–604. [CrossRef] [PubMed]
2. Kanehisa, M.; Goto, S.; Furumichi, M.; Tanabe, M.; Hirakawa, M. KEGG for representation and analysis of molecular networks involving diseases and drugs. *Nucleic Acids Res.* **2010**, *38* (Suppl. 1), D355–D360. [CrossRef] [PubMed]
3. Apweiler, R.; Bairoch, A.; Wu, C.H.; Barker, W.C.; Boeckmann, B.; Ferro, S.; Gasteiger, E.; Huang, H.; Lopez, R.; Magrane, M. UniProt: The universal protein knowledgebase. *Nucleic Acids Res.* **2004**, *32* (Suppl. 1), D115–D119. [CrossRef] [PubMed]
4. Bult, C.J.; Eppig, J.T.; Kadin, J.A.; Richardson, J.E.; Blake, J.A. The Mouse Genome Database (MGD): Mouse biology and model systems. *Nucleic Acids Res.* **2008**, *36* (Suppl. 1), D724–D728. [CrossRef] [PubMed]

5. Wishart, D.S.; Knox, C.; Guo, A.C.; Shrivastava, S.; Hassanali, M.; Stothard, P.; Chang, Z.; Woolsey, J. DrugBank: A comprehensive resource for in silico drug discovery and exploration. *Nucleic Acids Res.* **2006**, *34*, D668–D672. [CrossRef] [PubMed]
6. Nelson, S.J.; Zeng, K.; Kilbourne, J.; Powell, T.; Moore, R. Normalized names for clinical drugs: RxNorm at 6 years. *J. Am. Med. Inform. Assoc.* **2011**, *18*, 441–448. [CrossRef]
7. Gaulton, A.; Bellis, L.J.; Bento, A.P.; Chambers, J.; Davies, M.; Hersey, A.; Light, Y.; McGlinchey, S.; Michalovich, D.; Al-Lazikani, B. ChEMBL: A large-scale bioactivity database for drug discovery. *Nucleic Acids Res.* **2012**, *40*, D1100–D1107. [CrossRef]
8. Günther, S.; Kuhn, M.; Dunkel, M.; Campillos, M.; Senger, C.; Petsalaki, E.; Ahmed, J.; Urdiales, E.G.; Gewiess, A.; Jensen, L.J. SuperTarget and Matador: Resources for exploring drug-target relationships. *Nucleic Acids Res.* **2008**, *36* (Suppl. 1), D919–D922. [CrossRef]
9. Zhu, F.; Han, B.; Kumar, P.; Liu, X.; Ma, X.; Wei, X.; Huang, L.; Guo, Y.; Han, L.; Zheng, C. Update of TTD: Therapeutic target database. *Nucleic Acids Res.* **2010**, *38* (Suppl. 1), D787–D791. [CrossRef]
10. Hewett, M.; Oliver, D.E.; Rubin, D.L.; Easton, K.L.; Stuart, J.M.; Altman, R.B.; Klein, T.E. PharmGKB: The pharmacogenetics knowledge base. *Nucleic Acids Res.* **2002**, *30*, 163–165. [CrossRef]
11. Degtyarenko, K.; de Matos, P.; Ennis, M.; Hastings, J.; Zbinden, M.; McNaught, A.; Alcantara, R.; Darsow, M.; Guedj, M.; Ashburner, M. ChEBI: A database and ontology for chemical entities of biological interest. *Nucleic Acids Res.* **2008**, *36*, D344–D350. [CrossRef]
12. Avram, S.; Bologa, C.G.; Holmes, J.; Bocci, G.; Wilson, T.B.; Nguyen, D.-T.; Curpan, R.; Halip, L.; Bora, A.; Yang, J.J. DrugCentral 2021 supports drug discovery and repositioning. *Nucleic Acids Res.* **2021**, *49*, D1160–D1169. [CrossRef]
13. Barabasi, A.L.; Gulbahce, N.; Loscalzo, J. Network medicine: A network-based approach to human disease. *Nat. Rev. Genet.* **2011**, *12*, 56–68. [CrossRef]
14. Xue, H.; Li, J.; Xie, H.; Wang, Y. Review of drug repositioning approaches and resources. *Int. J. Biol. Sci.* **2018**, *14*, 1232. [CrossRef]
15. Moreau, Y.; Tranchevent, L.-C. Computational tools for prioritizing candidate genes: Boosting disease gene discovery. *Nat. Rev. Genet.* **2012**, *13*, 523–536. [CrossRef]
16. Yang, Y.; Adelstein, S.J.; Kassis, A.I. Target discovery from data mining approaches. *Drug Discov. Today* **2009**, *14*, 147–154. [CrossRef]
17. Hurle, M.; Yang, L.; Xie, Q.; Rajpal, D.; Sanseau, P.; Agarwal, P. Computational drug repositioning: From data to therapeutics. *Clin. Pharmacol. Ther.* **2013**, *93*, 335–341. [CrossRef]
18. Cheng, D.; Knox, C.; Young, N.; Stothard, P.; Damaraju, S.; Wishart, D.S. PolySearch: A web-based text mining system for extracting relationships between human diseases, genes, mutations, drugs and metabolites. *Nucleic Acids Res.* **2008**, *36*, W399–W405. [CrossRef]
19. Julfayev, E.S.; McLaughlin, R.J.; Tao, Y.P.; McLaughlin, W.A. KB-Rank: Efficient protein structure and functional annotation identification via text query. *J. Struct. Funct. Genomics* **2012**, *13*, 101–110. [CrossRef]
20. Wu, C.; Gudivada, R.C.; Aronow, B.J.; Jegga, A.G. Computational drug repositioning through heterogeneous network clustering. *BMC Syst. Biol.* **2013**, *7* (Suppl. 5), S6. [CrossRef]
21. Chen, Y.-A.; Tripathi, L.P.; Mizuguchi, K. TargetMine, an integrated data warehouse for candidate gene prioritisation and target discovery. *PLoS ONE* **2011**, *6*, e17844. [CrossRef] [PubMed]
22. Emig, D.; Ivliev, A.; Pustovalova, O.; Lancashire, L.; Bureeva, S.; Nikolsky, Y.; Bessarabova, M. Drug target prediction and repositioning using an integrated network-based approach. *PLoS ONE* **2013**, *8*, e60618. [CrossRef] [PubMed]
23. Martınez, V.; Navarro, C.; Cano, C.; Blanco, A. Network-based drug-disease relation prioritization using ProphNet. In Proceedings of the International Work-Conference on Bioinformatics and Biomedical Engineering, IWBBIO 2013, Granada, Spain, 18–20 March 2013.
24. Shameer, K.; Glicksberg, B.S.; Hodos, R.; Johnson, K.W.; Badgeley, M.A.; Readhead, B.; Tomlinson, M.S.; O'Connor, T.; Miotto, R.; Kidd, B.A. Systematic analyses of drugs and disease indications in RepurposeDB reveal pharmacological, biological and epidemiological factors influencing drug repositioning. *Brief. Bioinform.* **2017**, *19*, 656–678. [CrossRef] [PubMed]
25. Himmelstein, D.S.; Lizee, A.; Hessler, C.; Brueggeman, L.; Chen, S.L.; Hadley, D.; Green, A.; Khankhanian, P.; Baranzini, S.E. Systematic integration of biomedical knowledge prioritizes drugs for repurposing. *eLife* **2017**, *6*, e26726. [CrossRef] [PubMed]
26. Tranchevent, L.-C.; Capdevila, F.B.; Nitsch, D.; De Moor, B.; De Causmaecker, P.; Moreau, Y. A guide to web tools to prioritize candidate genes. *Brief. Bioinform.* **2011**, *12*, 22–32. [CrossRef]
27. Schriml, L.M.; Mitraka, E.; Munro, J.; Tauber, B.; Schor, M.; Nickle, L.; Felix, V.; Jeng, L.; Bearer, C.; Lichenstein, R. Human Disease Ontology 2018 update: Classification, content and workflow expansion. *Nucleic Acids Res.* **2019**, *47*, D955–D962. [CrossRef]
28. Piñero, J.; Ramírez-Anguita, J.M.; Saüch-Pitarch, J.; Ronzano, F.; Centeno, E.; Sanz, F.; Furlong, L.I. The DisGeNET knowledge platform for disease genomics: 2019 update. *Nucleic Acids Res.* **2020**, *48*, D845–D855. [CrossRef]
29. Barrell, D.; Dimmer, E.; Huntley, R.P.; Binns, D.; O'Donovan, C.; Apweiler, R. The GOA database in 2009—An integrated Gene Ontology Annotation resource. *Nucleic Acids Res.* **2009**, *37* (Suppl. 1), D396–D403. [CrossRef]
30. Huntley, R.P.; Sawford, T.; Martin, M.J.; O'Donovan, C. Understanding how and why the Gene Ontology and its annotations evolve: The GO within UniProt. *Gigascience* **2014**, *3*, 1. [CrossRef]
31. Gelman, A. Scaling regression inputs by dividing by two standard deviations. *Stat. Med.* **2008**, *27*, 2865–2873. [CrossRef]
32. Wei, W.-Q.; Cronin, R.M.; Xu, H.; Lasko, T.A.; Bastarache, L.; Denny, J.C. Development and evaluation of an ensemble resource linking medications to their indications. *J. Am. Med. Inform. Assoc.* **2013**, *20*, 954–961. [CrossRef]

33. Team, R.C. *R: A Language and Environment for Statistical Computing*; R Foundation for Statistical Computing: Vienna, Austria, 2013; ISBN 3-900051-07-0: 2014.
34. Almeida, A.M.d.; Castel-Branco, M.M.; Falcao, A.; R Core. Linear regressionR: A language and environment for calibration lines revisited: Weighting schemes for bioanalytical methods statistical computing. Vienna, Austria; 2014. *J. Chromatogr. B* **2002**, *774*, 215–222. [CrossRef]
35. Pushpakom, S.; Iorio, F.; Eyers, P.A.; Escott, K.J.; Hopper, S.; Wells, A.; Doig, A.; Guilliams, T.; Latimer, J.; McNamee, C. Drug repurposing: Progress, challenges and recommendations. *Nat. Rev. Drug Discov.* **2019**, *18*, 41–58. [CrossRef]
36. Groesser, L.; Herschberger, E.; Ruetten, A.; Ruivenkamp, C.; Lopriore, E.; Zutt, M.; Langmann, T.; Singer, S.; Klingseisen, L.; Schneider-Brachert, W. Postzygotic HRAS and KRAS mutations cause nevus sebaceous and Schimmelpenning syndrome. *Nat. Genet.* **2012**, *44*, 783–787. [CrossRef]
37. Green, T.E.; MacGregor, D.; Carden, S.M.; Harris, R.V.; Hewitt, C.A.; Berkovic, S.F.; Penington, A.J.; Scheffer, I.E.; Hildebrand, M.S. Identification of a recurrent mosaic KRAS variant in brain tissue from an individual with nevus sebaceous syndrome. *Mol. Case Stud.* **2021**, *7*, a006133. [CrossRef]
38. Maegawa, G.H.; Tropak, M.; Buttner, J.; Stockley, T.; Kok, F.; Clarke, J.T.; Mahuran, D.J. Pyrimethamine as a potential pharmacological chaperone for late-onset forms of GM2 gangliosidosis. *J. Biol. Chem.* **2007**, *282*, 9150–9161. [CrossRef]
39. Osher, E.; Fattal-Valevski, A.; Sagie, L.; Urshanski, N.; Sagiv, N.; Peleg, L.; Lerman-Sagie, T.; Zimran, A.; Elstein, D.; Navon, R. Effect of cyclic, low dose pyrimethamine treatment in patients with Late Onset Tay Sachs: An open label, extended pilot study. *Orphanet J. Rare Dis.* **2015**, *10*, 45. [CrossRef]
40. Leal, A.F.; Benincore-Flórez, E.; Solano-Galarza, D.; Garzón Jaramillo, R.G.; Echeverri-Peña, O.Y.; Suarez, D.A.; Alméciga-Díaz, C.J.; Espejo-Mojica, A.J. GM2 gangliosidoses: Clinical features, pathophysiological aspects, and current therapies. *Int. J. Mol. Sci.* **2020**, *21*, 6213. [CrossRef]
41. Balasubramanian, A.; Manzano, M.; Teramoto, T.; Pilankatta, R.; Padmanabhan, R. High-throughput screening for the identification of small-molecule inhibitors of the flaviviral protease. *Antivir. Res.* **2016**, *134*, 6–16. [CrossRef]
42. Nitsche, C. Strategies towards protease inhibitors for emerging flaviviruses. In *Dengue and Zika: Control and Antiviral Treatment Strategies*; Springer: Berlin/Heidelberg, Germany, 2018; pp. 175–186.
43. Food, U.; Administration, D.; FDA Online Label Repository. FDA Label Search-Package Code. Available online: http://labels.fda.gov/packagecode.cfm (accessed on 3 August 2022).
44. Kellar, D.; Register, T.; Lockhart, S.N.; Aisen, P.; Raman, R.; Rissman, R.A.; Brewer, J.; Craft, S. Intranasal insulin modulates cerebrospinal fluid markers of neuroinflammation in mild cognitive impairment and Alzheimer's disease: A randomized trial. *Sci. Rep.* **2022**, *12*, 1346. [CrossRef]
45. Imamura, Y.; Okuzumi, A.; Yoshinaga, S.; Hiyama, A.; Furukawa, Y.; Miyasaka, T.; Hattori, N.; Nukina, N. Quantum-dot-labeled synuclein seed assay identifies drugs modulating the experimental prion-like transmission. *Commun. Biol.* **2022**, *5*, 636. [CrossRef] [PubMed]
46. Twohig, D.; Nielsen, H.M. α-synuclein in the pathophysiology of Alzheimer's disease. *Mol. Neurodegener.* **2019**, *14*, 23. [CrossRef] [PubMed]
47. Matthews, D.C.; Mao, X.; Dowd, K.; Tsakanikas, D.; Jiang, C.S.; Meuser, C.; Andrews, R.D.; Lukic, A.S.; Lee, J.; Hampilos, N. Riluzole, a glutamate modulator, slows cerebral glucose metabolism decline in patients with Alzheimer's disease. *Brain* **2021**, *144*, 3742–3755. [CrossRef] [PubMed]
48. Czapski, G.A.; Strosznajder, J.B. Glutamate and GABA in microglia-neuron cross-talk in Alzheimer's disease. *Int. J. Mol. Sci.* **2021**, *22*, 11677. [CrossRef] [PubMed]
49. Okamoto, M.; Gray, J.D.; Larson, C.S.; Kazim, S.F.; Soya, H.; McEwen, B.S.; Pereira, A.C. Riluzole reduces amyloid beta pathology, improves memory, and restores gene expression changes in a transgenic mouse model of early-onset Alzheimer's disease. *Transl. Psychiatry* **2018**, *8*, 153. [CrossRef]
50. Rygiel, K. Can angiotensin-converting enzyme inhibitors impact cognitive decline in early stages of Alzheimer's disease? An overview of research evidence in the elderly patient population. *J. Postgrad. Med.* **2016**, *62*, 242. [CrossRef]
51. Ryan, D.K.; Karhunen, V.; Su, B.; Traylor, M.; Richardson, T.G.; Burgess, S.; Tzoulaki, I.; Gill, D. Genetic Evidence for Protective Effects of Angiotensin-Converting Enzyme Against Alzheimer Disease But Not Other Neurodegenerative Diseases in European Populations. *Neurol. Genet.* **2022**, *8*, e200014. [CrossRef]
52. Qaradakhi, T.; Gadanec, L.K.; McSweeney, K.R.; Tacey, A.; Apostolopoulos, V.; Levinger, I.; Rimarova, K.; Egom, E.E.; Rodrigo, L.; Kruzliak, P. The potential actions of angiotensin-converting enzyme II (ACE2) activator diminazene aceturate (DIZE) in various diseases. *Clin. Exp. Pharmacol. Physiol.* **2020**, *47*, 751–758. [CrossRef]
53. Rodríguez-Puertas, R. ACE2 activators for the treatment of COVID 19 patients. *J. Med. Virol.* **2020**, *92*, 1701–1702. [CrossRef]
54. Villa, C.; Rivellini, E.; Lavitrano, M.; Combi, R. Can SARS-CoV-2 infection exacerbate Alzheimer's disease? An overview of shared risk factors and pathogenetic mechanisms. *J. Pers. Med.* **2022**, *12*, 29. [CrossRef]
55. Duan, R.; Xue, X.; Zhang, Q.-Q.; Wang, S.-Y.; Gong, P.-Y.; Yan, E.; Jiang, T.; Zhang, Y.-D. ACE2 activator diminazene aceturate ameliorates Alzheimer's disease-like neuropathology and rescues cognitive impairment in SAMP8 mice. *Aging* **2020**, *12*, 14819. [CrossRef]
56. Rajapaksha, I.G.; Mak, K.Y.; Huang, P.; Burrell, L.M.; Angus, P.W.; Herath, C.B. The small molecule drug diminazene aceturate inhibits liver injury and biliary fibrosis in mice. *Sci. Rep.* **2018**, *8*, 10175. [CrossRef]

57. Li, Y.Y.; Jones, S.J. Drug repositioning for personalized medicine. *Genome Med.* **2012**, *4*, 27. [CrossRef]
58. Chong, C.R.; Sullivan, D.J. New uses for old drugs. *Nature* **2007**, *448*, 645–646. [CrossRef]
59. Sanseau, P.; Koehler, J. Editorial: Computational methods for drug repurposing. *Brief. Bioinform.* **2011**, *12*, 301–302. [CrossRef]
60. Paik, H.; Chung, A.-Y.; Park, H.-C.; Park, R.W.; Suk, K.; Kim, J.; Kim, H.; Lee, K.; Butte, A.J. Repurpose terbutaline sulfate for amyotrophic lateral sclerosis using electronic medical records. *Sci. Rep.* **2015**, *5*, 8580. [CrossRef]
61. Xu, H.; Aldrich, M.C.; Chen, Q.; Liu, H.; Peterson, N.B.; Dai, Q.; Levy, M.; Shah, A.; Han, X.; Ruan, X. Validating drug repurposing signals using electronic health records: A case study of metformin associated with reduced cancer mortality. *J. Am. Med. Inform. Assoc.* **2015**, *22*, 179–191. [CrossRef]
62. Duran-Frigola, M.; Mateo, L.; Aloy, P. Drug repositioning beyond the low-hanging fruits. *Curr. Opin. Syst. Biol.* **2017**, *3*, 95–102. [CrossRef]
63. Xie, L.; Xie, L.; Kinnings, S.L.; Bourne, P.E. Novel computational approaches to polypharmacology as a means to define responses to individual drugs. *Annu. Rev. Pharmacol. Toxicol.* **2012**, *52*, 361–379. [CrossRef]
64. Taboureau, O.; Nielsen, S.K.; Audouze, K.; Weinhold, N.; Edsgärd, D.; Roque, F.S.; Kouskoumvekaki, I.; Bora, A.; Curpan, R.; Jensen, T.S. ChemProt: A disease chemical biology database. *Nucleic Acids Res.* **2011**, *39* (Suppl. 1), D367–D372. [CrossRef]
65. Ekins, S.; Williams, A.J. Finding promiscuous old drugs for new uses. *Pharm. Res.* **2011**, *28*, 1785–1791. [CrossRef] [PubMed]
66. Sardana, D.; Zhu, C.; Zhang, M.; Gudivada, R.C.; Yang, L.; Jegga, A.G. Drug repositioning for orphan diseases. *Brief. Bioinform.* **2011**, *12*, 346–356. [CrossRef] [PubMed]
67. Ekins, S.; Williams, A.J.; Krasowski, M.D.; Freundlich, J.S. In silico repositioning of approved drugs for rare and neglected diseases. *Drug Discov. Today* **2011**, *16*, 298–310. [CrossRef] [PubMed]
68. Smith, R.B. Repositioned drugs: Integrating intellectual property and regulatory strategies. *Drug Discov. Today Ther. Strateg.* **2012**, *8*, 131–137. [CrossRef]
69. Dudley, J.T.; Deshpande, T.; Butte, A.J. Exploiting drug–disease relationships for computational drug repositioning. *Brief. Bioinform.* **2011**, *12*, 303–311. [CrossRef] [PubMed]
70. Sam, E.; Athri, P. Web-based drug repurposing tools: A survey. *Brief. Bioinform.* **2019**, *20*, 299–316. [CrossRef]
71. Xie, L.; Xie, L.; Bourne, P.E. Structure-based systems biology for analyzing off-target binding. *Curr. Opin. Struct. Biol.* **2011**, *21*, 189–199. [CrossRef]
72. Jin, G.; Wong, S.T. Toward better drug repositioning: Prioritizing and integrating existing methods into efficient pipelines. *Drug Discov. Today* **2014**, *19*, 637–644. [CrossRef]
73. Martínez, V.; Navarro, C.; Cano, C.; Fajardo, W.; Blanco, A. DrugNet: Network-based drug–disease prioritization by integrating heterogeneous data. *Artif. Intell. Med.* **2015**, *63*, 41–49. [CrossRef]
74. Kissa, M.; Tsatsaronis, G.; Schroeder, M. Prediction of drug gene associations via ontological profile similarity with application to drug repositioning. *Methods* **2015**, *74*, 71–82. [CrossRef]
75. Brown, A.S.; Patel, C.J. A review of validation strategies for computational drug repositioning. *Brief. Bioinform.* **2018**, *19*, 174–177. [CrossRef]
76. Yamaguchi, S.; Kaneko, M.; Narukawa, M. Approval success rates of drug candidates based on target, action, modality, application, and their combinations. *Clin. Transl. Sci.* **2021**, *14*, 1113–1122. [CrossRef]
77. Zhou, Z.; Chen, S.; Zhao, H.; Wang, C.; Gao, K.; Guo, Y.; Shen, Z.; Wang, Y.; Wang, H.; Mei, X. Probucol inhibits neural cell apoptosis via inhibition of mTOR signaling pathway after spinal cord injury. *Neuroscience* **2016**, *329*, 193–200. [CrossRef]

Article

KinFams: De-Novo Classification of Protein Kinases Using CATH Functional Units

Tolulope Adeyelu [1,2,†], Nicola Bordin [1,†], Vaishali P. Waman [1,†], Marta Sadlej [1], Ian Sillitoe [1], Aurelio A. Moya-Garcia [3,4,*] and Christine A. Orengo [1,*]

1 Institute of Structural and Molecular Biology, University College London, London WC1E 6BT, UK
2 Department of Comparative Biomedical Science, Louisiana State University, Baton Rouge, LA 70803, USA
3 Departamento de Biología Molecular y Bioquímica, Universidad de Málaga, 29071 Málaga, Spain
4 Laboratorio de Biología Molecular del Cáncer, Centro de Investigaciones Médico-Sanitarias (CIMES), Universidad de Málaga, 29071 Málaga, Spain
* Correspondence: amoyag@uma.es (A.A.M.-G.); c.orengo@ucl.ac.uk (C.A.O.)
† These authors contributed equally to this work.

Abstract: Protein kinases are important targets for treating human disorders, and they are the second most targeted families after G-protein coupled receptors. Several resources provide classification of kinases into evolutionary families (based on sequence homology); however, very few systematically classify functional families (FunFams) comprising evolutionary relatives that share similar functional properties. We have developed the FunFam-MARC (Multidomain ARchitecture-based Clustering) protocol, which uses multi-domain architectures of protein kinases and specificity-determining residues for functional family classification. FunFam-MARC predicts 2210 kinase functional families (KinFams), which have increased functional coherence, in terms of EC annotations, compared to the widely used KinBase classification. Our protocol provides a comprehensive classification for kinase sequences from >10,000 organisms. We associate human KinFams with diseases and drugs and identify 28 druggable human KinFams, i.e., enriched in clinically approved drugs. Since relatives in the same druggable KinFam tend to be structurally conserved, including the drug-binding site, these KinFams may be valuable for shortlisting therapeutic targets. Information on the human KinFams and associated 3D structures from AlphaFold2 are provided via our CATH FTP website and Zenodo. This gives the domain structure representative of each KinFam together with information on any drug compounds available. For 32% of the KinFams, we provide information on highly conserved residue sites that may be associated with specificity.

Keywords: protein kinases; functional families; KinFams; KinBase classification

Citation: Adeyelu, T.; Bordin, N.; Waman, V.P.; Sadlej, M.; Sillitoe, I.; Moya-Garcia, A.A.; Orengo, C.A. KinFams: De-Novo Classification of Protein Kinases Using CATH Functional Units. *Biomolecules* **2023**, *13*, 277. https://doi.org/10.3390/biom13020277

Academic Editors: Cameron Mura and Lei Xie

Received: 8 December 2022
Revised: 24 January 2023
Accepted: 26 January 2023
Published: 2 February 2023

1. Introduction

Protein kinases are enzymes involved in multiple cellular pathways. They catalyse the transfer of phosphate from a phosphate donor to the hydroxyl groups of acceptor molecules which can either be protein substrates, lipids or small molecules. Most kinases use ATP as their phosphate donor, however some use other donors, such as GTP, ADP, inorganic pyrophosphate (PPi) and others [1,2]. Through this phosphorylation process, the targets are covalently modified leading to the regulation of biological processes, such as the control of metabolism, transcription processes, cell division and movement, programmed cell death and several other signal transduction events in the cell. About 2% of the human genome encodes for protein kinases [2]. They are the second largest enzyme family and the fifth largest family of genes in humans, following zinc finger proteins, G-protein coupled receptors, immunoglobulins, and proteases [3]. Protein kinases can be broadly classified as either tyrosine kinases or serine/threonine kinases based on the specificity of the substrate they phosphorylate.

The protein kinase catalytic domain is structurally conserved and comprises around 250 to 300 amino acid residues [4]. It contains two lobes (N- and -C) connected through a flexible hinge region with the active site in a cleft between the lobes, which together acts as a functional unit (see Figure 1). The smaller N-lobe contains the highly conserved C-helix [5]. The larger C-lobe is mainly α-helical and contains the helices called E and G in its conserved core. Other important structural motifs are the phosphate-binding loop and the activation loop (A-loop), which bind ATP and the peptide substrate, respectively [6]. Kinases display remarkable diversity in their primary sequences, substrate specificity, structure and the pathways associated with them. However, they share a great degree of similarity in their 3D structure and especially in their catalytic site where the ATP-binding cavity is found [7,8]. ATP binds in the cleft between the N and C lobes and therefore most kinase inhibitors interact with this region to perturb the binding of ATP.

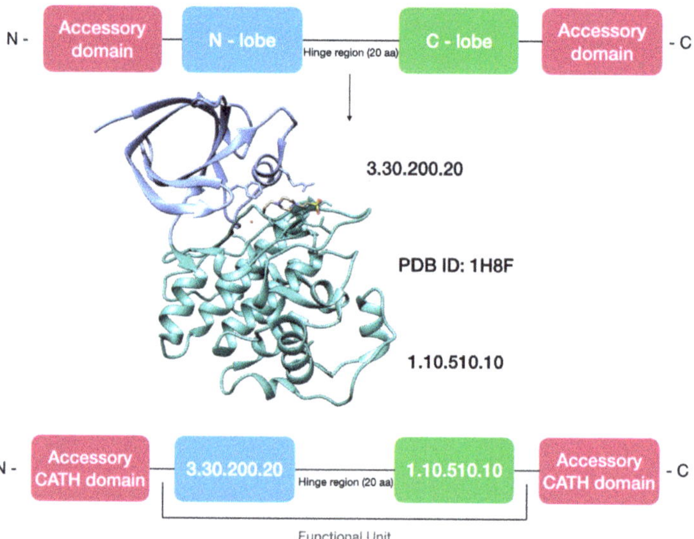

Figure 1. Schematic representing the generation of kinase functional units from the separate kinase domains in CATH. The 3D structure is shown using PDB ID:1H8F. The kinase N-lobe domain (blue box) is classified in the CATH 3.30.200.20 superfamily, while the C-lobe domain is classified in the CATH 1.10.510.10 superfamily.

Most kinase family classification systems derive from the seminal work by Hanks and Hunter [7] that uses the amino acid sequences of the catalytic domains, and which divides kinases into groups, families and subfamilies. In 1997, the Bourne group built on this work and included an additional dataset of ~1600 kinase sequences from the SwissProt and PIR resources [9]. They made their data available through the Protein Kinase Resource (PKR), which comprises nine groups, 81 families and 238 subfamilies [9]. This was one of the very first resources to make the kinase classification data available online together with structural annotations from the Protein Databank (PDB) and disease information from the OMIM database [9].

The currently most widely used standard classification system was later developed by Manning and colleagues in 2002 and made available via the KinBase resource [10]. Members within a KinBase group have a broad substrate site specificity; members within a family are grouped together based on sequence similarity and their biological function. Some of the families in KinBase are further subdivided into subfamilies based on finer sequence-level and functional similarity. To date, KinBase classifies protein kinases from 15 organisms, into 14 groups, 240 families and 339 subfamilies (according to the latest KinBase version 2014;

kinase.com, accessed on 24 January 2023). The kinomes from the following 15 organisms are classified in KinBase–*H. sapiens, M. musculus, C. elegans, D. melanogaster, S. cerevisiae, D. discoideum* and *T. thermophila, A. queenslandica, M. brevicollis, C. cinerea, G. lamblia, L. major, T. vaginalis* and *S. moellendorffii*.

Several other studies subsequently used or expanded these kinase classification schemes (See Table 1). The Barton group used a multilevel hidden Markov model (HMM) library to map sequences from SwissProt (version 2004) for *H. sapiens, M. musculus, D. melanogaster, C. elegans, S. cerevisiae,* and *D. discoideum* and from 21 other additional eukaryotic species [11,12]. This data was made available through the Kinomer database, which provides only group-level classification based on KinBase version 2008 [13].

Table 1. The summary of existing kinase classification schemes and associated resources. Hanks and Hunter (1995) developed the first kinase classification scheme, which was expanded in 1997 by the Bourne group (Protein Kinase Resource) and Manning group (KinBase). KinBase (version 2014, shown in bold). Other groups subsequently applied these classification schemes to map sequences from other additional species. For example, KinG has mapped sequences from over 200 organisms to families in Protein Kinase Resource (PKR). PrOKiNO has mapped kinase sequences from 1321 species to families from KinBase (2012). More recently, CATH-KinFams, described in this study, maps sequences from 13,981 species to 2210 functional families, using the novel FunFam-MARC protocol. [a] KinFams consist of alignments of kinase domain sequences from 13,981 species (UniProt release 2018_02). [b] Using HMMs built from these KinFams [a], we detected hits to an additional 20,494 organisms (total 34,475) from the latest UniProt release (2022_03).

Year	Name of Family Classification/ Database	Number of Groups/Families/ Subfamilies	Organisms/ Version of Uniprot or Swissprot Used	Website	Reference
1995	Hanks and Hunter	5 groups, 55 subfamilies	Model organisms	Not available	[7,8]
1997	PKR—Protein kinase resource	9 groups, 81 families, 238 subfamilies	SwissProt (2004)	http://pkr.sdsc.edu/html/index.shtml (not unavailable), accessed on 24 January 2023	[9]
2002	KinBase	KinBase 2014 version 14 Groups, 240 families, 339 Subfamilies	15 organisms	http://www.kinase.com/kinbase, accessed on 24 January 2023	[10]
2004, 2010	KinG database	PKR (as above)	>2000 organisms UniProt (2019)	http://king.mbu.iisc.ernet.in/, accessed on 24 January 2023	[14]
2007, 2009	Kinomer v.1	8 groups from KinBase (2008)	43 eukaryotic organisms	http://www.compbio.dundee.ac.uk/kinomer/index.html, accessed on 24 January 2023	[13]
2011, 2015	PrOKiNO	KinBase (2012 version) 14 groups, 273 Families and 359 Subfamilies	1321 organisms UniProt (2021)	http://vulcan.cs.uga.edu/prokino/about/prokino, accessed on 24 January 2023	[15]
2022	KinFams (CATH v4.3)	2210 KinFams	[a] 13,981 organisms (from UniProt 2018) [b] 34,475 organisms (from UniProt 2022)	https://www.cathdb.info/, accessed on 24 January 2023	[16]

In 2004, the Srinivasan group developed the KinG database [14], the first database to include sequences from bacteria (total 27 species), archaea (total eight species) and plant species (*Arabidopsis thaliana*). These were mapped to families from Bourne's Protein Kinase Resource (PKR, 9). Srinivasan and co-workers showed how information on other domains tethered to the kinase catalytic domain revealed outliers in classical kinase classifications that could be used to refine the classification [17]. They considered the composition of the kinase accessory domains and the organisation of these domains. Classification was refined manually using an alignment-free method to detect the similarity between sequences by assessing short amino acid sequence patterns and structural features outside the catalytic domain. Using this approach, they were able to detect outliers called "hybrid kinases" that had sequence regions associated with the catalytic domains matching a particular subfamily but regions outside the catalytic domain matching a different subfamily [17]. The standard classification approach using only the catalytic domain sequences would not have been adequate to capture these cases. KinG currently holds information on >2000 organisms (including eukaryotes, viruses and prokaryotes) and allows searches for kinases based on domain combinations [14].

Other integrated resources also exist [18–20]. In 2011, Ghosal et al. [15] developed the protein kinase ontology (ProKinO) framework for human kinases, which now also provides family annotations for 1321 species, by mapping sequences to KinBase (version 2012) and integrating data from COSMIC, UniProt and Reactome [15]. This framework has been used in the analysis of cancer-associated mutations [18] and recently to annotate dark kinases (i.e., experimentally uncharacterised) in humans [21,22]. A similar resource, KinHub, also provides annotations specific to human kinases (http://www.kinhub.org, accessed on 24 January 2023). KIDFamMap [19] provides a platform for accessing the kinase conformational types and functions to gain biological insights into the selectivity of human kinase inhibitors and mechanisms of action.

These resources provide a rich source of annotations of existing families in KinBase or Protein Kinase Resource (PKR) (see Table 1), however most of them are based on KinBase and are not completely up to date with sequences from all organisms in UniProt (see Table 1). Since the Manning group developed KinBase, there has been a significant expansion in protein kinase sequences deposited in UniProt (https://www.uniprot.org/help/downloads [23], accessed on 24 January 2023) and other public sequence repositories. Whilst other large resources, such as Pfam ([24], https://pfam.xfam.org, accessed on 24 January 2023) and PANTHER ([25], http://www.pantherdb.org/, acceseed on 24 January 2023) classify these proteins into evolutionary families, they do not explicitly classify them into distinct functional families, i.e., comprising evolutionary relatives sharing similar functional properties.

The CATH classification currently classifies the kinase functional unit into two separate domains corresponding to the N-lobe and the C-lobe, as they are distinct globular regions. Since both are required to provide the function, we have generated a new category of superfamily in CATH, corresponding to the kinase 'functional unit', which concatenates one or more domains contributing to the functional role of the protein. Subsequently our CATH-FunFam (functional family) resource uses automated approaches including agglomerative clustering and an entropy-based protocol, to segregate functionally distinct groups by implicit identification of specificity-determining positions (SDPs) and other functional sites [26]. CATH-FunFams have been endorsed in-silico [26,27] and by blind independent assessment in CAFA, in which CAFA-FunFams were recently highly ranked for prediction of molecular function [28].

In this study, we report the classification of kinase functional families (CATH-KinFams), from all kinase sequences available in UniProt (version 2018_02) using an improved FunFam classification method (FunFam-MARC). Our automated approach has allowed us to update information on sequences deposited in UniProt since the development of Protein Kinase Resource (PKR), KinBase, and other related kinase resources, and to identify new families (and subfamilies) and their relationship to families defined in Manning's KinBase classification.

The CATH FunFam-based protocol explicitly exploits information on the multi-domain architecture (MDAs) of protein kinases. Our automated classification protocol identifies a total of 2210 CATH-KinFams, the majority of which are observed to have high functional purity in terms of EC annotations. Since mutations in protein kinases have been recorded in several diseases especially cancer and kinases are a major therapeutic target, we also analyse our human-associated KinFams in the context of disease information and drugs, based on a CATH FunFam-based protocol developed earlier [29]. Our KinFam classification is currently the most comprehensive in terms of functional families and species and the data is available for download on Zenodo (https://zenodo.org/record/7575924, accessed on 24 January 2023), the CATH FTP site (ftp://orengoftp.biochem.ucl.ac.uk/kinfams, accessed on 24 January 2023) and will be made available via the CATH-FunVar website (https://funvar.cathdb.info/ [16], accessed on 24 January 2023),. The multi-domain-based functional family classification method designed for classification of the kinases, can be readily extended to other important classes of enzymes and drug targets.

2. Materials and Methods

2.1. Generating CATH-KinFams

CATH typically classifies the functional unit in protein kinases into two separate domains corresponding to the N- and C- lobes (or domains). These are represented as the CATH superfamilies 3.30.200.20 (N-domain) and 1.10.510.10 (C-domain), respectively (https://www.cathdb.info/, accessed on 24 January 2023). As the majority of the protein kinase inhibitors act at the hinge region between these two domains, we have created a new level within CATH to classify such 'functional units'. This will clearly be valuable for enzymes and other proteins where the functional unit straddles more than one domain. In the context of the kinases, not only will it enable us to better understand the relationships between different kinases, but it will be essential for understanding kinase–drug interactions and enabling drug repurposing. The concept of a functional unit is illustrated in Figure 1 (illustrated using PDB ID: 1H8F).

2.2. Updating Kinase Domain Sequences in the CATH Family Classification and Generating the Kinase Functional Unit

The CATH kinase superfamilies were updated to include the most recent version of UniProt (UniProt release 2018_02). This was achieved by scanning UniProt sequences against the library of HMMs built from all CATH structural representatives using HMMer3 [30]. CATH-resolve-hits [31] was then used to identify significant matches to the kinase N- and C- domain superfamilies (CATH superfamilies '1.10.510.10' and '3.30.200.20', respectively) and to other CATH domains. Kinase functional units were constructed for each protein kinase by concatenating the domain sequences from the N- and C- lobe domains. We allowed a linker (up to 20 residues long) between the two lobes to ensure that we covered the complete kinase hinge region. The multi-domain architectures (MDA) (i.e., the order of the domains along the protein sequence, including the kinase functional unit and the additional domain partners) were determined using the CATH resolve-hits (CRH) protocol [31]. CRH uses an optimisation algorithm to resolve matches to the CATH HMM libraries and obtain a set of non-overlapping domain annotations for the sequence.

2.3. Running the FunFam-MARC Algorithm

FunFam-MARC (multidomain architecture-based clustering) is a suite of protocols, as summarised in Figure 2. It first partitions the set of kinase functional unit sequences into subsets of sequences having the same MDA (i.e., the same domains in the same order in the protein sequence). Within each MDA partition, the sequences are clustered into 90% sequence identity clusters (S90) using CD-HIT [32]. These CD-HIT clusters are the starting point for the next step in FunFam-MARC which applies GeMMA [33] (see Figure 2a), a method for deriving a tree of sequence relationships in a protein superfamily. In the first step of GeMMA, S90 clusters are annotated with experimentally characterised GO terms (e.g., TAS) obtained using the UniProt API [34]. Since FunFam-MARC is computationally expensive, clusters having no experimental GO annotations are discarded. HHsuite is used to generate HMMs for each S90 cluster [35]. Subsequently, GeMMA applies agglomerative clustering by performing all against all HMM comparisons between clusters and then progressively merging clusters with the highest scores (see Figure 2a). This generates an input tree for FunFHMMER [27], a method that cuts the tree into clusters of functionally similar sequences. FunFHMMer traverses the tree from leaves to the root, cutting the tree where the branches comprise clusters with significant differences in function determining residues. These are identified by using GroupSim [36]; see Figure 2b) which detects differences in conservation patterns between equivalent residues in the combined multiple sequence alignment of the two clusters being considered. Functional determinants, i.e., Specificity-Determining Positions (SDPs) are identified as residues which are differentially conserved between FunFam clusters.

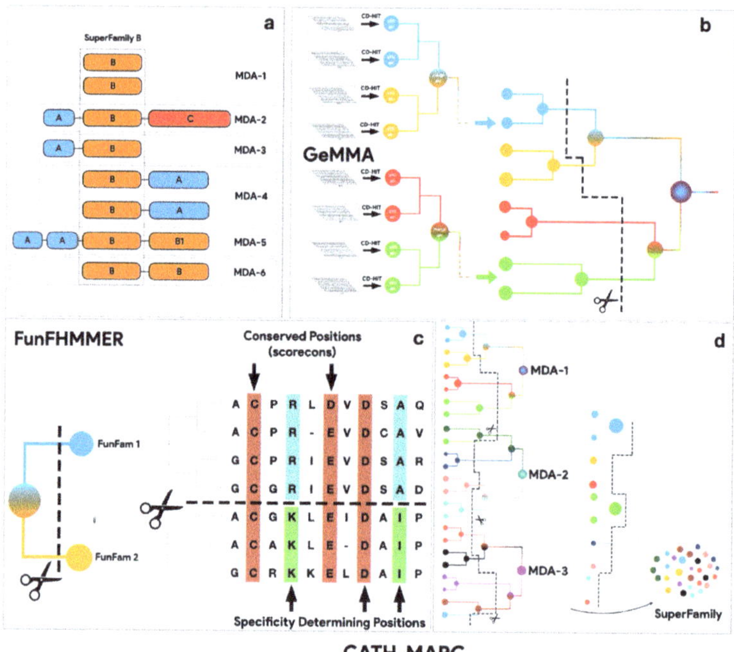

Figure 2. FunFam-MARC protocol. (**a**). FunFam-MARC approach based on multi domain architectures, (**b**). Overview of GeMMA/FunFHMMER protocol, (**c**). Example of FunFHMMER detection of specificity determining positions, (**d**). Multiple iterations of GeMMA/FunFHMMER with MDA pooling.

Once all MDAs have been processed, FunFam clusters from each MDA partition are pooled and form the starting clusters for a final run of GeMMA and FunFHMMer. Following this final iteration of tree building and segregation into clusters, the resulting clusters are the final kinase FunFams (subsequently referred to as KinFams). Finally, we scan the sequences from the experimentally uncharacterised S90 clusters against the final kinase FunFam HMMs to determine how close they are to functionally characterised FunFams to help guide the functional characterisation of these clusters. The FunFam-MARC protocol is illustrated in Figure 2.

2.4. Assessing the Functional Coherence of KinFams and KinBase Classifications Using the Enzyme Classification

We assessed the functional coherence of the Kinase FunFams (KinFams) and Kinase families by examining the agreement in experimental EC-annotations between sequence relatives in a given KinFam. That is, we determined whether relatives in each KinFam had the same or similar Enzyme Classification (EC) numbers. This is an established approach previously used to validate CATH FunFams [26,37]. The enzyme classification is a 4-digit numerical classification scheme based on the chemical reactions of enzymes [38]. The first digit describes the general type of reaction the enzyme undergoes; the second digit is the subclass, reflecting the type of bond breakage or formation taking place; the third digit represents the sub-subclass, which provides information on the chemical group involved in the enzymatic reaction; and the fourth level indicates the substrate specificity of the enzyme. The enzyme classification numbers of members in each FunFam were compared both at the 3-digit (EC3) and 4-digit (EC4) levels. The number of different EC codes among the relatives within a KinFam gives a measure of the functional purity of that kinase functional family.

For each KinFam, we calculated the information content of the multiple sequence alignment (MSA). This is captured as a diversity of position score (DOPs score) using Scorecons [39]. A DOPs score above 70 is a good indicator of a high diversity in the sequences. For FunFams with sufficient information content (DOPs > 70), Scorecons was also used to calculate the residue conservation at each position in the MSA. Previous analyses have shown that highly conserved residues in a FunFam are enriched in known functional residues (e.g., catalytic, ligand binding or protein interface residues) [26]. Thirty-two percent of the KinFams have a high DOPs score (>70). FunFams with low DOPs either contain very few sequences (<6) or are very species specific and lack diverse sequences.

As KinBase only provides sequences using their internal naming scheme, in order to extract Enzyme Commission codes (EC) we mapped KinBase entries to UniProt using BLASTP [40] to retrieve matches with 100% sequence identity.

2.5. Mapping of the CATH KinFams and KinBase Classifications

To compare the predicted CATH KinFams with the curated KinBase family classification, sequences from each KinBase family and subfamily were scanned against the KinFams-HMMs library using HMMER3 [24] with an e-value cutoff of 1×10^{-18} to give a mapping between the classifications.

2.6. Mapping Drug Information from ChEMBL to Human KinFams

We previously developed a protocol [29] which associated domain families with drugs, by calculating the over-representation of drug targets within domain families. To identify druggable KinFams associated with human protein kinases, we adopted a similar approach: an FDA-approved kinase-inhibitor drug dataset was extracted from ChEMBL release 30 [41], https://www.ebi.ac.uk/chembl/, accessed on 2 November 2022). A drug was considered as a small molecule with therapeutic application, with direct binding to a single protein (ASSAY-TYPE = "B"), with a maximum phase of development = "4", which indicates that the drug has been approved. Those with weak activity were filtered out by only considering a drug-target activity stronger than 1mM and a pChEMBL value of 6. The pChEMBL value is the measure of the half-maximal potency/affinity on a negative logarithmic scale. The anatomical therapeutic code (ATC-code) was used to select drugs that are protein kinase inhibitors. The ATC code classifies drugs into different groups at different levels (https://www.whocc.no/atc_ddd_index/, accessed on 24 January 2023). The code "L01E" corresponds to antineoplastic drugs which are protein kinase inhibitors.

2.7. Obtaining 3D Structures (PDB and AlphaFold2) for Human-KinFams

For all the sequences in human associated KinFams, we extracted the kinase domains from the PDB (https://www.rcsb.org/, accessed on 23 November 2022 [42]) or from the AlphaFold Protein Structure Database (https://alphafold.ebi.ac.uk/, accessed on

23 November 2022 [43]), as a 3D-model based on the sequence region of the functional unit in the UniProt sequence. We removed all AlphaFold2 models that did not fulfil the internal quality criteria established in Bordin et al. 2022 [44], which filters models based on below- pLDDT score > 70, more than 3 secondary structural elements, less than 65% of residues not in secondary structures, less than 30% of residues in long unordered regions, core packing and globularity.

We examined the extent of the structural diversity within kinase functional families within each human KinFam, by doing all-against-all structure comparisons of domain structures using the Sequential Structure Alignment Program (SSAP) [45].

3. Results and Discussion

3.1. Updating the CATH Kinase Superfamily

Following the update of the CATH kinase domain superfamilies with sequences from UniProt (version release 2018_02) and concatenating N- and C- lobe sequences, 330,085 kinase functional unit sequences were obtained. As reported by Martin et al. [17], many kinases are multi-domain proteins and there is considerable diversity in their architectures, i.e., in the nature and order of domains in the protein sequence. Sequence distribution by MDA, is shown in Figure 3. There are 245 MDAs (out of 6958) comprising 100 sequences or more (see Figure 3).

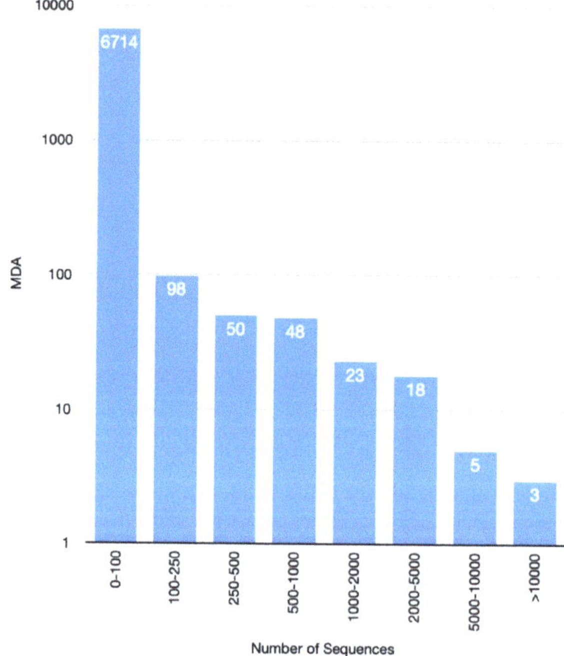

Figure 3. CATH kinase sequence distribution by multi–domain architecture (MDA).

The majority of MDAs are associated with small numbers of sequences whilst the largest 100 MDAs comprise 86.7% of the total sequences. The topmost populated domain architectures, in terms of number of sequences, are illustrated in Figure 4. It is worth noting that ~50% of the kinase sequences possess only the fused canonical N-C architecture (3.30.200.20-1.10.510.10) (Figure 4a).

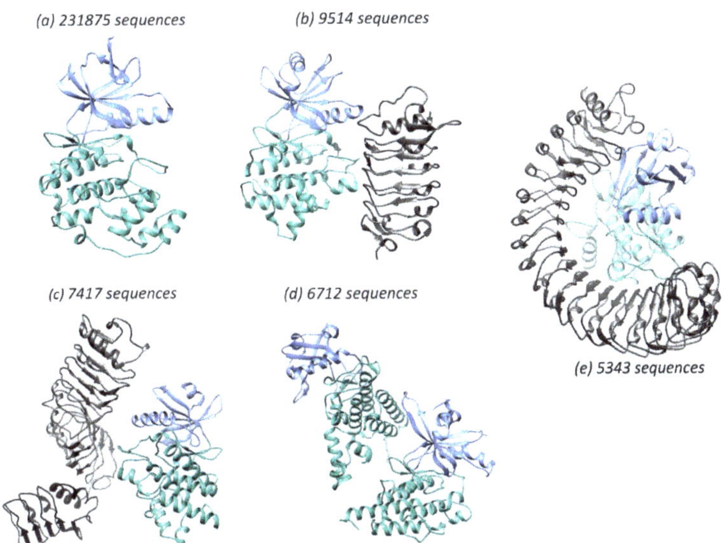

Figure 4. The most populated kinase multidomain architectures containing >5000 kinase sequences are shown. N-lobe is shown in blue, and the C-lobe is shown in cyan. The accessory domain is shown in grey. The domain architectures are illustrated using Alphafold2 structures from the following UniProt entries- (**a**) L7I0P6, (**b**) A0A445ETT0, (**c**) A0A178WEY8, (**d**) A0A444WN80 and (**e**) Q6XAT2.

Within each of the 245 different MDA groups, kinase sequences were first clustered using CD-HIT at a 90% sequence similarity, resulting in 12,392 starting clusters across all MDAs. There were 39 clusters without experimental GO annotation that were not included in the FunFam generation but subsequently scanned against FunFams to identify the closest GO-annotated FunFams. The FunFam-MARC protocol was applied (see Methods) generating 2210 Kinase FunFams referred to as KinFams. Our KinFams contained a total of 330,085 sequences, a more than 40-fold increase over the number of sequences currently provided by KinBase.

The majority of KinFams are organism-specific while a few KinFams represent sequences from more than 250 species (Figure 5).

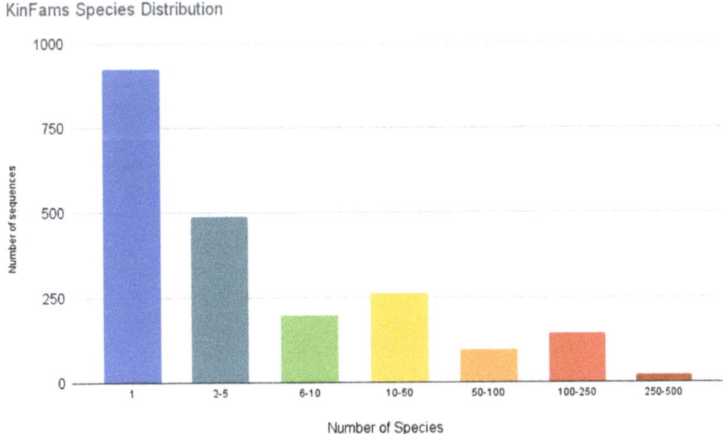

Figure 5. Species distribution associated with KinFams.

3.2. Assessing the Functional Coherence of the KinFams

We analysed a subset of 543 KinFams comprising a total of 124,000 sequences (37.6% of the total kinase sequences), as these KinFams had one or more relatives with an experimentally characterised EC classification. We determined the number of EC terms in each KinFam, considering enzyme classification (EC) numbers at both the 3-digit and 4-digit EC-levels. Kinase sequences fall into 24 EC4 classes. Figure 6a,b shows the number of KinFams that fall into the different EC3 and EC4 terms assigned to kinase sequences. Figure 6c,d shows the number of EC terms assigned to KinFams.

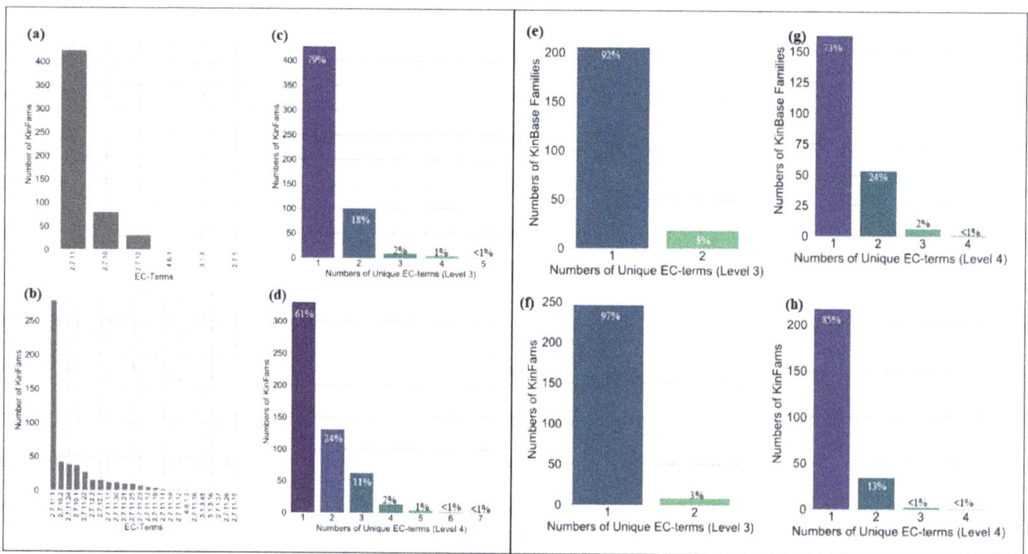

Figure 6. Distribution of EC-terms at levels 3 (**a**) and 4 (**b**) found in the KinFams for the complete set of 330,085 sequences classified in KinFams. Numbers of KinFams with one or more EC at level 3 (**c**) and EC at level 4 (**d**) in KinFams. It can be seen from Figure 6c,d that the majority of KinFams are associated with one EC3 and one EC4 term. For a subset of sequences in KinFams that map to KinBase, the right panel of the figure compares the numbers of unique EC terms at level 3 for KinBase (**e**) and KinFams (**f**) and level 4 for KinBase (**g**) and KinFams (**h**).

We compared the functional coherence of our KinFams classification with the KinBase classification using the same approach of considering the number of unique EC terms in each KinBase family and subfamily. We compared with KinBase because this resource was manually curated using experimental annotations for the sequences and is one of the most widely used and highly cited kinase classifications available. To make this comparison, we mapped KinBase sequences to KinFams to identify equivalent sequence sets (see Methods). It can be seen from Figure 6 that KinFams are more functionally coherent than KinBase subfamilies; the majority (85%) having only one EC4 term, compared to 73% for the KinBase classification. At the EC3 level, both KinFams and KinBase classifications have most families annotated with only one EC3 term 97% of relatives in KinFams and 92% of relatives in the KinBase classification have only one EC3 annotation.

The improvement in EC functional coherence in KinFams was associated with the splitting of some KinBase subfamilies by the FunFam-MARC protocol. Figure 7 shows that 163 KinBase families and subfamilies have one-to-one mapping with KinFams, while 342 are split into two or more KinFams.

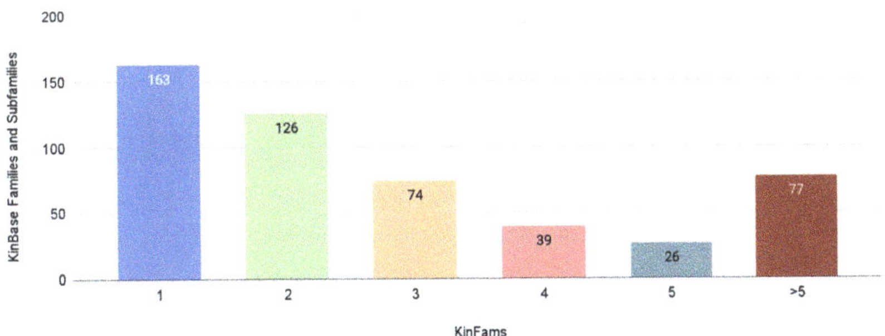

Figure 7. Mapping of KinBase to KinFams. The figure illustrates how many KinBase families and subfamilies are split into one or more KinFams by the FunFam-MARC protocol.

For certain KinBase groups and families, a further level of subclassification in subfamilies is not available. For a subset of these KinBase groups and families, KinFams is able to capture finer granularity in function by expanding the number of sub-families.

The group-wise expansion in the number of sub-families in KinBase due to KinFams is shown in Figure 8. The highest expansion (~5-fold) of family space in KinFams is observed for the KinBase 'Other' group. For the other kinase groups, the expansion varies from about 1.5-fold in case of two groups (Atypical and PKL) to about 3-fold expansion in the case of six groups (CMGC, TLK, CAMK, TK, AGC, STE and CK1). No expansion is seen in the case of the RGC group.

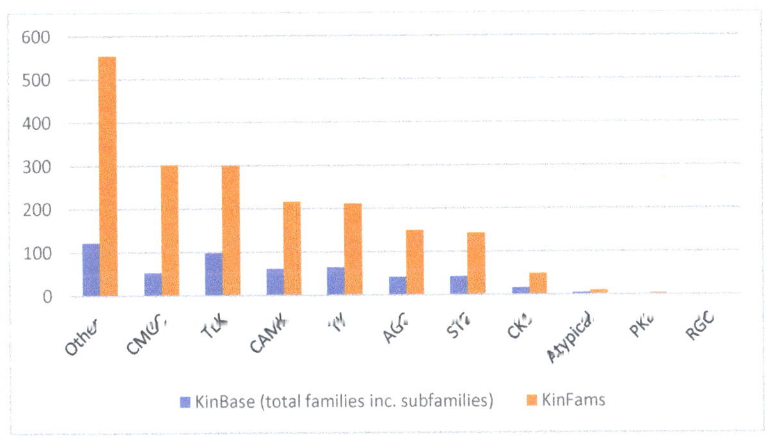

Figure 8. Group-wise expansion of subfamilies in KinFams, as compared to KinBase.

Whilst our EC analyses of the KinBase classifications suggested that a majority (73%) are likely to be functionally coherent, only a small proportion of KinBase sequences (11%) are experimentally annotated and therefore the subclassification of KinBase families by our KinFam protocol could reflect the detection of differences in Specificity-Determining Positions (SDPs). Below, we provide some examples illustrating the ability of our strategy to detect functional differences in relatives within KinBase families based on conservation of SDPs.

3.3. Example Illustrating KinFam Sub-Classification of the KinBase JAK Family

The TK:Jak KinBase family includes the genes coding for the JAK1, JAK2, JAK3 and TYK2 proteins. These proteins comprise two distinct types of kinases, one of which is known to be catalytic and the other is reported as a pseudokinase (i.e., involved in non-catalytic, regulatory functions) [46]. For example, human JAK1 (UniProt ID: P23458) proteins are all annotated as EC 2.7.10.2. They comprise two kinases: the non-catalytic kinase (residues 583-855) and a catalytic kinase (residues 875-1153). Our protocol correctly subclassifies these into two distinct KinFams, namely KinFam-101 (catalytic) and KinFam-104 (non-catalytic, i.e., pseudokinase).

Our SDP analyses (see Figure 9) clearly indicate a variation between these two KinFams in several crucial sites of the kinase: the key HRD motif in the catalytic loop (RD is substituted with GN in the non-catalytic kinase) and within the DFG motif of the activation loop (F to P in the non-catalytic domain). Moreover, the C-helix E925, in the active kinase, which is in contact with the key active site K908 of the B3-strand (in the N-lobe), is equivalent to A/T638 in the pseudokinase. The salt bridge between the glutamic acid and lysine is crucial for the formation of the activated conformation of the kinase, as well as binding ATP, and the mutation might be partially responsible for the inactivity of the pseudokinase.

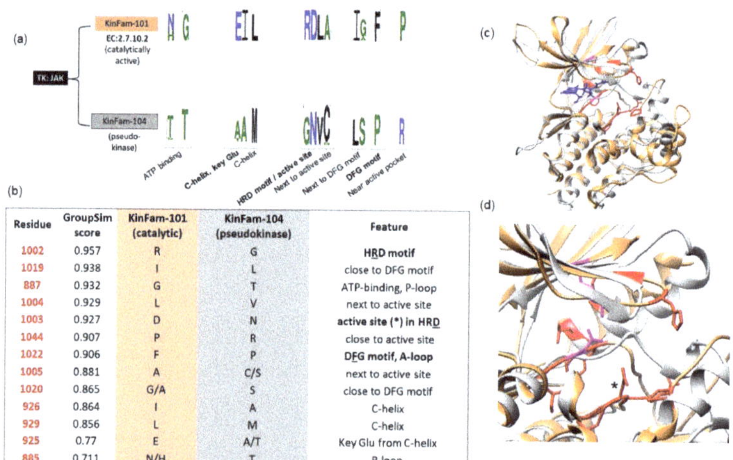

Residue	GroupSim score	KinFam-101 (catalytic)	KinFam-104 (pseudokinase)	Feature
1002	0.957	R	G	**HRD motif**
1019	0.938	I	L	close to DFG motif
887	0.932	G	T	ATP-binding, P-loop
1004	0.929	L	V	next to active site
1003	0.927	D	N	**active site (*) in HRD**
1044	0.907	P	R	close to active site
1022	0.906	F	P	**DFG motif, A-loop**
1005	0.881	A	C/S	next to active site
1020	0.865	G/A	S	close to DFG motif
926	0.864	I	A	C-helix
929	0.856	L	M	C-helix
925	0.77	E	A/T	Key Glu from C-helix
885	0.711	N/H	T	P-loop

Figure 9. Specificity-determining positions (SDPs) predicted using CATH-KinFams: an example using the TK: JAK family. (**a**) TK: JAK family from KinBase is subdivided into two KinFams using CATH, each representing distinct kinases (catalytic and non-catalytic). (**b**) List of top-ranked SDPs (red) that are specific to each KinFam: 101 and 104: SDPs occur at/near the active site (within HRD motif in catalytic loop), at the DFG motif of the activation loop (residue numbering is shown according to the active kinase of JAK1, KinFam-101, PDB:6W8L). (**c**) Superposition of structures of the representatives of the catalytically active (orange, PDB: 6W8L, domain 875-1153) and pseudokinase (grey, AF_P23458, domain:583-855) of JAK1 (UniProt id: P23458). SDPs are shown in red. Ligand molecule from PDB:6W8L (namely R4S, which binds at the ATP-binding site), is shown in blue. (**d**) Close-up view of SDPs and their location within the catalytic and activation loops. Active site residue (D1003) in HRD motif is shown as asterisk. The other two active site residues are D1022 (from DFG motif) and K908 in the N-lobe (shown in magenta).

The impacts of these mutations have been previously discussed in the literature [47], in a study which also highlights the lack of crucial autophosphorylation sites in the A-loop of the pseudokinase. Our SDP analysis, based on the KinFam classification, identifies further possible sites responsible for the inactivity of the pseudokinase, such as those involved in ATP binding (next to the DFG motif and the active site), as well as sites near the active site pocket: proline 1044, in contact with the active site, is mutated to an arginine in the pseudokinase, which may prohibit the ATP or phosphorylation substrate from entering into the active site pocket. The position of the SDPs is shown on Figure 9 below.

3.4. KinFam Subclassification of the KinBase HIPK Subfamily

HIPK (homeodomain-interacting protein kinase) is a subfamily belonging to KinBase family DYRK and the group CMGC. The HIPK subfamily comprises co-repressors that differentially interact homeodomain transcription factors [48].

The KinBase subfamily HIPK is divided by the FunFam-MARC protocol into two KinFams-10 and 319. KinFam-10 contains vertebrate HIPK1, HIPK2 and HIPK3 proteins (which share more than 90% sequence identity with each other). These are primarily present in the nucleus and expressed in all tissues [49]. By contrast, KinFam-319 consists solely of HIPK4 proteins, which occur in cytoplasm and are expressed mainly in testis and brain. The classification of HIPK4 into a distinct KinFam, is consistent with the fact that the HIPK4 protein is known to be a distant member of the KinBase HIPK family (sharing only 50% sequence identity with other HIPK1-3). In contrast to other HIPKs [1–3], HIPK4 occurs in the cytoplasm and lacks a nuclear localisation sequence and homeobox-interacting domain [49–52]. Additionally, in vitro studies confirmed that HIPK4 plays a unique role in regulating phosphorylation of manchette protein RIMBP3 during spermiogenesis [53]. A recent genome-wide microarray study suggested that HIPK4 does not primarily act through transcriptional control (unlike other HIPKs1-3), and that HIPK4 is essential for acrosome–acroplaxome function and male fertility [54]. The growing evidence from various experimental studies thus supports a distinct functional role of HIKP4 and endorses assignment to a distinct KinFam (KinFam-10), compared to the other HIPK1-3 proteins (KinFam-319).

Our SDP analysis shows that the majority of differentially conserved residues occur within the 'activation loop', that harbors the tyrosine residue required for autophosphorylation of HIPKs (Figure 10). This is particularly interesting because catalytic activity and subcellular localization of HIPKs is observed to be dependent on tyrosine autophosphorylation in the activation loop [49].

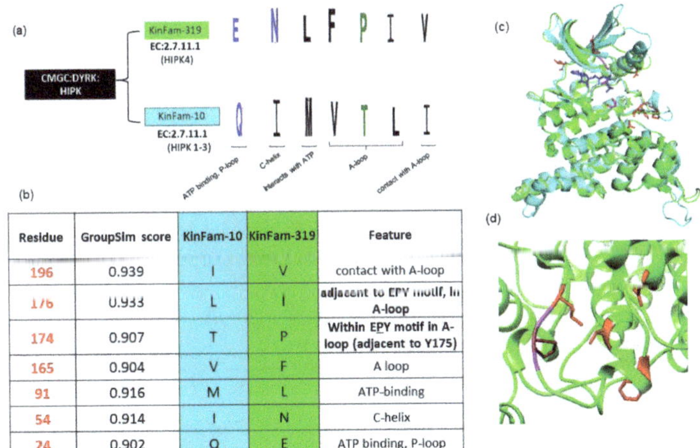

Residue	GroupSim score	KinFam-10	KinFam-319	Feature
196	0.939	I	V	contact with A-loop
176	0.933	L	I	adjacent to EPY motif, in A-loop
174	0.907	T	P	Within EPY motif in A-loop (adjacent to Y175)
165	0.904	V	F	A loop
91	0.916	M	L	ATP-binding
54	0.914	I	N	C-helix
24	0.902	Q	E	ATP binding, P-loop

Figure 10. Specificity-determining positions (SDPs) predicted using CATH-KinFams: an example using the HIPK family. (**a**) HIPK family from KinBase is subdivided into two KinFams using CATH, each represents a distinct set of HIPK proteins (**b**) list of top-ranked SDPs (in red) that are specific to each KinFams-10 (HIPKs 1-3, cyan) and KinFam-319 (HIPK4, green). The majority of SDPs occur at and near the activation loop. SDPs (red) are numbered and mapped according to the AlphaFold2 af_Q8NE63_model. (**c**) Superposition of representative structures from KinFam-10 (PDB: 6P5S, green) and KInFam-319 (af_Q8NE63_model, HIPK4, cyan), respectively. Active site residues (K40, D136, D158) are shown in magenta; Ligand molecule (namely 3NG), which binds at the ATP binding site is shown in blue. (**d**) Close-up view of SDPs (in red) and their location within the ATP-binding site and activation loops.

The activation loop of the DYRK family has a characteristic YxY element, whose second tyrosine is auto phosphorylated for kinase activation [55]. This motif is known to be altered to STY and EPY in HIPK1-3 and HIPK4, respectively. Interestingly, most of the SDPs are observed at and near this tyrosine-containing motif (See Figure 10). Additionally, E24Q substitution is observed in the P-loop lining the ATP-binding pocket, which also forms an interaction with an active site residue in the N-lobe. In summary, we identified additional SDPs within the activation loop, which are likely to be associated with distinct functional phenotypes in the HIPKs and which can suggest further investigation using experimental studies.

3.5. Merging of KinBase Groupings by KinFams

In some cases, the FunFam-MARC protocol merges distinct KinBase groupings into a single KinFam. The majority (82%) of the KinFams (73% of the sequences) map to a single KinBase family or subfamily. However, 18% of KinFams (comprising 4% of the total kinase sequences in KinFams) contain sequences from two or more KinBase subfamilies, whilst 11% of KinFams (2% of sequences) merge sequences from KinBase families and 4% of KinFams (0.3% of sequences) merge sequences from KinBase Groups (Figure 11, Table S3). This suggests that KinFams may sometimes miss subtle variations, for example between closely related species. It may also reflect the fact that the KinBase manual curation exploited other information besides sequence data, e.g., tissue specificity.

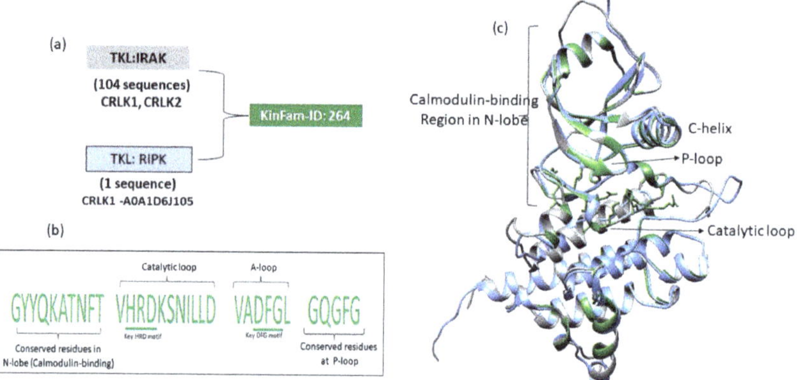

Figure 11. An example illustrating the merging of members of distinct families by the FunFam-MARC protocol. (**a**) KinFam-264 comprises 104 sequences from the KinBase TKL-IRAK family including calcium/calmodulin-regulated receptor-like kinases from plants (CRLK1 and CRLK2, e.g., UniProt ID: Q9FIU5). KinFam-264 merges a singleton sequence from TKL-RIPK, i.e., CRLK1 from *Zea Mays* [UniProt ID: A0A1D6J105]. (**b**) Closer inspection of conserved sites (shown in green) identified by Scorecons [39] indicates that many (91%) of the highly conserved residues (sites with Scorecons $\geq$ 90) in the larger group were shared by the singleton sequence. Conserved sites that are in the key functional regions are indicated in the figure-b. (**c**) The conserved sites are depicted using alphafold2 structures from TKL: RIPK (UniProt ID: A0A1D6J105, blue), and from the representative from TKL: IRAK (Q9FIU5, grey). The key regions are annotated. The majority of conserved sites are located in the N-lobe (known to harbor the calmodulin-binding site), the catalytic loop, the activation loop and the substrate binding site.

We examined some of these cases and observed that most of the time, the protocol was merging a single sequence with a much larger set of KinBase sequences and that many of the highly conserved residues in the larger group were shared by the singleton sequence (see Figure 11). Our protocol exploits information on differentially conserved positions to segregate functionally distinct relatives. However, when one of the FunFams is very small (i.e., having few sequences) it can be difficult to determine the highly conserved positions unless the sequences are from very distant species.

3.6. Increase in Kinase Family Space in KinFams Relative to KinBase

Our scans bring in protein kinases from all kingdoms and cover a total of 34,475 unique taxa (i.e., species). There is a 5-fold increase in the coverage of human kinase sequences relative to KinBase (2666 human domains in KinFams vs. 530 in KinBase). Out of 1,660,849 UniProt sequences assigned to KinFams, 47,359 (~3%), were annotated as putative or uncharacterised proteins in UniProt, so classification in KinFams is providing putative functions for these proteins based on the GO experimental annotations for the matched KinFam.

Our KinFam classification identifies many more functional subfamilies than KinBase. Whilst some of these families may relate to a finer subclassification of KinBase families based on SDPs, some are likely to be novel families (see also below). Sequences from a more recent version of UniProt (release 2022_03) were scanned against the KinFam HMMs and sequences with an e-value below 1e-18 (threshold chosen to ensure functional similarity) were denoted as matches (see Materials and Methods), resulting in 1,790,576 matches from 1,660,849 UniProt entries, since some proteins contain more than one kinase domain.

A total of 505 (out of 579) KinBase families (i.e., 208 families and 297 subfamilies) map to 969 KinFams (out of 2210 KinFams). The remaining 74 KinBase families were not mapped to any KinFams as they were small or single sequence families, and the sequences are no longer maintained by UniProt. A further 1215 out of 2210 KinFams, are putative novel families comprising sequences not classified in KinBase. However, these KinFams appear to be functionally close to a KinBase family, i.e., they match the HMM for that family with an E-value of 10^{-18}. This threshold has been suggested in previous studies to be associated with the functional similarity in catalytic mechanism and may be associated with some similarity in specificity. The remaining KinFams (26/2210) are outside the 1×10^{-18} (sequence similarity space) from a KinBase family and are therefore more likely to be completely novel families. Where experimental functional annotations are available for these potentially novel families, they are provided in Supplementary Table S1.

3.7. Identifying Druggable KinFams

Since human kinases have been implicated in several diseases including cancer, we mapped clinically approved drugs to human KinFams. Sixty-one out of 246 human KinFams have relatives that are associated with drugs and diseases using data from ChEMBL version 30 [41] for drugs and from UniProt-Disease (https://www.uniprot.org/help/involvement_in_disease, accessed on 2 November 2022) for diseases.

Kinases represent the second most targeted superfamily after the GPCRs, and they have the ability to provide novel usage of drugs to families associated with diseases which may help in repurposing available drugs. Therefore, we focused on identifying further druggable KinFams by means of a statistical analysis we used previously [29]. This connects drug with protein families, based on the statistical overrepresentation of the targets of the drug among the relatives of a protein family. We identified 28 druggable KinFams (Figure 12), that were associated with 47 drugs (BH False discovery q-value < 0.05) (See details in Supplementary Table S2).

Our analysis of the druggable KinFam shows a multi-drug association of drug compounds with the numbers of drugs associated with KinFams, ranging from 1–7. Most of the approved drugs associated with the KinFams are antineoplastic drugs (i.e., they prevent the growth of new tissues that may become cancerous). For example, the KinFam ("kinases_4.3-FF-000030") is associated with the drugs ceritinib (CHEMBL2403108; used for the treatment of non-small cell lung cancer), ponatinib (CHEMBL1171837; developed for the treatment of chronic myeloid leukemia) and nintedanib (CHEMBL502835; used for some types of non-small-cell lung cancers).

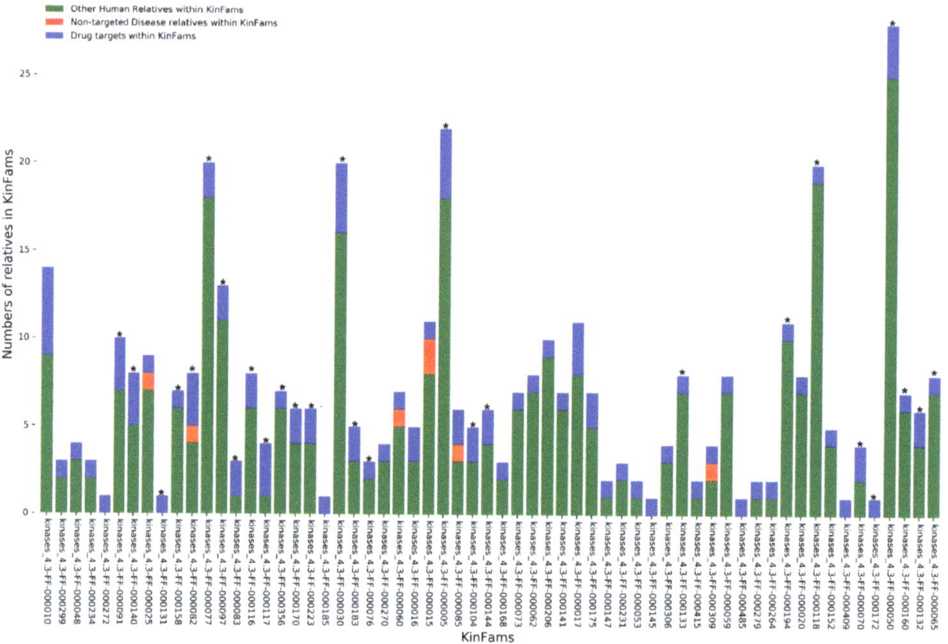

Figure 12. KinFams association with drug targets, i.e., 28 druggable KinFams based on overrepresentation of drug targets in the KinFam (shown as asterisk).

We have previously shown that relatives within the same druggable CATH FunFams are structurally conserved and have high conservation in the drug binding site [29]. Therefore, knowledge of the drug binding site in one or more of the relatives in a druggable KinFam may be useful for inheriting drug binding information to other relatives [56].

We already have experimental Protein Databank (PDB) structures for some relatives in some of the human KinFams. For the remaining sequences we extracted good quality models (see Methods Section 2.7) from the AlphaFold2 portal [43,57], where available. There are 246 human KinFams (comprising a total of 1379 sequences). Using the Alphafold2 and PDB domains associated with all sequences in these KinFams we performed and all-vs-all structural comparisons of relatives in each KinFam using SSAP [4]. We observed that for 80% of the human KinFams (75% of druggable KinFams) there is a very high average structural similarity between the relatives (RMSD < 3 Å, SSAP structure similarity > 90 (out of 100). Therefore, these KinFams may be particularly valuable for inheriting information on the drug binding pocket.

Druggable KinFams in which relatives share considerable similarity in structure, may also be valuable to consider when narrowing down therapeutic targets for a disease condition. This could be further substantiated by carrying out molecular dynamic simulations to establish potential binding energies for the various drugs associated with a druggable KinFam. Furthermore, relatives in the KinFams could be further explored to suggest possible side-effects of the drugs. However, this is beyond the scope of this current paper.

4. Conclusions and Future Directions

We developed the FunFam-MARC protocol which considers the multi-domain architecture of protein kinases and specificity determining residues to classify kinases into 2210 distinct kinase functional families (KinFams). KinFams are observed to have a higher functional purity in terms of EC annotations than families in the widely used canonical KinBase classification. This is due to the fact that we subclassified many (67%) of the

KinBase families into two or more distinct KinFams. Although some apparently pure KinBase families may be split unnecessarily, given the lack of experimental annotation in most KinBase families, it is difficult to determine the extent to which we might over-split families. For example, in mammals, there may be differences in kinase interactions and thus functional specificities of kinases in different tissues, leading to changes in SDPs and giving rise to new KinFams. By contrast we rarely merge KinBase families.

A major advantage of our KinFam classification is the functional coherence of our functional families, which ensures that relatives can be easily aligned to give robust multiple sequence alignments that can be further analysed to detect highly conserved residues likely to be associated with the specificity of the kinase.

Our classification approach is based purely on protein sequence information and does not take account of any experimental information on the oligomerization state, or known functional properties of the proteins, including substrate specificity, activity or subcellular localisation. Whilst these annotations are publicly available for some kinases, they are not comprehensive and therefore currently not sufficient for large-scale automated classifications, similar to our KinFams resource. However, previous analyses have shown that our CATH-FunFam protocol tends to implicitly capture residues differentially conserved between relatives associated with different multidomain compositions or oligomerization states, i.e., residues involved in domain-domain or protein-protein interfaces [26,27]. FunFams can also capture residues involved in promiscuous or moonlighting functions of the enzyme [58]. Furthermore, since KinFams are built from sequences in UniProt, it is possible to use the Uniprot ID to obtain a range of additional structural and functional annotations available from other resources (e.g., GO, PDB, REACTOME) to examine the similarity of these properties across a KinFam.

We provide a catalogue of protein kinase families (KinFams) comprising sequences available in UniProt version 2018_02. We also provide information on the predicted multidomain composition of each kinase sequence with information on CATH superfamily assignment for each domain so that users can determine all of the partner domains in the kinase, beside the functional unit. We also provide hidden Markov models (HMMs) generated for each of the KinFams using HMMer3 [30,59]. The comprehensive mapping of UniProt sequences to KinFams revealed that our kinase family space covers all available taxa in UniProt (release 2022) including eukaryotes, prokaryotes and viruses.

We demonstrated the application of our previously developed protocol [29] to find druggable families in the set of human protein kinases. Some of the structurally uncharacterised human (30%) KinFams have AlphaFold2 models of very good quality (i.e., pLDDT > 90). For some of these (80.8%), the high structural similarity between relatives is further evidence of a high functional similarity and suggests that drug binding characteristics can be inherited across relatives. The high-quality models will also be valuable for determining whether disease associated mutations lie on or close to functional sites and could be a modifying function or whether they are buried in the protein and the mutation could be destabilising the protein. In a small proportion of cases, we merge two or more KinBase families, but our FunFam-MARC protocol rarely merges KinBase groups.

We have provided information on the KinFams via our CATH-FunVar (Functional Variation) resource [16]. This was previously established to display cancer associated CATH FunFams enriched in driver mutations. Whilst we have provided an initial set of predicted structures for human KinFams, we also aim to bring in alphfold2 domains of high quality (pLDDT $\geq$ 90) for all UniProt sequences in KinFams. The human KinFam classification will also be made available through CATH-FunVar. Thirty-two percent of KinFams have high information content (DOPs > 70) for which we provide information on conserved residues. This data can aid the characterisation of functional sites involved in the specificity and mechanism of the kinase. Where possible, we provided information on the EC purity of the KinFam as measured based on available experimental EC annotations for the relatives. We also provided information on the KinBase families that map to the KinFam to highlight KinFams where we merged KinBase families. This will allow users to derive multiple sequence alignments for these families and verify the degree of likely functional coherence across the family, by analysing highly conserved sites shared by the majority of the relatives.

Our KinFam classification was generated to test the ability of our FunFam-MARC protocol to identify functionally distinct families in a highly populated evolutionary superfamily. Only sequence data has been used to generate the classification.

In future, we will explore the value of using the predicted structural data now available from the AlphaFold2 portal to improve our FunFam classification protocol. We hope that our KinFam data will facilitate the study of this diverse and medically relevant superfamily and help guide other kinase classification schemes and the experimental targeting of kinases that are predicted to have novel specificities.

Supplementary Materials: The following supporting information can be downloaded at: https://www.mdpi.com/article/10.3390/biom13020277/s1, Table S1: Kinfams outside E-18, Table S2: 28 druggable KinFams in humans, Table S3: KinFams to KinBase mapping.

Author Contributions: C.A.O. and A.A.M.-G. designed the idea. N.B., T.A., V.P.W., M.S. and I.S. performed experiments. T.A., N.B. and V.P.W. contributed equally to the manuscript. N.B., T.A., V.P.W., A.A.M.-G., C.A.O. and M.S. analysed the data. All authors wrote the manuscript. All authors have read and agreed to the published version of the manuscript.

Funding: Tolulope Adeyelu is supported for his research through the Nigeria Federal Scholarship Board. Vaishali P. Waman and Nicola Bordin are funded by the Wellcome Trust grant [221327/Z/20/Z]. Ian Sillitoe is funded by BBSRC [BB/R014892/1]. Aurelio A. Moya García is funded by a grant from Junta de Andalucía with the European Regional Development Fund [UMA18-FEDERJA-102].

Institutional Review Board Statement: Not applicable.

Informed Consent Statement: Not applicable.

Data Availability Statement: The data generated in this study is made available through Zenodo (https://zenodo.org/record/7575924, accessed on 24 January 2023) and the CATH FTP (ftp://orengoftp.biochem.ucl.ac.uk/kinfams, accessed on 24 January 2023). Additional information on EC codes, GO terms and links to UniProt will be made available on CATH-FunVar (https://funvar.cathdb.info/, accessed on 24 January 2023).

Conflicts of Interest: The authors declare no conflict of interest. The funders had no role in the design of the study; in the collection, analyses, or interpretation of data; in the writing of the manuscript, or in the decision to publish the results.

References

1. Fabbro, D.; Cowan-Jacob, S.W.; Moebitz, H. Ten things you should know about protein kinases: IUPHAR Review 14. *Br. J. Pharmacol.* **2015**, *172*, 2675–2700. [CrossRef] [PubMed]
2. Milanesi, L.; Petrillo, M.; Sepe, L.; Boccia, A.; D'Agostino, N.; Passamano, M.; di Nardo, S.; Tasco, G.; Casadio, R.; Paolella, G. Systematic analysis of human kinase genes: A large number of genes and alternative splicing events result in functional and structural diversity. *BMC Bioinform.* **2005**, *6* (Suppl. S4), S20. [CrossRef] [PubMed]
3. Roskoski, R. Classification of small molecule protein kinase inhibitors based upon the structures of their drug-enzyme complexes. *Pharmacol. Res.* **2016**, *103*, 26–48. [CrossRef] [PubMed]
4. Hanks, S.K.; Hunter, T. The eukaryotic protein kinase superfamily: Kinase (catalytic) domain structure and classification. *FASEB J.* **1995**, *9*, 576–596. [CrossRef] [PubMed]
5. Taylor, S.S.; Kornev, A.P. Protein kinases: Evolution of dynamic regulatory proteins. *Trends Biochem. Sci.* **2011**, *36*, 65–77. [CrossRef]
6. Kobe, B.; Kemp, B.E. Chapter 74—Principles of Kinase Regulation. In *Handbook of Cell Signaling*, 2nd ed.; Academic Press: Cambridge, MA, USA, 2010; pp. 559–563. [CrossRef]
7. Hanks, S.; Quinn, A.; Hunter, T. The protein kinase family: Conserved features and deduced phylogeny of the catalytic domains. *Science* **1988**, *241*, 42–52. [CrossRef]
8. Dreher, J.; Baumann, K. Comparison of ATP binding sites using structure-based similarity methods and molecular interaction fields. *J. Cheminform.* **2011**, *3* (Suppl. S1), P34. [CrossRef]
9. Smith, C.M.; Shindyalov, I.N.; Veretnik, S.; Gribskov, M.; Taylor, S.S.; Ten Eyck, L.F.; Bourne, P.E. The protein kinase resource. *Trends Biochem. Sci.* **1997**, *22*, 444–446. [CrossRef]

10. Manning, G.; Whyte, D.B.; Martinez, R.; Hunter, T.; Sudarsanam, S. The Protein Kinase Complement of the Human Genome. *Science* **2002**, *298*, 1912–1934. [CrossRef]
11. Miranda-Saavedra, D.; Barton, G.J. Classification and functional annotation of eukaryotic protein kinases. *Proteins* **2007**, *68*, 893–914. [CrossRef]
12. Bairoch, A.; Boeckmann, B.; Ferro, S.; Gasteiger, E. Swiss-Prot: Juggling between evolution and stability. *Brief. Bioinform.* **2004**, *5*, 39–55. [CrossRef]
13. Martin, D.M.; Miranda-Saavedra, D.; Barton, G.J. Kinomer v. 1.0: A database of systematically classified eukaryotic protein kinases. *Nucleic Acids Res.* **2009**, *37*, D244–D250. [CrossRef]
14. Krupa, A.; Abhinandan, K.R.; Srinivasan, N. KinG: A database of protein kinases in genomes. *Nucleic Acids Res.* **2004**, *32*, D153–D155. [CrossRef] [PubMed]
15. Gosal, G.; Kochut, K.J.; Kannan, N. ProKinO: An ontology for integrative analysis of protein kinases in cancer. *PLoS ONE* **2011**, *6*, e28782. [CrossRef]
16. Sillitoe, I.; Bordin, N.; Dawson, N.; Waman, V.P.; Ashford, P.; Scholes, H.M.; Pang, C.S.M.; Woodridge, L.; Rauer, C.; Sen, N.; et al. CATH: Increased structural coverage of functional space. *Nucleic Acids Res.* **2021**, *49*, D266–D273. [CrossRef] [PubMed]
17. Martin, J.; Anamika, K.; Srinivasan, N. Classification of Protein Kinases on the Basis of Both Kinase and Non-Kinase Regions. *PLoS ONE* **2010**, *5*, e12460. [CrossRef] [PubMed]
18. McSkimming, D.I.; Dastgheib, S.; Talevich, E.; Narayanan, A.; Katiyar, S.; Taylor, S.S.; Kochut, K.; Kannan, N. ProKinO: A unified resource for mining the cancer kinome. *Hum. Mutat.* **2015**, *36*, 175–186. [CrossRef] [PubMed]
19. Chiu, Y.Y.; Lin, C.T.; Huang, J.W.; Hsu, K.C.; Tseng, J.H.; You, S.R.; Yang, J.M. KIDFamMap: A database of kinase-inhibitor-disease family maps for kinase inhibitor selectivity and binding mechanisms. *Nucleic Acids Res.* **2013**, *41*, D430–D440. [CrossRef]
20. Eid, S.; Turk, S.; Volkamer, A.; Rippmann, F.; Fulle, S. KinMap: A web-based tool for interactive navigation through human kinome data. *BMC Bioinform.* **2017**, *18*, 16. [CrossRef]
21. Huang, L.C.; Taujale, R.; Gravel, N.; Venkat, A.; Yeung, W.; Byrne, D.P.; Eyers, P.A.; Kannan, N. KinOrtho: A method for mapping human kinase orthologs across the tree of life and illuminating understudied kinases. *BMC Bioinform.* **2021**, *22*, 446. [CrossRef]
22. Soleymani, S.; Gravel, N.; Huang, L.-C.; Yeung, W.; Bozorgi, E.; Bendzunas, N.G.; Kochut, K.J.; Kannan, N. Dark kinase annotation, mining and visualization using the Protein Kinase Ontology. *bioRxiv* **2022**. [CrossRef]
23. UniProt Consortium. UniProt: The universal protein knowledgebase in 2021. *Nucleic Acids Res.* **2021**, *49*, D480–D489. [CrossRef] [PubMed]
24. Mistry, J.; Chuguransky, S.; Williams, L.; Qureshi, M.; Salazar, G.A.; Sonnhammer, E.L.L.; Tosatto, S.C.; Paladin, L.; Raj, S.; Richardson, L.J.; et al. Pfam: The protein families database in 2021. *Nucleic Acids Res.* **2021**, *49*, D412–D419. [CrossRef]
25. Thomas, P.D.; Ebert, D.; Muruganujan, A.; Mushayahama, T.; Albou, L.; Mi, H. PANTHER: Making genome-scale phylogenetics accessible to all. *Protein Sci.* **2022**, *31*, 8–22. [CrossRef] [PubMed]
26. Das, S.; Lee, D.; Sillitoe, I.; Dawson, N.L.; Lees, J.G.; Orengo, C.A. Functional classification of CATH superfamilies: A domain-based approach for protein function annotation. *Bioinformatics* **2015**, *31*, 3460–3467. [CrossRef]
27. Das, S.; Sillitoe, I.; Lee, D.; Lees, J.G.; Dawson, N.L.; Ward, J.; Orengo, C.A. CATH FunFHMMer web server: Protein functional annotations using functional family assignments. *Nucleic Acids Res.* **2015**, *43*, W148–W153. [CrossRef]
28. Zhou, N.; Jiang, Y.; Bergquist, T.R.; Lee, A.J.; Kacsoh, B.Z.; Crocker, A.W.; Lewis, K.A.; Georghiou, G.; Nguyen, H.N.; Hamid, N.; et al. The CAFA challenge reports improved protein function prediction and new functional annotations for hundreds of genes through experimental screens. *Genome Biol.* **2019**, *20*, 244. [CrossRef]
29. Moya-García, A.; Adeyelu, T.; Kruger, F.A.; Dawson, N.L.; Lees, J.G.; Overington, J.P.; Orengo, C.; Ranea, J.A.G. Structural and Functional View of Polypharmacology. *Sci. Rep.* **2017**, *7*, 10102. [CrossRef]
30. Finn, R.D.; Clements, J.; Eddy, S.R. HMMER web server: Interactive sequence similarity searching. *Nucleic Acids Res.* **2011**, *39*, W29–W37. [CrossRef]
31. Lewis, T.E.; Sillitoe, I.; Lees, J.G. Cath-resolve-hits: A new tool that resolves domain matches suspiciously quickly. *Bioinformatics* **2019**, *35*, 1766–1767. [CrossRef]
32. Fu, L.; Niu, B.; Zhu, Z.; Wu, S.; Li, W. CD-HIT: Accelerated for clustering the next generation sequencing data. *Bioinformatics* **2012**, *28*, 3150–3152. [CrossRef] [PubMed]
33. Lee, D.A.; Rentzsch, R.; Orengo, C. GeMMA: Functional subfamily classification within superfamilies of predicted protein structural domains. *Nucleic Acids Res.* **2010**, *38*, 720–737. [CrossRef] [PubMed]
34. Nightingale, A.; Antunes, R.; Alpi, E.; Bursteinas, B.; Gonzales, L.; Liu, W.; Luo, J.; Qi, G.; Turner, E.; Martin, M.-J. The Proteins API: Accessing key integrated protein and genome information. *Nucleic Acids Res.* **2017**, *45*, W539–W544. [CrossRef]
35. Steinegger, M.; Meier, M.; Mirdita, M.; Vöhringer, H.; Haunsberger, S.J.; Söding, J. HH-suite3 for fast remote homology detection and deep protein annotation. *BMC Bioinform.* **2019**, *20*, 473. [CrossRef] [PubMed]
36. Capra, J.A.; Singh, M. Characterization and prediction of residues determining protein functional specificity. *Bioinformatics* **2008**, *24*, 1473–1480. [CrossRef] [PubMed]
37. Littmann, M.; Bordin, N.; Heinzinger, M.; Schütze, K.; Dallago, C.; Orengo, C.; Rost, B. Clustering FunFams using sequence embeddings improves EC purity. *Bioinformatics* **2021**, *37*, 3449–3455. [CrossRef] [PubMed]
38. McDonald, A.G.; Tipton, K.F. Enzyme nomenclature and classification: The state of the art. *FEBS J.* **2021**. [CrossRef] [PubMed]
39. Valdar, W.S. Scoring residue conservation. *Proteins* **2002**, *48*, 227–241. [CrossRef]

40. Altschul, S.F.; Gish, W.; Miller, W.; Myers, E.W.; Lipman, D.J. Basic local alignment search tool. *J. Mol. Biol.* **1990**, *215*, 403–410. [CrossRef]
41. Gaulton, A.; Hersey, A.; Nowotka, M.; Bento, A.P.; Chambers, J.; Mendez, D.; Mutowo, P.; Atkinson, F.; Bellis, L.J.; Cibrián-Uhalte, E.; et al. The ChEMBL database in 2017. *Nucleic Acids Res.* **2017**, *45*, D945. [CrossRef]
42. Berman, H.M.; Westbrook, J.; Feng, Z.; Gilliland, G.; Bhat, T.N.; Weissig, H.; Shindyalov, I.N.; Bourne, P.E. The Protein Data Bank. *Nucleic Acids Res.* **2000**, *28*, 235–242. [CrossRef]
43. Varadi, M.; Anyango, S.; Deshpande, M.; Nair, S.; Natassia, C.; Yordanova, G.; Yuan, D.; Stroe, O.; Wood, G.; Laydon, A.; et al. AlphaFold Protein Structure Database: Massively expanding the structural coverage of protein-sequence space with high-accuracy models. *Nucleic Acids Res.* **2022**, *50*, D439–D444. [CrossRef] [PubMed]
44. Bordin, N.; Sillitoe, I.; Nallapareddy, V.; Rauer, C.; Lam, S.D.; Waman, V.P.; Sen, N.; Heinzinger, M.; Littmann, M.; Kim, S.; et al. AlphaFold2 reveals commonalities and novelties in protein structure space for 21 model organisms. *bioRxiv* **2022**. [CrossRef]
45. Taylor, W.R.; Orengo, C.A. Protein structure alignment. *J. Mol. Biol.* **1989**, *208*, 1–22. [CrossRef]
46. Min, X.; Ungureanu, D.; Maxwell, S.; Hammarén, H.; Thibault, S.; Hillert, E.-K.; Ayres, M.; Greenfield, B.; Eksterowicz, J.; Gabel, C.; et al. Structural and Functional Characterization of the JH2 Pseudokinase Domain of JAK Family Tyrosine Kinase 2 (TYK2). *J. Biol. Chem.* **2015**, *290*, 27261–27270. [CrossRef] [PubMed]
47. Lupardus, P.J.; Ultsch, M.; Wallweber, H.; Kohli, P.B.; Johnson, A.R.; Eigenbrot, C. Structure of the pseudokinase-kinase domains from protein kinase TYK2 reveals a mechanism for Janus kinase (JAK) autoinhibition. *Proc. Natl. Acad. Sci. USA* **2014**, *111*, 8025–8030. [CrossRef] [PubMed]
48. Kim, Y.H.; Choi, C.Y.; Lee, S.-J.; Conti, M.A.; Kim, Y. Homeodomain-interacting protein kinases, a novel family of co-repressors for homeodomain transcription factors. *J. Biol. Chem.* **1998**, *273*, 25875–25879. [CrossRef] [PubMed]
49. Van der Laden, J.; Soppa, U.; Becker, W. Effect of tyrosine autophosphorylation on catalytic activity and subcellular localisation of homeodomain-interacting protein kinases (HIPK). *Cell Commun. Signal.* **2015**, *13*, 3. [CrossRef]
50. Arai, S.; Matsushita, A.; Du, K.; Yagi, K.; Okazaki, Y.; Kurokawa, R. Novel homeodomain-interacting protein kinase family member, HIPK4, phosphorylates human p53 at serine 9. *FEBS Lett.* **2007**, *581*, 5649–5657. [CrossRef]
51. He, Q.; Shi, J.; Sun, H.; An, J.; Huang, Y.; Sheikh, M.S. Characterization of Human Homeodomain-interacting Protein Kinase 4 (HIPK4) as a Unique Member of the HIPK Family. *Mol. Cell. Pharmacol.* **2010**, *2*, 61–68.
52. Rinaldo, C.; Siepi, F.; Prodosmo, A.; Soddu, S. HIPKs: Jack of all trades in basic nuclear activities. *Biochim. Biophys. Acta* **2008**, *1783*, 2124–2129. [CrossRef] [PubMed]
53. Liu, X.; Zang, C.; Wu, Y.; Meng, R.; Chen, Y.; Jiang, T.; Wang, C.; Yang, X.; Guo, Y.; Situ, C.; et al. Homeodomain-interacting protein kinase HIPK4 regulates phosphorylation of manchette protein RIMBP3 during spermiogenesis. *J. Biol. Chem.* **2022**, *298*, 102327. [CrossRef] [PubMed]
54. Crapster, J.A.; Rack, P.G.; Hellmann, Z.J.; Le, A.D.; Adams, C.M.; Leib, R.D.; E Elias, J.; Perrino, J.; Behr, B.; Li, Y.; et al. HIPK4 is essential for murine spermiogenesis. *Elife* **2020**, *9*, e50209. [CrossRef] [PubMed]
55. Kaltheuner, I.H.; Anand, K.; Moecking, J.; Düster, R.; Wang, J.; Gray, N.S.; Geyer, M. Abemaciclib is a potent inhibitor of DYRK1A and HIP kinases involved in transcriptional regulation. *Nat. Commun.* **2021**, *12*, 6607. [CrossRef] [PubMed]
56. Zheng, X.; Gan, L.; Wang, E.; Wang, J. Pocket-based drug design: Exploring pocket space. *AAPS J.* **2013**, *15*, 228–241. [CrossRef]
57. Jumper, J.; Evans, R.; Pritzel, A.; Green, T.; Figurnov, M.; Ronneberger, O.; Tunyasuvunakool, K.; Bates, R.; Žídek, A.; Potapenko, A.; et al. Highly accurate protein structure prediction with AlphaFold. *Nature* **2021**, *596*, 583–589. [CrossRef]
58. Das, S.; Khan, I.; Kihara, D.; Orengo, C. Exploring Structure–Function Relationships in Moonlighting Proteins. In *Moonlighting Proteins: Novel Virulence Factors in Bacterial Infections*, 1st ed.; Henderson, B., Ed.; John Wiley & Sons, Inc.: Hoboken, NJ, USA, 2017; pp. 21–43.
59. Eddy, S.R. A new generation of homology search tools based on probabilistic inference. *Genome Inform.* **2009**, *23*, 205–211.

Article

Mutational Signatures as Sensors of Environmental Exposures: Analysis of Smoking-Induced Lung Tissue Remodeling

Yoo-Ah Kim †, Ermin Hodzic †, Bayarbaatar Amgalan, Ariella Saslafsky, Damian Wojtowicz and Teresa M. Przytycka *

National Center for Biotechnology Information, National Library of Medicine, National Institutes of Health, Bethesda, MD 20894, USA
* Correspondence: przytyck@ncbi.nlm.nih.gov
† These authors contributed equally to this work.

Abstract: Smoking is a widely recognized risk factor in the emergence of cancers and other lung diseases. Studies of non-cancer lung diseases typically investigate the role that smoking has in chronic changes in lungs that might predispose patients to the diseases, whereas most cancer studies focus on the mutagenic properties of smoking. Large-scale cancer analysis efforts have collected expression data from both tumor and control lung tissues, and studies have used control samples to estimate the impact of smoking on gene expression. However, such analyses may be confounded by tumor-related micro-environments as well as patient-specific exposure to smoking. Thus, in this paper, we explore the utilization of mutational signatures to study environment-induced changes of gene expression in control lung tissues from lung adenocarcinoma samples. We show that a joint computational analysis of mutational signatures derived from sequenced tumor samples, and the gene expression obtained from control samples, can shed light on the combined impact that smoking and tumor-related micro-environments have on gene expression and cell-type composition in non-neoplastic (control) lung tissue. The results obtained through such analysis are both supported by experimental studies, including studies utilizing single-cell technology, and also suggest additional novel insights. We argue that the study provides a proof of principle of the utility of mutational signatures to be used as sensors of environmental exposures not only in the context of the mutational landscape of cancer, but also as a reference for changes in non-cancer lung tissues. It also provides an example of how a database collected with the purpose of understanding cancer can provide valuable information for studies not directly related to the disease.

Keywords: mutational signatures; smoking; lung cancers; APOBEC; immune response to smoking; cell-type composition; goblet cells; ciliated cells; basal cells

Citation: Kim, Y.-A.; Hodzic, E.; Amgalan, B.; Saslafsky, A.; Wojtowicz, D.; Przytycka, T.M. Mutational Signatures as Sensors of Environmental Exposures: Analysis of Smoking-Induced Lung Tissue Remodeling. *Biomolecules* **2022**, *12*, 1384. https://doi.org/10.3390/biom12101384

Academic Editors: Cameron Mura and Lei Xie

Received: 5 August 2022
Accepted: 15 September 2022
Published: 27 September 2022

1. Background

Over the last few decades, the scientific community has continued to collect large quantities of biomedical data, typically organized in specialized databases. One such effort, The Cancer Genome Atlas (TCGA), a landmark cancer genomics program, includes data on over 20,000 primary cancer and matched normal samples, spanning 33 cancer types. As research questions continue to evolve, such historical data, combined with new computational approaches, remain fundamental for generating and testing new hypotheses and suggesting new experimental analyses.

Many lung diseases, including cancer, are associated with environmental factors, such as smoking or air pollution. Prolonged exposure to these factors often leads to chronic changes in lung structure and function. However, interactions between such environmental exposures and molecular-level changes in lung function are not fully understood. The amounts of environmental exposures are difficult to measure, making it challenging to quantify their impacts. In some cases, individuals might even be unaware of being exposed to harmful elements. Even when a sustained exposure can be established, as it is the case in smoking, the level of the exposure is often under-reported [1]. Furthermore, cigarette smoke contains a mixture of chemicals [2], and many factors, such as cigarette type, strength, and smoking habits, also contribute to the net exposure to individual factors. To bypass this challenge, studies typically resort to using binary classification—ever smoker vs. never smoker (e.g., [3])—even though continuous measurements could be more informative. The impact of cigarette smoking might also be indirect. For example, it is known that cigarette smoking is one of sources of chronic inflammation [4], which might in turn lead to chronic obstructive pulmonary disease (COPD) or cancer. Cigarette smoking has been also linked to differences in response to immunotherapy [5–11]. Thus, in order to better understand the process of emergence of lung diseases, it is important to develop computational approaches, which, while leveraging existing data, can help to untangle the impact of various factors on molecular changes in lung tissue. The emerging concept of mutational signatures can offer an interesting opportunity to uncover hidden relations between cellular level changes and a certain class of external exposures.

Smoking, and many other environmental exposures, are known to be mutagenic. The effects of such mutagenic exposures have been studied extensively in the context of cancer [12,13] and recent studies leveraged the idea of mutational signatures—characteristic mutation patterns imprinted on DNA molecules by specific mutagens [14–18]. Mutational signatures are typically defined based on a partition of mutations into mutation categories. Most studies utilize mutation categories defined based on six types of single nucleotide substitutions (C > A, C > G, C > T, T > A, T > C, and T > G), considered in the context of the $5'$ and $3'$ flanking nucleotides, yielding 96 mutation categories (e.g., TCC > TAC, and CAG > CTG). Given such categories, mutational signatures are defined as multinomial distributions of mutation counts over these categories. Following the pioneering paper of Alexandrov et al. [17], several computational methods have been proposed to infer such signatures based on large cancer datasets. The Catalogue of Somatic Mutations in Cancer (COSMIC) contains a reference set of signatures defined using the 96 mutation categories mentioned above. COSMIC signatures have been broadly explored and many, but not all, have been linked to specific mutagenic processes. A decomposition of somatic mutations in a tumor genome into COSMIC signatures and mutation counts attributed to each signature (signature exposure) can provide patient-specific information about mutagenic factors contributing to the somatic mutations in the tumor (reviewed in [19–22]).

Mutational signatures can be easily inferred from bulk genome sequencing of tumor samples. Although the influence of environmental factors, such as smoking, is not restricted to tumors but also affects the whole organism, mutations in non-cancer cells are not common and are difficult to capture by bulk sequencing since such cells are not related by common ancestry from a tumor-initiating cell. Since mutagenic processes caused by environment-related mutagens are exogenous for both cancer and non-cancer samples, signature exposures inferred based on cancer mutation data can be used to estimate the strength of the corresponding environmental factors acting on non-cancer cells as well. However, while considering environmental processes through the lenses of mutational signatures provides unique opportunities, it also comes with its set of challenges. Some environmental factors, including cigarette smoke, are mixtures of many potentially harmful components. While some such components might be uniquely associated with smoking, others might be present in other contexts as well. In addition, even if the sample itself is non-neoplastic, it should not be ignored that the sample donor was a cancer patient. Conveniently, in many cancer types, tumor growth is correlated with a specific mutational signature (SBS1), allowing for pinpointing correlations that could be due the disease's status rather than environment. Finally, the etiologies of many mutational signatures are not fully understood, and not all chemicals impacting cell function are mutagenic but might instead co-occur with mutagenic exposures. Thus, as in any association-based analysis, additional studies might be required to obtain mechanistic explanations of the uncovered associations (see Section 3).

In this paper, we explore the application of the TCGA data for providing a better understanding of the relation between smoking (and other external processes) and molecular-level changes in the lung. Utilizing mutational signatures, derived from cancer lung tissue, and gene expression, derived from the corresponding normal (non-neoplastic control) samples, we hypothesize that such data can inform about the impact of environmental processes on the function of normal lung tissue (Figure 1). We take two complementary computational approaches in our analysis; First, utilizing an approach developed in a previous study [23], we analyzed the relation between patients' exposures to mutational signatures and gene expression in control samples. Next, recognizing that chronic changes might be related to cellular reprogramming on the tissue level, we utilized methods to decompose bulk samples into cell type proportions to uncover correlations of signature exposures with changes in epithelial and immune cell type proportions. Our study demonstrates the usefulness of such a joint analysis, recapitulating much of the known associations obtained by previous studies (including results obtained using single cell analyses) and providing additional novel insights. It provides a proof of principle of the utility of mutational signatures to be used as sensors of environmental exposures not only in the context of the mutational landscape of cancer, but also as a measurement of important exogenous influences on non-cancer tissues.

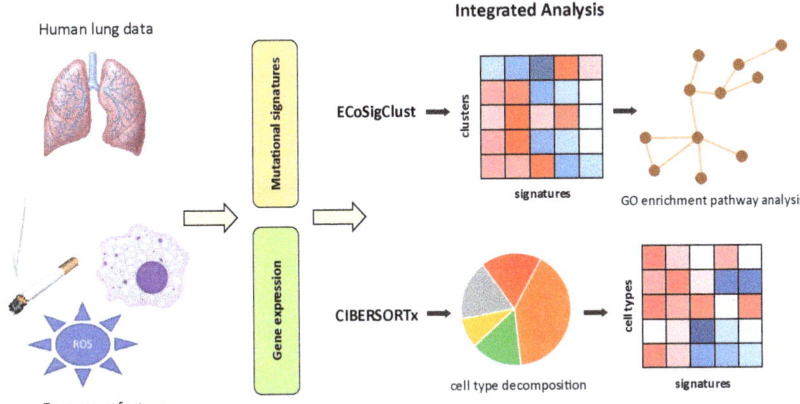

Figure 1. The analysis overview of the impact of smoking and exogenous processes on non-neoplastic lung tissue.Given a tumor and control sample from the same patient, the tumor sample is used as a readout of mutational signatures, while the control sample is used as a readout of changes in gene expression in non-neoplastic control as a function of mutation signature exposure. The combined analysis of mutational signatures and gene expression with ECoSigClust uncovers functional changes in gene expression (**upper panel**), while the analysis of these signatures leveraging CIBERSORTx uncovers changes in cellular composition (**lower panel**) and sheds light on their correlation with exposures to exogenous processes.

2. Results

2.1. Properties of Mutational Signatures Observed in LUAD Patients

Smoking is a widely recognized risk factor in the emergence of lung diseases. It is also one of the primary mutagens contributing to the emergence of lung adenocarcinoma (LUAD). Previous studies have identified a specific mutational signature (SBS4) that is uniquely associated with smoking [15] and is not observed in non-smokers [24]. This signature is very similar to the mutational signature induced in vitro by exposing cells to a known tobacco smoke carcinogen benzo[a]pyrene, and was shown to correlate with pack years of smoking [15]. This provides strong evidence that SBS4 is a direct consequence of tobacco carcinogens and presents a unique opportunity to study the relation between environmental exposures, represented by mutational signatures from tumor sequencing, and gene expression from control samples. Importantly, even in the context of LUAD—a cancer type that is related to smoking—information on the smoking status is often missing. Quantification of the signature exposure allows to bypass this issue, providing an unbiased estimate of exposure to smoking. We utilized TCGA LUAD mutation data to infer mutational signatures in individual cancer patients, as described in the Methods section.

In addition to the presence of the SBS4 mutational signature in TCGA LUAD data, the genomes of LUAD patients also harbor five additional COSMIC mutational signatures—SBS1, SBS2, SBS5, SBS13, and SBS40 (Methods). Three LUAD signatures—SBS1, SBS5, and SBS40, are often referred to as "clock-like" signatures, as their strength is positively correlated with patients' age in many (but not all) cancer types. However, no such correlation is observed in LUAD patients (Figure S1B). Such loss of correlation with age suggests the existence of other factors that accelerate (or otherwise modify) the accumulation of naturally occurring mutations.

Out of the three clock-like signatures, SBS1 is the best understood. It is assumed to arise due to a spontaneous or enzymatic deamination of 5-methylcytosine during replication. Thus, SBS1 is gained during cell division and its accumulation is accelerated in tumor. Consequently, the exposure of this signature is used to estimate the timing of the tumor initiating event [25]. Consistent with this interpretation, we found that in LUAD, SBS1 is highly associated with primary tumor grade (p-value $< 4.8 \times 10^{-5}$, Figure S1B).

SBS5 is present in nearly all cancer types but its etiology is less understood. As it is frequently correlated with smoking [26], including in LUAD (Figure S1B), it might be the result of exposure to environmental causes occurring with smoking, but also broadly present in other, smoking-independent, contexts. One potential cause might be the accumulation of mutations due to reactive oxygen species (ROS) that play an important role in environment-related mutagenesis, and are prominently associated with smoking [27,28]. SBS5 has also been previously linked to the NER DNA repair pathway [29], but the exact mechanism remains unknown.

The accumulation of SBS40 mutations with age in some cancer types suggests that it might also be related to environmental factors. This is a recently defined signature, characterized by a relatively uniform distribution of mutation types, similarly to SBS5. This renders its contribution uncertain [18]. In the TCGA LUAD dataset, the signature strength of SBS40 is correlated with the signature strength of SBS4 (Figure S1). Thus, we consider these two signatures together in our analysis.

The two remaining signatures, SBS2 and SBS13, are attributed to mutations introduced by the AID/APOBEC family of cytidine deaminases enzymes. The activity of these enzymes is often related to innate immune response [30]. For example, the strength of these signatures has been shown to correlate with the expression of immune-related genes and pathways [23].

The cause of the over-activity of APOBECs in LUAD is yet to be established, but Alexandrov et al. speculated that the cellular machinery underlying SBS2 and SBS13 can be activated by tobacco smoking, perhaps as a smoking-related inflammatory response [15]. Indeed, it has been observed that cigarette smoke incites a potent inflammatory reaction in the airways and alveoli [31], and, in LUAD data, SBS13 exposure is correlated with smoking status (Figure S1B). However, it is also possible that the immune response is related to the fact that the sample was taken from a cancer patient, even if it is from a non-neoplastic lung. In what follows, we will attempt to shed more light on this distinction.

In summary, the mutational signatures observed in LUAD can be divided into three groups: smoking-associated (SBS4, SBS5, SBS40), immune-related (SBS2, SBS13) and the tumor growth-related signature (SBS1).

2.2. *Pathway-Based Analysis and Relation between Signature Exposures and Gene Expression in Control Samples*

First, we asked if mutational signatures can reveal how smoking and other mutagenic processes identified in LUAD interact with gene expression in non-cancer control samples. In an attempt to understand the impact of external mutagens on molecular processes in cells, we utilized the approach developed in a previous study [23] and identified clusters of genes whose expression is correlated with different combinations of signatures (Figure 2a and Table S1). More specifically, we selected genes whose expression is significantly correlated with the strength of at least one mutational signature ($p < 0.05$), and clustered the genes based on their correlation patterns with mutational signatures. We refer to this clustering procedure as `ECoSigClust` (**e**xpression **co**rrelated **sig**nature **clust**ering).

Gene ontology (GO) enrichment analysis of the clusters obtained by `ECoSigClust` revealed that the clusters are enriched with specific GO terms, providing insights into the interactions between signatures and molecular pathways. In addition, we analyzed the cluster assignment of known markers of specific lung cell types. Guided by the observations from this analysis, we further explored the association between exposure to exogenous processes and changes in cell-type composition in the lung in the following Section (Section 2.3).

2.2.1. Exposure to Smoking Signature is Correlated with Increased Inflammatory Response in Non-Cancer Lung Tissue and Elevated Expression of the PD-L1 Immune Checkpoint Gene

The cluster with the strongest positive correlation with the smoking-specific signature SBS4 (and thus with SBS40), which we call *smoking-specific cluster* (CL5, Figure 2a), includes 837 genes, enriched with the cytokine-mediated signaling pathway ($p < 10^{-13}$), inflammatory response ($p < 10^{-13}$) and cell activation ($p < 10^{-14}$, Table S2). This is consistent with previous observations that the exposure of epithelial cells to smoking triggers pro-inflammatory response and increases the release of pro-inflammatory cytokines and chemokines [28,32], many of which are included in the cluster. For example, the cluster includes several chemokines (CCL2, CCL3, CCL4, CCL7, and CCL11), and pro-inflammatory cytokines (Interleukin 1α (IL1A), and tumor necrosis factor (TNF)) (Table S2). Interestingly, the smoking cluster includes MUC5AC, the canonical marker of mucus-producing secretory goblet cells [33,34], suggesting a relation between smoking and goblet cell population. We investigate the relation further in Section 2.3.

Another notable gene in the cluster is GPR15, a chemoattractant receptor for lymphocytes. The expression of GPR15 was previously found to be up-regulated in smokers [35].

Cluster 5 contains the PD-L1 (CD274) gene. The up-regulation of PD-L1 is believed to allow cancers to evade the host immune system. Thus, immune checkpoint inhibitors of PD-L1 are promising tools for cancer immunotherapy [36,37]. The fact that the association of expression of PD-L1 with smoking is observed in non-cancer lung tissue, and is not related to tumor growth (no correlation with SBS1), is of particular importance. Indeed, a recent experimental study demonstrated that cigarette smoke and the carcinogen benzo(a)pyrene (BaP) induce PD-L1 expression on lung epithelial cells [11].

Finally, Cluster 5 also includes the APOBEC3B gene, which is known to induce mutations related to the emergence of mutational signatures (SBS2 and SBS13). The fact that APOBEC3B belongs to the smoking cluster, rather than a cluster associated with signatures SBS2 and SBS13, suggests that over-activity of this APOBEC enzyme is likely to be triggered by an inflammatory response to smoking [15]. As for negative correlations, we observe that the exposure of these two signatures (SBS2 and SBS13) is negatively correlated with Clusters 8 and 9, both of which are enriched with cell differentiation and morphogenesis. This negative correlation suggests that smoking may shift the overall epithelial function away from a diversity of cell types with specialized functions, toward a consensus increase in mucus secretion, proliferation, and response to stress.

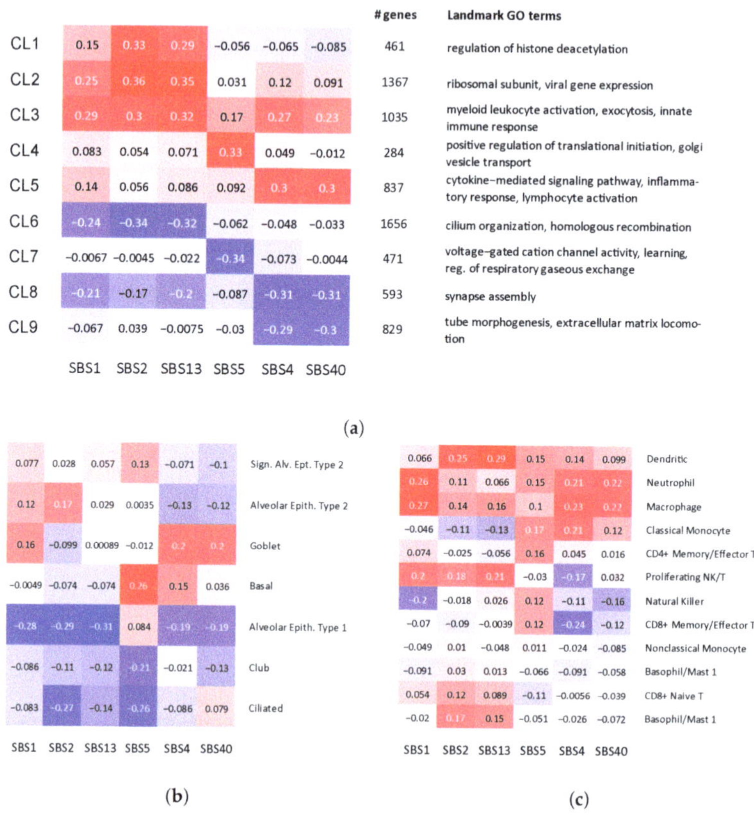

Figure 2. Correlation between mutational signatures and cell type composition and gene expression. (**a**) ECoSigClust clusters, based on the correlation between mutational signatures and gene expression. Genes having a significant correlation with at least one mutational signature ($p < 0.05$) are included in the clustering. The heat map shows the mean correlation between signature and expression among all genes in the cluster (left). For each cluster, the number of genes and representative GO terms enriched in the cluster genes are also shown. (**b**,**c**) Correlation between mutational signatures and cell composition. Bulk expression counts are decomposed into different cell types using CIBERSORTx, and Spearman correlation coefficients are shown for (**b**) epithelial cells and (**c**) immune cells, separately.

2.2.2. Strength of SBS5, a Signature Correlated with Smoking but Not Unique to This Mutagen, Is Correlated with Changes in Ciliogenesis

The exposure to signature SBS5 is overall correlated with smoking in many cancers, including LUAD. However, as discussed before, this signature is not unique to smokers. The exposure to this signature is negatively correlated with Cluster 7, which is enriched with genes related to voltage gated cation channel activity and neurotransmitter receptor complex. It is known that these channels are targets of a number of naturally occurring toxins and therapeutic agents, as well as environmental toxicants [38], including nicotine [39]. In addition, the cluster also contains known early transcriptional drivers of ciliogenesis, such as MYB and TP73 (Table S1), consistent with the reports that smoking blocks early ciliogenesis [40,41]. The results discussed in Section 2.3 provide further insights into the relation of this signature and changes in the population of ciliated cells in lung.

2.2.3. Relation between the Strengths of APOEBEC-Related Signatures and Gene Expression

The two APOBEC signatures (SBS2 and SBS13) are positively correlated with the expression of genes in Clusters 1–3, and negatively correlated with Cluster 6. We note that correlation of Clusters 2 and 3 with SBS1 suggests a possible relation to tumor growth, so it is not clear to which extent the activity of this cluster is related to smoking and to which extent the changes in the immune system are triggered by tumor growth. Interestingly, Cluster 2 also includes SFTPB and SFTPC, the genes responsible for encoding pulmonary-associated surfactants secreted by the alveolar cells of the lung and maintaining the stability of pulmonary tissue by reducing the surface tension of fluids that coat the lung (Table S1). Interestingly, we found that the expression of the APOBEC3C gene is correlated with the expression of the immune checkpoint gene PD-1 (PDCD1) (p-value < 0.0051). The APOBEC3C gene is a member of Cluster 2, suggesting that, in contrast to PD-L1, PD-1 might be stimulated by immune response. Out of the three clusters with positive correlation, Cluster 1 correlated with APOBEC signatures most specifically. GO enrichment analysis of this cluster reveals a relation with the regulation of histone deacetylation (Table S2). While a general relation between immune response and histone deacetylation has been well appreciated [42], the association with APOBEC activity remains to be investigated. GO enrichment analysis of Cluster 6, showing negative correlation with APOBEC signatures, found that this cluster is significantly enriched with cilium. Cluster 6 also includes TUBB1, a marker of ciliated cells. This suggests a link between the number of ciliated cells and APOBEC activity.

2.3. *Mutational Signatures Reveal Relation between Exposure to Exogenous Processes and a Remodeling of Cell-Type Composition in Lung*

The signature-dependent expression changes of MUC5AC, a canonical marker of mucus producing secretory goblet cells, as well as other markers discussed in the previous section, suggest a relation between exposures of mutational signatures and changes in the cell-type composition. Indeed, previous studies reported that exposure to smoking leads to the reprogramming of cell-type composition in lungs [3,43]. Thus, we asked whether examining the relation between the exposures of mutational signatures and cell-type composition can identify such trends and potentially provide additional insights.

To investigate the relation between cell-type composition and mutational signatures, we decomposed the bulk expression data using CIBERSORTx [44] and estimated the cell composition in each sample (see Section 4). Considering epithelial and immune cells separately, we then computed the correlation coefficients between the proportions of cell types (within epithelial and immune cell types, respectively) and the strengths of mutational signatures (Figure 2b,c), which revealed several changes in both epithelial and immune cell type composition correlated with mutational signature activities.

Among epithelial cell types, the proportion of goblet cells is positively correlated with smoking signatures (SBS4, SBS40), while SBS5 has the strongest correlation with Basal cells (Figure 2b). This is consistent with the previous observation that the exposure to cigarette smoke increases the number of mucous-secreting goblet cells and thus can lead to goblet cell hyperplasia, mucus hypersecretion and promote inflammatory responses [45,46].

The correlation pattern of goblet cells is similar to the pattern of smoking cluster in Figure 2a, supporting the hypothesis that the inflammatory responses are generated by epithelial cells with altered cytokine-mediated signaling pathways in response to smoking exposure. Previous studies found that bronchial epithelial cells exposed to cigarette smoke produced a dose-dependent increase in the expression of MUC5AC, IL8 (also called CXCL8) and TNFα genes [47], all of which belong to the smoking cluster.

Interestingly, exposure of SBS4 and SBS5 is correlated with an increased proportion of Basal cells. Basal cells are located below the surface epithelial cell layer and serve as progenitor cells from which ciliated, secretory, and goblet cells differentiate.

Consistent with the results of the previous section, the proportion of ciliated cells has negative correlations with SBS2 and SBS5. The major function of airway ciliated cells is to mediate the propulsion of mucus gel. Thus, a proper balance between goblet and ciliated cells is required for the correct functioning of lungs. Previous studies indicated that this balance might be disturbed by smoking [3,43] and our results confirm this view, but additionally reveal a contribution of APOBEC-related processes captured by SBS2. Interestingly, the reduction in the number of ciliated cells is also associated with SBS2, suggesting a potential relation of the reduction in the ciliated cell number to APOBEC and immune response, which warrants further investigations.

As for immune cells, we observe that innate immune cells, such as dentritic cells, macrophages, and neutrophiles, have overall positive correlation across all mutational signatures (Figure 2c), including the tumor growth-related SBS1. Interestingly the exposure of smoking signature (SBS4) is associated with reduction in CD8+ cells, suggesting an immunosuppressive effect. A similar effect was previously observed in HNSCC cancer [48].

3. Conclusions

Exposure of individuals to environmental factors, such as smoking, might lead to molecular changes within cells and the reprogramming of cellular tissue composition. Such changes might be relevant to human health. Yet, the relations between environmental exposures and the above-mentioned changes are not well understood. One of the challenges in studies of the impact of environmental factors on cellular changes is related to the fact that historical exposure to environmental factors is often difficult to quantify. However, many such adverse environmental factors are mutagenic and leave characteristic mutational signatures.

In this paper, we explored whether a joint analysis of mutational signatures and gene expression of non-cancer samples can provide insights into the impact of mutagenic factors on the expression of genes, pathways, and cellular composition in non-neoplastic lung tissue.

Currently, mutational signatures are the most readily accessible for cancer patients by sequencing tumor samples. We reasoned that even if the signatures are inferred from mutations in cancer cells, exogenous environmental factors act on both the cancer and non-cancer cells. Therefore, in this study, we performed a combined analysis of mutational signatures, obtained from cancer genomes, and gene expression from control samples. The fact that a specific mutational signature, SBS1, is known to be correlated with tumor growth, allowed us to identify relations that might be due to tumor growth response in non-neoplastic lung tissue rather than a direct effect of smoking.

Our signature-based analysis uncovered many interesting insights on how smoking can impact the activities of genes, pathways, and tissue composition in lung. The results of our studies are in good agreement with current knowledge, providing confidence in our approach; see Table 1. Furthermore, our results provide additional insights that were not accessible with previous approaches. For example, previous studies demonstrated that smoking can decrease ciliated cells and increase goblet cells in their proportion [3,49]. By analyzing correlations with mutational signature values rather than binary smoking status, our analysis further revealed that the decrease in the ciliated cells proportion is related to the exposure of the SBS5 signature—a signature known to be correlated with smoking but also occurring in contexts not related to smoking.

The interplay between smoking and immune system that we uncovered is also consistent with current knowledge, although the correlation of SBS1 with one of the two immune related clusters suggests that some of the immune response in the control lung tissue could be contributed by an immune response to cancer.

Knowledge of mutational processes acting on a patient's genome might also help to develop personalized therapies. For example, signature SBS3 indicates homologous recombination deficiency (HRD), and since the patients with HRD are known to benefit from PARP inhibitor therapy [50], the presence of this signature can be used as a marker for PARP inhibitor therapy [51]. Furthermore, APOBEC signatures have been associated with sensitivity to ataxia telangiectasia and Rad3-related kinase (ATR) in some cancer cell lines, suggesting a potential for targeted therapy [52–54]. Interestingly, some studies indicated that a smoking history can have an effect on the efficacy of immune checkpoint inhibitors [55]. Our signature-based analysis points to several different mechanisms that, in addition to high tumor mutation burden, can contribute to this effect. First, we found that the strength of smoking signatures is correlated with the expression of the immune checkpoint gene PD-L1, which might promote immune escape. Next, smoking is associated with a reduction in the proportion of CD8+cells, which can kill transformed tumor cells. Finally, the expression of important immune checkpoint gene PD-1 appears to be (indirectly) associated with APOBEC signatures. These examples illustrate an increasing role that mutational signatures play in identifying treatment options.

Overall, we show that looking at the expression changes through the lenses of mutational signatures provides a new and powerful stepping stone for studying the impact of environmental factors on individual's health, disease susceptibility, and progression. The smoking-associated mutational signature allowed for an unbiased inference of smoking status, key information that is often missing in collected data. In fact, the analysis provided here would have been under-powered if we restricted the study to control samples with reported smoking status only. Finally, cigarette smoke includes a complex mixture of potentially harmful factors, and both direct and indirect as well as mutational signatures based analyses allow for separating at least some of these factors. However, our analysis has also some limitations. Most importantly, the current understating of the mechanisms of many mutational signatures is incomplete, which can limit the interpretability of our association-based results. In addition, as with any association-based approach, additional experiments and knowledge are required to provide mechanistic explanations of the observed dependencies. Finally, while it is easy to obtain mutational signatures from tumor samples, such an approach is indirect, and it would be desirable to measure the mutations directly in the sample of interest. In future, large-scale single-cell sequencing is likely to enable the robust analysis of mutational signatures in non-cancer tissue.

Despite these limitations, our study shows that the utility of mutational signatures can go beyond cancer studies and shed light on the role of environmental mutagens in chronic molecular level changes in the organism. It also provides an example of how a database collected with the purpose of understanding cancer can provide valuable information for studies not directly related to the disease.

Table 1. Results of the analysis of the relation between mutational signatures and gene expression in the context of previous studies.

Observation From Mutational Signatures	Supporting Literature
Cluster 2:	
ABOBEC signatures are associated with expression of SFTPB and SFTPC	novel observation
APOBEC might indirectly trigger the expression PD-1	[56]
Cluster 5:	
Smoking triggers pro-inflammatory response and cytokines signaling	[28,32]
Smoking increases MUC5AC expression	[34]
Smoking increases PD-L1 expression	[36]
Smoking increases GPR15 expression	[35]
Cluster 6:	
APOBEC is associated with a reduction in cilium organization	novel observation
Cell-type composition:	
ABOBEC signatures are associated with a reduction in CD8+ cells	[48]
Smoking is associated with increase of goblet cells	[45,46,57],
Smoking is associated with decrease of ciliated cells	[3,40,41]
APOBEC is associated with decrease of ciliated cells	novel observation

4. Methods

4.1. Mutational Signatures

We downloaded the TCGA LUAD (lung adenocarcinoma) exome mutation spectra and the exome COSMIC reference mutational signatures, provided by Alexandrov et al. [18], from Synapse (accession numbers: syn11801889 and syn11726602, respectively). We utilized the data from 48 patients with known gene expression data for both cancer and control lung tissue. The statistics on this cohort are provided in Table S3. To determine the predominant signatures being active in LUAD samples, we started with the initial sample exposures to mutational signatures from [18] (version 3.1, June 2020, Synapse accession number: syn11804065). The list of active signatures was refined to remove any rare signatures; namely, we keep only signatures that were present in at least 5% of samples and were responsible for at least 1% of mutations. Next, using such a list of active mutational signatures in LUAD (SBS1, SBS2, SBS4, SBS5, SBS13, SBS40, and SBS45), we determined their sample-specific exposures using the quadratic programming (QP) approach available in the R package—SignatureEstimation [58]. Signature SBS45 was omitted from the analyses presented in this study, as this signature is likely an artifact due to the 8-oxo-guanine introduced during sequencing (see COSMIC Mutational Signatures website: https://cancer.sanger.ac.uk/signatures/ (accessed on 14 September 2022)).

4.2. Expression Data

TCGA LUAD RNAseq expression data were obtained from the Genomic Data Commons Data Portal (https://portal.gdc.cancer.gov/ (accessed on 14 September 2022)) on 5 June 2020. HTseq counts were normalized and variance-stabilizing transformed (vst) using DESeq2 [59]. Only donors that had both gene expression and mutational signature exposures were kept, which resulted in 48 normal samples and 466 tumor samples used in this study.

4.3. Clustering

To identify expression-based pathways that are associated with signatures, we used `ECoSigClust` developed for our previous analysis [23]. Specifically, we first computed Spearman correlation coefficients of the expression level and mutation counts for each pair of genes and mutational signatures. We then selected the genes exhibiting significant correlation with at least one of the mutational signatures; the expression of a gene is considered significantly correlated with a signature if nominal $p < 0.05$. This procedure selected 7533 genes. We then clustered the genes based on their correlation patterns using a consensus K-means algorithm; running K-means clustering 100 times with random start, varying k from 5 to 50, and subsequently running hierarchical clustering with the consensus matrix from 100 runs of the K-means algorithm. To determine the optimal cluster number, three different clustering validation metrics—Silhouette Index, Calinski–Harabasz Index, and Davies–Bouldin Index—were used, measuring compactness within clusters and separation between clusters slightly differently. The chosen number of clusters $k = 9$ was based on these metrics (Figure S2) and was kept small for the interpretability of each cluster. GO enrichment analysis was performed using the hypergeometric test for each cluster with all genes included in the clustering as the background to assess the differences among the clusters. The list of genes and enrichment analysis results for all clusters are provided in Tables S1 and S2.

4.4. Cell Composition Analysis with CIBERSORTx

HTseq raw counts in bulk expression data for the normal samples from TCGA LUAD dataset were used for the analysis. For each gene, the counts in every sample were normalized by the total sum of counts in that sample, multiplied by 1,000,000. The genes without at least one normalized count with a value greater than 1 were discarded. The *Human Lung Cell Atlas* (HLCA) [60] single-cell reference data containing 42 distinct cell types was obtained in the form of counts from synapse (accession number: syn21560511). As per

CIBERSORTx guidelines, the same normalization procedure was used on the single-cell reference data and used as input to CIBERSORTx to impute the cell proportions of the 42 given cell types in the bulk TCGA-Lung expression data.

For two subsets of cell types—epithelial and immune cell types, we computed the Spearman correlation of each imputed cell type's fraction with the exposures of Signatures 1, 2, 4, 5, 13, and 40. The strength of the correlation and the resulting heatmaps are shown in Figure 2.

Supplementary Materials: The following supporting information can be downloaded at: https://www.mdpi.com/article/10.3390/biom12101384/s1, Table S1: Gene membership in clusters of genes whose expression is correlated with different combinations of signatures. Table S2: GO enrichment analysis of expression clusters from Table 1. Table S3: Statics on the cohort used in these study. Four stages refer to the extent of patient's cancer, the mean cigarettes in per day (CPD), mean age in days, gender and number of samples whose corresponding information is available. The mean values were computed over the samples whose corresponding CPD and Days are available. The last column is the number of patients with the given information. The complete information can be obtained from the TCGA data portal. Figure S1: Spearman correlations (above) and corresponding *p*-values (below) represent the pairwise associations in control samples. Figure S2: Evaluation of clustering for varying *k*'s (the number of clusters) using different metrics.

Author Contributions: Conceptualization, Y.-A.K. and T.M.P.; methodology, Y.-A.K. and E.H.; validation, Y.-A.K. and T.M.P.; formal analysis, all authors; data curation, D.W. and A.S.; writing—original draft preparation, Y.-A.K., E.H. and T.M.P.; writing—review and editing, E.H. and T.M.P.; supervision, T.M.P.; project administration, T.M.P. All authors have read and agreed to the published version of the manuscript.

Funding: This research was supported by the Intramural Research Program of the National Library of Medicine, NIH.

Institutional Review Board Statement: Not applicable.

Informed Consent Statement: Not applicable.

Data Availability Statement: `ECoSigClust` is available at https://github.com/ncbi/ECoSigClust (accessed on 14 September 2022).

Conflicts of Interest: The authors declare no conflict of interest.

References

1. Gorber, S.C.; Schofield-Hurwitz, S.; Hardt, J.; Levasseur, G.; Tremblay, M. The accuracy of self-reported smoking: A systematic review of the relationship between self-reported and cotinine-assessed smoking status. *Nicotine Tob. Res.* **2009**, *11*, 12–24. [CrossRef] [PubMed]
2. Hecht, S.S. Tobacco Smoke Carcinogens and Lung Cancer. *JNCI J. Natl. Cancer Inst.* **1999**, *91*, 1194–1210. [CrossRef] [PubMed]
3. Smith, J.C.; Sausville, E.L.; Girish, V.; Yuan, M.L.; Vasudevan, A.; John, K.M.; Sheltzer, J.M. Cigarette Smoke Exposure and Inflammatory Signaling Increase the Expression of the SARS-CoV-2 Receptor ACE2 in the Respiratory Tract. *Dev. Cell* **2020**, *53*, 514–529. [CrossRef] [PubMed]
4. Lee, J.; Taneja, V.; Vassallo, R. Cigarette Smoking and Inflammation. *J. Dent. Res.* **2011**, *91*, 142–149. [CrossRef] [PubMed]
5. Cortellini, A.; Giglio, A.D.; Cannita, K.; Cortinovis, D.L.; Cornelissen, R.; Baldessari, C.; Giusti, R.; D'Argento, E.; Grossi, F.; Santoni, M.; et al. Smoking status during first-line immunotherapy and chemotherapy in NSCLC patients: A case–control matched analysis from a large multicenter study. *Thorac. Cancer* **2021**, *12*, 880–889. [CrossRef]
6. Li, J.J.; Karim, K.; Sung, M.; Le, L.W.; Lau, S.C.; Sacher, A.; Leighl, N.B. Tobacco exposure and immunotherapy response in PD-L1 positive lung cancer patients. *Lung Cancer* **2020**, *150*, 159–163. [CrossRef]
7. Norum, J.; Nieder, C. Tobacco smoking and cessation and PD-L1 inhibitors in non-small cell lung cancer (NSCLC): A review of the literature. *ESMO Open* **2018**, *3*, e000406. [CrossRef]
8. Sun, L.Y.; Cen, W.J.; Tang, W.T.; Long, Y.K.; Yang, X.H.; Ji, X.M.; Yang, J.J.; Zhang, R.J.; Wang, F.; Shao, J.Y.; et al. Smoking status combined with tumor mutational burden as a prognosis predictor for combination immune checkpoint inhibitor therapy in non-small cell lung cancer. *Cancer Med.* **2021**, *10*, 6610–6617. [CrossRef]
9. Desrichard, A.; Kuo, F.; Chowell, D.; Lee, K.W.; Riaz, N.; Wong, R.J.; Chan, T.A.; Morris, L.G.T. Tobacco Smoking-Associated Alterations in the Immune Microenvironment of Squamous Cell Carcinomas. *JNCI J. Natl. Cancer Inst.* **2018**, *110*, 1386–1392. [CrossRef]

10. Lafuente-Sanchis, A.; Zúñiga, Á.; Estors, M.; Martínez-Hernández, N.J.; Cremades, A.; Cuenca, M.; Galbis, J.M. Association of PD-1 , PD-L1 , and CTLA-4 Gene Expression and Clinicopathologic Characteristics in Patients with Non–Small-Cell Lung Cancer. *Clin. Lung Cancer* **2017**, *18*, e109–e116. [CrossRef]
11. Wang, G.Z.; Zhang, L.; Zhao, X.C.; Gao, S.H.; Qu, L.W.; Yu, H.; Fang, W.F.; Zhou, Y.C.; Liang, F.; Zhang, C.; et al. The Aryl hydrocarbon receptor mediates tobacco-induced PD-L1 expression and is associated with response to immunotherapy. *Nat. Commun.* **2019**, *10*, 1125. [CrossRef] [PubMed]
12. Basu, A. DNA Damage, Mutagenesis and Cancer. *Int. J. Mol. Sci.* **2018**, *19*, 970. [CrossRef] [PubMed]
13. Poon, S.; McPherson, J.R.; Tan, P.; Teh, B.; Rozen, S.G. Mutation signatures of carcinogen exposure: Genome-wide detection and new opportunities for cancer prevention. *Genome Med.* **2014**, *6*, 24. [CrossRef] [PubMed]
14. Alexandrov, L.B.; Nik-Zainal, S.; Wedge, D.C.; Aparicio, S.; Behjati, S.; Biankin, A.V.; Bignell, G.R.; Bolli, N.; Borg, A.; Børresen-Dale, A.L.; et al. Signatures of mutational processes in human cancer. *Nature* **2013**, *500*, 415–421. [CrossRef]
15. Alexandrov, L.B.; Ju, Y.S.; Haase, K.; Van Loo, P.; Martincorena, I.; Nik-Zainal, S.; Totoki, Y.; Fujimoto, A.; Nakagawa, H.; Shibata, T.; et al. Mutational signatures associated with tobacco smoking in human cancer. *Science* **2016**, *354*, 618–622. [CrossRef]
16. Nik-Zainal, S.; Alexandrov, L.B.; Wedge, D.C.; Van Loo, P.; Greenman, C.D.; Raine, K.; Jones, D.; Hinton, J.; Marshall, J.; Stebbings, L.A.; et al. Mutational processes molding the genomes of 21 breast cancers. *Cell* **2012**, *149*, 979–993. [CrossRef]
17. Alexandrov, L.B.; Nik-Zainal, S.; Wedge, D.C.; Campbell, P.J.; Stratton, M.R. Deciphering Signatures of Mutational Processes Operative in Human Cancer. *Cell Rep.* **2013**, *3*, 246–259. [CrossRef]
18. Alexandrov, L.B.; Kim, J.; Haradhvala, N.J.; Huang, M.N.; Tian Ng, A.W.; Wu, Y.; Boot, A.; Covington, K.R.; Gordenin, D.A.; Bergstrom, E.N.; et al. The repertoire of mutational signatures in human cancer. *Nature* **2020**, *578*, 94–101. [CrossRef]
19. Kim, Y.A.; Leiserson, M.D.; Moorjani, P.; Sharan, R.; Wojtowicz, D.; Przytycka, T.M. Mutational Signatures: From Methods to Mechanisms. *Annu. Rev. Biomed. Data Sci.* **2021**, *4*, 189–206. [CrossRef]
20. Koh, G.; Degasperi, A.; Zou, X.; Momen, S.; Nik-Zainal, S. Mutational signatures: Emerging concepts, caveats and clinical applications. *Nat. Rev. Cancer* **2021**, *21*, 619–637. [CrossRef]
21. Helleday, T.; Eshtad, S.; Nik-Zainal, S. Mechanisms underlying mutational signatures in human cancers. *Nat. Rev. Genet.* **2014**, *15*, 585–598. [CrossRef] [PubMed]
22. Ma, J.; Setton, J.; Lee, N.Y.; Riaz, N.; Powell, S.N. The therapeutic significance of mutational signatures from DNA repair deficiency in cancer. *Nat. Commun.* **2018**, *9*, 3292. [CrossRef] [PubMed]
23. Kim, Y.A.; Wojtowicz, D.; Sarto Basso, R.; Sason, I.; Robinson, W.; Hochbaum, D.S.; Leiserson, M.D.M.; Sharan, R.; Vadin, F.; Przytycka, T.M. Network-based approaches elucidate differences within APOBEC and clock-like signatures in breast cancer. *Genome Med.* **2020**, *12*, 52. [CrossRef] [PubMed]
24. Landi, M.T.; Synnott, N.C.; Rosenbaum, J.; Zhang, T.; Zhu, B.; Shi, J.; Zhao, W.; Kebede, M.; Sang, J.; Choi, J.; et al. Tracing Lung Cancer Risk Factors Through Mutational Signatures in Never-Smokers. *Am. J. Epidemiol.* **2020**, *190*, 962–976. [CrossRef] [PubMed]
25. Gerstung, M.; Jolly, C.; Leshchiner, I.; Dentro, S.C.; Gonzalez, S.; Rosebrock, D.; Mitchell, T.J.; Rubanova, Y.; Anur, P.; Yu, K.; et al. The evolutionary history of 2,658 cancers. *Nature* **2020**, *578*, 122–128. [CrossRef]
26. Alexandrov, L.B.; Jones, P.H.; Wedge, D.C.; Sale, J.E.; Campbell, P.J.; Nik-Zainal, S.; Stratton, M.R. Clock-like mutational processes in human somatic cells. *Nat. Genet.* **2015**, *47*, 1402–1407. [CrossRef]
27. Waris, G.; Ahsan, H. Reactive oxygen species: Role in the development of cancer and various chronic conditions. *J. Carcinog.* **2006**, *5*, 14. [CrossRef]
28. Strzelak, A.; Ratajczak, A.; Adamiec, A.; Feleszko, W. Tobacco Smoke Induces and Alters Immune Responses in the Lung Triggering Inflammation, Allergy, Asthma and Other Lung Diseases: A Mechanistic Review. *Int. J. Environ. Res. Public Health* **2018**, *15*, 1033. [CrossRef]
29. Kim, J.; Mouw, K.W.; Polak, P.; Braunstein, L.Z.; Kamburov, A.; Tiao, G.; Kwiatkowski, D.J.; Rosenberg, J.E.; Van Allen, E.M.; D'Andrea, A.D.; et al. Somatic ERCC2 mutations are associated with a distinct genomic signature in urothelial tumors. *Nat. Genet.* **2016**, *48*, 600–606. [CrossRef]
30. Vieira, V.C.; Soares, M.A. The Role of Cytidine Deaminases on Innate Immune Responses against Human Viral Infections. *BioMed Res. Int.* **2013**, *2013*, 683095. [CrossRef]
31. O'Callaghan, D.S.; O'Donnell, D.; O'Connell, F.; O'Byrne, K.J. The role of inflammation in the pathogenesis of non-small cell lung cancer. *J. Thorac. Oncol.* **2010**, *5*, 2024–2036. [CrossRef] [PubMed]
32. Qiu, F.; Liang, C.L.; Liu, H.; Zeng, Y.Q.; Hou, S.; Huang, S.; Lai, X.; Dai, Z. Impacts of cigarette smoking on immune responsiveness: Up and down or upside down? *Oncotarget* **2017**, *8*, 268–284. [CrossRef] [PubMed]
33. Nomi, K.; Hayashi, R.; Ishikawa, Y.; Kobayashi, Y.; Katayama, T.; Quantock, A.J.; Nishida, K. Generation of functional conjunctival epithelium, including goblet cells, from human iPSCs. *Cell Rep.* **2021**, *34*, 108715. [CrossRef] [PubMed]
34. Shao, M.X.G.; Nakanaga, T.; Nadel, J.A. Cigarette smoke induces MUC5AC mucin overproduction via tumor necrosis factor-α-converting enzyme in human airway epithelial (NCI-H292) cells. *Am. J. Physiol.-Lung Cell. Mol. Physiol.* **2004**, *287*, L420–L427. [CrossRef] [PubMed]
35. Bilsborough, J.; Viney, J.L. GPR15: A tale of two species. *Nat. Immunol.* **2015**, *16*, 137–139. [CrossRef]
36. Akinleye, A.; Rasool, Z. Immune checkpoint inhibitors of PD-L1 as cancer therapeutics. *J. Hematol. Oncol.* **2019**, *12*, 92. [CrossRef]
37. Ai, L.; Chen, J.; Yan, H.; He, Q.; Luo, P.; Xu, Z.; Yang, X. Research Status and Outlook of PD-1/PD-L1 Inhibitors for Cancer Therapy. *Drug Des. Dev. Ther.* **2020**, *14*, 3625–3649. [CrossRef]

38. Atchison, W.D. Effects of toxic environmental contaminants on voltage-gated calcium channel function: From past to present. *J. Bioenerg. Biomembr.* **2003**, *35*, 507–532. [CrossRef]
39. Schuller, H.M. Is cancer triggered by altered signalling of nicotinic acetylcholine receptors? *Nat. Rev. Cancer* **2009**, *9*, 195–205. [CrossRef]
40. Goldfarbmuren, K.C.; Jackson, N.D.; Sajuthi, S.P.; Dyjack, N.; Li, K.S.; Rios, C.L.; Plender, E.G.; Montgomery, M.T.; Everman, J.L.; Bratcher, P.E.; et al. Dissecting the cellular specificity of smoking effects and reconstructing lineages in the human airway epithelium. *Nat. Commun.* **2020**, *11*, 2485. [CrossRef]
41. Nemajerova, A.; Kramer, D.; Siller, S.S.; Herr, C.; Shomroni, O.; Pena, T.; Gallinas Suazo, C.; Glaser, K.; Wildung, M.; Steffen, H.; et al. TAp73 is a central transcriptional regulator of airway multiciliogenesis. *Genes Dev.* **2016**, *30*, 1300–1312. [CrossRef] [PubMed]
42. Licciardi, P.V.; Karagiannis, T.C. Regulation of Immune Responses by Histone Deacetylase Inhibitors. *ISRN Hematol.* **2012**, *2012*, 690901. [CrossRef]
43. Schamberger, A.C.; Staab-Weijnitz, C.A.; Mise-Racek, N.; Eickelberg, O. Sci RepCigarette smoke alters primary human bronchial epithelial cell differentiation at the air-liquid interface. *Sci. Rep.* **2015**, *5*, 8163. [CrossRef] [PubMed]
44. Newman, A.M.; Steen, C.B.; Liu, C.L.; Gentles, A.J.; Chaudhuri, A.A.; Scherer, F.; Khodadoust, M.S.; Esfahani, M.S.; Luca, B.A.; Steiner, D.; et al. Determining cell type abundance and expression from bulk tissues with digital cytometry. *Nat. Biotechnol.* **2019**, *37*, 773–782. [CrossRef] [PubMed]
45. Haswell, L.E.; Hewitt, K.; Thorne, D.; Richter, A.; Gaça, M.D. Cigarette smoke total particulate matter increases mucous secreting cell numbers in vitro: A potential model of goblet cell hyperplasia. *Toxicol. In Vitro* **2010**, *24*, 981–987. [CrossRef]
46. Lumsden, A.B.; McLean, A.; Lamb, D. Goblet and Clara cells of human distal airways: Evidence for smoking induced changes in their numbers. *Thorax* **1984**, *39*, 844–849. [CrossRef]
47. Damiá, A.d.e.D.; Gimeno, J.C.; Ferrer, M.J.; Fabregas, M.L.; Folch, P.A.; Paya, J.M. Arch BronconeumolA study of the effect of proinflammatory cytokines on the epithelial cells of smokers, with or without COPD. *Arch. Bronconeumol.* **2011**, *47*, 447–453. [CrossRef]
48. de la Iglesia, J.V.; Slebos, R.J.; Martin-Gomez, L.; Wang, X.; Teer, J.K.; Tan, A.C.; Gerke, T.A.; Aden-Buie, G.; van Veen, T.; Masannat, J.; et al. Effects of Tobacco Smoking on the Tumor Immune Microenvironment in Head and Neck Squamous Cell Carcinoma. *Clin. Cancer Res.* **2020**, *26*, 1474–1485. [CrossRef]
49. Kim, V.; Oros, M.; Durra, H.; Kelsen, S.; Aksoy, M.; Cornwell, W.D.; Rogers, T.J.; Criner, G.J. Chronic bronchitis and current smoking are associated with more goblet cells in moderate to severe COPD and smokers without airflow obstruction. *PLoS ONE* **2015**, *10*, e0116108. [CrossRef] [PubMed]
50. Bryant, H.E.; Schultz, N.; Thomas, H.D.; Parker, K.M.; Flower, D.; Lopez, E.; Kyle, S.; Meuth, M.; Curtin, N.J.; Helleday, T. Specific killing of BRCA2-deficient tumours with inhibitors of poly(ADP-ribose) polymerase. *Nature* **2005**, *434*, 913–917. [CrossRef]
51. Davies, H.; Glodzik, D.; Morganella, S.; Yates, L.R.; Staaf, J.; Zou, X.; Ramakrishna, M.; Martin, S.; Boyault, S.; Sieuwerts, A.M.; et al. HRDetect is a predictor of BRCA1 and BRCA2 deficiency based on mutational signatures. *Nat. Med.* **2017**, *23*, 517–525. [CrossRef] [PubMed]
52. Buisson, R.; Lawrence, M.S.; Benes, C.H.; Zou, L. APOBEC3A and APOBEC3B Activities Render Cancer Cells Susceptible to ATR Inhibition. *Cancer Res.* **2017**, *77*, 4567–4578. [CrossRef] [PubMed]
53. Wang, S.; Jia, M.; He, Z.; Liu, X.S. APOBEC3B and APOBEC mutational signature as potential predictive markers for immunotherapy response in non-small cell lung cancer. *Oncogene* **2018**, *37*, 3924–3936. [CrossRef] [PubMed]
54. Brady, S.W.; Gout, A.M.; Zhang, J. Therapeutic and prognostic insights from the analysis of cancer mutational signatures. *Trends Genet.* **2022**, *38*, 194–208. [CrossRef] [PubMed]
55. Chen, D.L.; Li, Q.Y.; Tan, Q.Y. Smoking history and the efficacy of immune checkpoint inhibitors in patients with advanced non-small cell lung cancer: A systematic review and meta-analysis. *J. Thorac. Dis.* **2021**, *13*, 220–231. [CrossRef] [PubMed]
56. Boichard, A.; Tsigelny, I.F.; Kurzrock, R. High expression of PD-1 ligands is associated with *Kataegis* Mutat. Signat. APOBEC3 Alterations. *OncoImmunology* **2017**, *6*, e1284719. [CrossRef]
57. Kim, V.; Jeong, S.; Zhao, H.; Kesimer, M.; Boucher, R.C.; Wells, J.M.; Christenson, S.A.; Han, M.K.; Dransfield, M.; Paine, R.; et al. Current smoking with or without chronic bronchitis is independently associated with goblet cell hyperplasia in healthy smokers and COPD subjects. *Sci. Rep.* **2020**, *10*, 20133. [CrossRef]
58. Huang, X.; Wojtowicz, D.; Przytycka, T.M. Detecting presence of mutational signatures in cancer with confidence. *Bioinformatics* **2018**, *34*, 330–337. [CrossRef]
59. Love, M.I.; Huber, W.; Anders, S. Moderated estimation of fold change and dispersion for RNA-seq data with DESeq2. *Genome Biol.* **2014**, *15*, 550. [CrossRef]
60. Travaglini, K.J.; Nabhan, A.N.; Penland, L.; Sinha, R.; Gillich, A.; Sit, R.V.; Chang, S.; Conley, S.D.; Mori, Y.; Seita, J.; et al. A molecular cell atlas of the human lung from single-cell RNA sequencing. *Nature* **2020**, *587*, 619–625. [CrossRef]

Article

A Machine Learning Approach to Identify the Importance of Novel Features for CRISPR/Cas9 Activity Prediction

Dhvani Sandip Vora [1], Yugesh Verma [1] and Durai Sundar [1,2,*]

1 Department of Biochemical Engineering and Biotechnology, Indian Institute of Technology Delhi, Hauz Khas, New Delhi 110016, India
2 Yardi School of Artificial Intelligence, Indian Institute of Technology Delhi, Hauz Khas, New Delhi 110016, India
* Correspondence: sundar@dbeb.iitd.ac.in

Abstract: The reprogrammable CRISPR/Cas9 genome editing tool's growing popularity is hindered by unwanted off-target effects. Efforts have been directed toward designing efficient guide RNAs as well as identifying potential off-target threats, yet factors that determine efficiency and off-target activity remain obscure. Based on sequence features, previous machine learning models performed poorly on new datasets, thus there is a need for the incorporation of novel features. The binding energy estimation of the gRNA-DNA hybrid as well as the Cas9-gRNA-DNA hybrid allowed generating better performing machine learning models for the prediction of Cas9 activity. The analysis of feature contribution towards the model output on a limited dataset indicated that energy features played a determining role along with the sequence features. The binding energy features proved essential for the prediction of on-target activity and off-target sites. The plateau, in the performance on unseen datasets, of current machine learning models could be overcome by incorporating novel features, such as binding energy, among others. The models are provided on GitHub (GitHub Inc., San Francisco, CA, USA).

Keywords: CRISPR/Cas9; genome editing; machine learning; SHAP values; binding energy; off-targets

Citation: Vora, D.S.; Verma, Y.; Sundar, D. A Machine Learning Approach to Identify the Importance of Novel Features for CRISPR/Cas9 Activity Prediction. *Biomolecules* **2022**, *12*, 1123. https://doi.org/10.3390/biom12081123

Academic Editors: Cameron Mura and Lei Xie

Received: 13 July 2022
Accepted: 10 August 2022
Published: 16 August 2022

1. Introduction

Clustered regularly interspersed short palindromic repeats (CRISPR) and its associated nuclease Cas9 constitute a versatile and reprogrammable genome editing mechanism that has been repurposed as a widely used tool [1–3]. The single guide RNA can be customised to target the DNA at any location by changing the 20 nucleotides "spacer". This spacer is designed to complement the "protospacer" region in the DNA, at which the Cas9 nuclease would create a double-stranded break [4]. A 3-nucleotide protospacer adjacent motif (PAM) is a prerequisite for probing and cleaving the target DNA by this two-component protein–RNA system [1]. The PAM site is generally of the form of NGG (where N is any nucleotide) for the *Streptococcus pyogenes*-derived Cas9 (SpCas9) protein [5,6]. The SpCas9 is a multidomain protein consisting of (i) three recognition domains that bind to the RNA and DNA strands, (ii) two nuclease domains to cleave each of the DNA strands, (iii) a PAM interaction domain, and (iv) an arginine-rich helix which acts as a linker [7]. Although this system is a facile and flexible genome editing tool, there are two critical design problems associated with this system: (i) designing a guide RNA with good activity at the intended target region and (ii) ensuring that the selected guide does not show activity at similar unintended sites, or in other words, has low off-target activity [8,9]. The presence of the Cas9 off-target activity has hindered clinical applications of Cas9, which is a significant area of focus for CRISPR/Cas9 study.

Great strides have been taken to understand the mechanism of action and, consequently, develop design rules to aid experimentalists in optimising guides for the intended applications. The field has benefited greatly over the past decade, majorly because of the development of multiple methods to detect Cas9 off-target activity in vitro and in situ within the cell [10–15]. Off-target detection techniques have enabled the identification of empirical rules that seem to drive off-target identification and activity by allowing analyses of various off-targets generated for multiple guides under different conditions [16–19].

The availability of an experimentally derived structure and sequence of target and off-target data has allowed computational studies to understand Cas9 activity. Many prediction algorithms have been proposed to achieve each of the tasks mentioned above, qualitative algorithms and scoring schemes to rank guides by on-target efficiency and off-target predictions [20,21]. Most algorithms are based on sequence features—number and position of mismatches (PAM proximal ends are less likely to tolerate mismatches, while the distal ends report more tolerance for mismatches) [17]. Many machine learning models have been built to predict the performance of guides and the prediction of their respective off-targets based on rules depending on the system's various sequence and structural features [17,22–25], yet there is a gap between the predictions and experimentally observed results. Popular machine learning models are based on features such as the sequence at the cut-site, the number of mismatches, experimentally validated efficiency and off-target activity of the guides. Recently, deep learning models have been reported, which are trained on large-scale datasets, and some have included novel features for validation; for example, DeepCRISPR, one of the earlier attempts at building a deep learning-based tool for prediction, introduced four epigenetic features apart from the sequence features [26]. DeepCpf1 is a convolution neural net (CNN) model, and CRISPcut is a rule-based model, both of which include chromatin accessibility as an additional feature to improve the prediction confidence [27,28]. CRISPcut and AttnToCrispr are prediction algorithms that also have included the cell-line information as features while predicting off-targets and on-target efficiency, respectively [28,29]. The addition of new and important features has, in each case, improved the model performance and confidence in the predictions. Recent studies have reported that DNA enthalpy (a proxy for the stability of the DNA duplex) and DNA-RNA duplex energy parameters play an essential role in predicting on-target efficiency and off-target activity [24,30]. This study presents two new features that prove to be important in future prediction algorithm designs: MMGBSA-based binding energy for (i) DNA and guide RNA, and (ii) Cas9 protein–nucleic acid recognition domain and the DNA-RNA hybrid.

2. Materials and Methods

2.1. Data Assembly

The data used for model training and validation were obtained from published methods of CRISPR/Cas9 off-target site prediction (CRISPcut) [28] and detection (CIRCLE-seq) [11,28] (SRA identifier SRP103697). The predictions obtained from CRISPcut, run with default parameters, for the 11 guide RNAs used in CIRCLE-seq were used to obtain a comprehensive list of potential off-target sites in the genome for the corresponding cell lines used in the CIRCLE-seq experiment. The experimentally validated off-target sites were called the positive dataset, while the predictions not validated experimentally were referred to as the negative dataset. All predictions obtained from CRISPcut were analysed for chromatin accessibility; only accessible sequences were selected since earlier studies have established the importance of this feature [31–33]. The data assembly and selection are summarised in Table S3. The cleavage efficiency obtained from the CIRCLE-seq dataset for all reported off-targets was normalised to fit a uniform scale. The features used for model training are detailed in Table S4.

2.2. Predictive Features

Multiple predictive features were calculated for each of the sequences—mismatch position, number of mismatches, mismatch in PAM, type of mismatch (transition, transversion or indel), cell line information, percentage GC for the protospacer, percentage GC in the seed region, chromosome number, DNA strand information and the two new proposed binding energy features. Two MMGBSA-based binding energy features were considered—

dG(REC3:hybrid) and dG(DNA:RNA). The dG(REC3:hybrid) was calculated between the REC3 domain of SpCas9 and the 20-nucleotide DNA-RNA hybrid. The binding energy of the 20-nucleotide RNA and target DNA strands was calculated as dG(DNA:RNA). The MMGBSA calculations were carried out using the Schrödinger Maestro suite's Prime utility after pre-processing and the restrained minimisation of the complexes [34,35].

2.3. MMGBSA Binding Energy Calculation

The structure used as a template was obtained from RCSB PDB (ID: 4UN3). The REC3 domain was selected (residues 447–718) along with the 20 nucleotides of the target DNA and the 20 nucleotides of the guide RNA. The PyMOL nucleic acid mutagenesis tool was used to create all target and off-target systems from the template [36]. The structures were imported in the Schrödinger Maestro suite and preprocessed, hydrogen bonds were optimised, and restrained minimisation was carried out before performing MMGBSA calculation using the Prime utility [34,37]. The energies of molecular mechanics when combined with the generalized Born and surface area continuum solvation (MMGBSA) is a popular approach to estimate the binding free energy between biomolecules. MMGBSA is an intermediate in both computational costs and accuracy, widely applied for various systems [38–40]. The free energy is calculated and summed over solvation energy, gas-phase energy and entropic contributions. The REC3 domain was chosen as the receptor and the DNA-RNA hybrid was used as the ligand for the dG(REC3:hybrid) feature; DNA was selected as the receptor for the dG(DNA:RNA) feature.

2.4. Mann–Whitney U Test

The Mann–Whitney U test, also called the Mann–Whitney–Wilcoxon test, is a non-parametric test to compare differences of a variable between two groups when the variable in question is not normally distributed. The test was performed on the dataset for both dG features, the values of which served as input for the test enabled by the Pingouin Python package (0.5.2) [41]. The common language effect size was calculated using a Python script. The output is a U statistic and p-value, which indicates whether the groups show stochastic equality or not. The test is also robust to outliers. The U test was used to determine if the dG values for the experimentally validated off-targets (positive) and the non-validated predictions (negative) were statistically different.

2.5. Machine Learning Model Implementation

Two machine learning models were implemented:

(1) A random forest regression model on a small fraction of the CIRCLE-seq dataset with the dependant variable as normalised cleavage frequencies following their normalisation;
(2) A random forest classification model on a fraction of the CIRCLE-seq and CRISPcut derived datasets with the dependant variable being whether the sequence is cleaved experimentally or not.

The regression model was to determine whether the binding energy features significantly impact the cleavage frequency of the off-target sequences. The classification model would help determine if the energy features play a role in differentiating experimentally unlikely predictions from experimentally validated off-target sequences. Since the dG values calculation was computationally intensive and time consuming, the dataset consisted of 186 positive examples and 126 negative examples. However, the sequences were collected manually to ensure sufficient diversity in cleavage frequency, the number of mismatches, and other sequence features that were previously reported as significant. The classification model was implemented to understand if the features were sufficient to differentiate between experimentally likely predictions and those that are not.

Multiple machine learning models were tested with varying parameters; the best performing models were reported. All models evaluated were implemented using the scikit-learn package in Python [42].

2.6. Sampling Data for Training

Initial training was performed on a 75% train set, and assessment of the model performance was measured on the 25% held-out test dataset. The best performing model architecture was selected. For analysis of feature importance, since the dataset was limited, training was carried out again with 5-fold cross validation to ensure that the unbalanced dataset was not a limiting factor for model performance. The 5-fold cross-validation was repeated to ensure the absence of bias for both models.

2.7. Assessing Model Performance

The regression model's performance was evaluated by comparing the mean squared error (MSE), mean absolute error (MAE) and the R-squared values, and the better performing model was selected for feature importance determination and feature ranking. The MAE and MSE measure the difference between the model predictions and actual observations; hence, the ideal score is 0. The R-squared value is a correlation coefficient measuring a linear correlation between two continuous variables. The variance weighted measure is an explanation of the variance in the model output, the best score being 1.

The classification model was assessed using its confusion matrix:

$$\mathrm{M} = \begin{bmatrix} TP & FN \\ FP & TN \end{bmatrix}$$

where *TP* stands for true positive, *FP* for false positive, *FN* for false negative and *TN* for true negative. The accuracy of a model is defined as

$$\text{Accuracy} = \frac{TP + TN}{TP + TN + FP + FN}$$

The recall is the measure of how many actual positives the model can capture, while the precision is how many of the predicted positives are correct. The precision–recall curve, a standard evaluation criterion for a classification model, is based on the following definitions:

$$\text{Recall} = \frac{TP}{TP + FN}$$

$$\text{Precision} = \frac{TP}{TP + FP}$$

The F1 score, or F-measure, is the harmonic mean of the precision and recall, conveying a balance between the two. It is defined as

$$\text{F1 score} = \frac{2 * Recall * Precision}{Recall + Precision}$$

2.8. Identifying Feature Importance

Interpreting the features that impact a machine learning model's outcome is important for enabling the predictions' validation. In the regression and classification models used, the feature set is small, and so is the dataset; hence, each feature's influence must be understood. Hence, Shapley additive explanations (SHAP) values were implemented using the shap library in Python [43]; the TreeExplainer utility was used to analyse the random forest regressor output and to describe the model output of the random forest classifier [44]. The shap method employs an explanatory model with feature weights to explain relative feature importance and is adapted from game theory. It is to be noted that shap values do not indicate causality.

3. Results

3.1. Data Assembly and Processing

The guide RNAs and their respective off-targets were obtained from the CIRCLE-seq data [11]. The data obtained from the prediction algorithm CRISPcut were checked for the number of sites predicted for each guide RNA input [28]. The number of sites predicted hold little correlation with the experimental sites (Figure S1a). However, when the chromatin accessible sites were selected and compared, a sufficient correlation was obtained between the number of sites predicted and the number of sites confirmed experimentally (Figure S1b). Moreover, since chromatin accessibility has been shown in earlier studies to be an important feature, sequences selected for the model were only from the accessible sequences' subset [31,45].

The sequences selected from the CIRCLE-seq (positive dataset) and CRISPcut predictions, but not found in the experimentally validated datasets (negative dataset), were selected manually to ensure that the other features, such as the number of mismatches, cleavage frequencies and cell lines, were sufficiently represented. The features included for the model prediction were calculated using Python scripts, except the binding energy features, which were calculated using the method described. The resulting dataset had 40 features and 312 data points.

To determine if the features were correlated with each other, correlation analysis was carried out, and the results are shown in Figure S2. No significant correlation between the features was observed. The correlation islands observed were between the cell lines that were one-hot encoded and are hence mutually exclusive. A high correlation was expected for the total mismatches and protospacer mismatches (referred to as number of mismatches, #mm); the same can be stated for total PAM mismatches and types of PAM mismatches—transversion or transition type. Hence, the features selected were unique and not redundant.

3.2. Statistical Analysis of the Binding Energy Features

To determine if the values of the binding energies, by themselves, could be used to differentiate between the positive and negative datasets, the Mann–Whitney U test was carried out to compare the values between the two sets (Supplementary Table S1). The Mann–Whitney U test is a non-parametric test to check if a feature's values are larger for one of the two populations being compared; it is the non-parametric equivalent of the unpaired *t* test.

The values of the two binding energy features were compared for the positive and negative datasets, where the H_0 hypothesis was that the values for the two groups are equal. Hence, the H_0 hypothesis's rejection indicated that the difference between randomly selected values of the features from both populations is big enough to be statistically significant (Table S1). The rank–biserial correlation coefficient indicated the difference between total amount of favourable and unfavourable evidence. The common language effect size is the probability that a random value from Group 1 is greater than a random value from Group 2.

The Mann–Whitney U (MWU) test indicated that the values of the two binding energy features—dG(REC3:hybrid) and dG(DNA:RNA)—have differing values for the positive and negative datasets (Table S1). Moreover, it is evident from the MWU test that a random value from the negative dataset is likely to be higher than a random value from the positive dataset. However, since the effect size values are low, the features cannot solely be used as a distinguishing factor for the negative and positive datasets. The difference in population means the calculation was not enough to reliably call these features distinguishing.

3.3. Regression Model Selection and Performance Assessment

Linear, quadratic, cubic, multi-layer perceptrons and random forest regressors were implemented with varying parameters and random states to determine the best performing model. The dependent variable was the cleavage frequency for the off-target sequences obtained from the CIRCLE-seq dataset. The performance measured in the *R*-squared value, mean absolute error, mean squared error and variance-weighted measure is summarised in Table 1. The random forest regressor was chosen based on its superior performance on the dataset, compared to the other models tested. The random forest algorithm is known for its ability to predict well on tabular data, as is the case here. The perceptron was also tested for multiple nodes in one and two hidden layers trained till convergence; however, it failed to outperform the random forest regressor.

Table 1. Summary of model performances. All values shown are for the test dataset.

Metrics	Regressor							
	Linear	Quadratic	Cubic	Decision Tree	SVR	MLP	XGBoost	Random Forest
Mean Absolute Error	0.19	1.23	0.51	0.21	0.19	0.19	0.17	0.06
Mean Squared Error	0.07	3.49	0.54	0.09	0.08	0.07	0.06	0.01
Root MSE	0.26	1.87	0.73	0.29	0.28	0.26	0.24	0.08
R-squared value	0.37	−32.26	−4.15	0.18	0.27	0.42	0.47	0.94
Variance weighted	0.38	−31.82	−4.13	0.24	0.27	0.47	0.55	0.94

The various model metrics listed in the first column are given for the regression models tested. For the random forest regressor, the metrics are comparatively much better than the other three. It was selected for feature importance analysis. SVR stands for support vector regressor. MSE stands for mean squared error. The values reported for each regressor is after the optimisation of individual models.

The best performing regression model, the random forest regressor, was initialized on various random states and number of trees (as shown in Figure S3). The model with the maximum R-squared and minimum mean absolute error (MAE) was selected for further analysis, following which 5-fold cross-validation was performed. The resulting mean squared error (MSE) remained at 0.05, standard deviation (STD) was 0.01, and the R^2 score was 0.92, indicating that the chosen model was robust.

3.4. Explaining Feature Importance for the Random Forest Regressor

The importance and magnitude of the impact of the features on the model output were explored in detail since the aim of the study was to establish the importance of the two features proposed, namely the energy of binding of the REC3 domain of Cas9 to the 20 nucleotide hybrid of the target DNA and guide RNA-dG(REC3:hybrid), and the binding energy of the 20 nucleotide DNA to the guide RNA strand-dG(DNA:RNA). The variable importance plot (Figure 1a) generated by implementing SHAP [43,44,46] lists the most important features in descending order. The ones on top contributed the most to the model output and hence, have high predictive capability.

The SHAP values also help determine the relationship of the features to the output. The SHAP variable importance plot (Figure 1b) ranked variables in descending order of importance, and the horizontal spread indicated the effect of the value and the corresponding higher or lower prediction. Each dot is a value for an instance in the data, and the colour indicates a higher or lower value for that instance. While distance (total mismatches in the sequence) and #mm (mismatches in the protospacer region) were redundant features and showed a similar impact on output, Figure 1 shows that the low binding energy of the DNA-RNA hybrid, dG(DNA:RNA), had a high impact on model output; while the binding energy of the Cas9 REC3 domain to the DNA-RNA hybrid, dG(REC3:hybrid) was negatively correlated with the model output. Figure 1 also indicates that the presence of mismatch at the 6th position played an important role in determining the model output.

The SHAP variable importance plot (Figure 2) takes three values: a base value, SHAP values, and the matrix of feature values. The base value was the average or expected model output, and the SHAP value of a feature and the value of the feature at that instance determined in which direction the features "push" the model output. The output value highlighted is the model output for this instance. The features in red direct the output higher, while those in blue push the predictions lower. The SHAP plot for three instances are shown; since each feature plays a different role for each instance, it is essential to consider the local as well as global relevance of the feature.

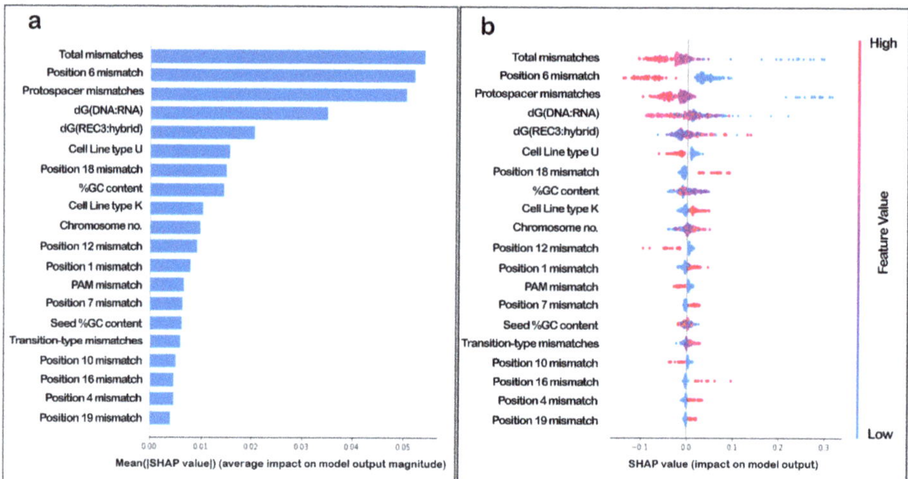

Figure 1. SHAP variable importance plots. (**a**) The plot arranges features in decreasing order of magnitude of impact on model output. (**b**) The features are listed in decreasing order of importance, the dots are coloured according to value (in a gradient from high to low, as red to blue) and the impact for each instance is plotted horizontally. The spread indicates impact on model output, and the colour indicates feature value for that output.

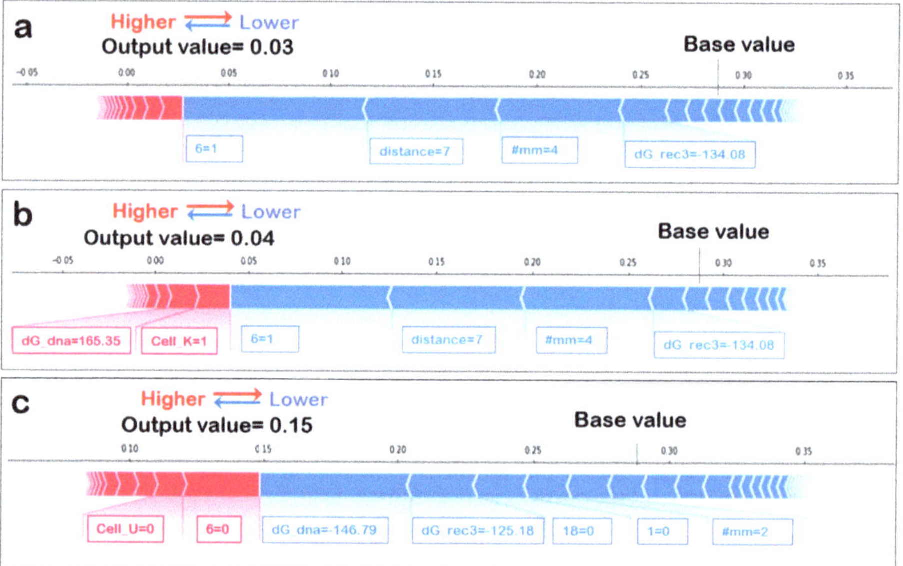

Figure 2. SHAP variable importance plot. The SHAP variable impact on outcome for singular datapoints are shown. Examples shown are explainer plots for dataset indices (**a**) 0, (**b**) 1 and (**c**) 2. The base value labelled in the figure in influenced by varying degrees by the features shown in the diagrams and the output value (shown in bold) was obtained. The features SHAP values are written alongside the features if it causes an increase in base value it is shown in red otherwise in blue.

The SHAP dependence plot (Figure 3) describes partial dependence between a feature selected, and the reference feature was chosen automatically by the script with which the chosen feature interacts the most. The dots mark each instance of the chosen variable, and the colour of the dots indicate the value of the reference feature for that instance. In Figure 3a,b, there is no clear trend between the two features; however, in Figure 3a the absence of a mismatch at position 4 and the lower values of dG(DNA:RNA) have a higher impact on the model output. Figure 3b shows that the partial dependence between the two features is not significant and no trend can be observed. The spread of the plot indicates the relationship between the two features. As in Figure 3c, the vertical dispersion at a particular value shows the interaction effect between the two features. Moreover, an approximately negative correlation exists between the variables, and a smaller Hamming distance (total mismatches in the off-target) would have more influence on the model output; it also corresponds with lower values of dG(DNA:RNA).

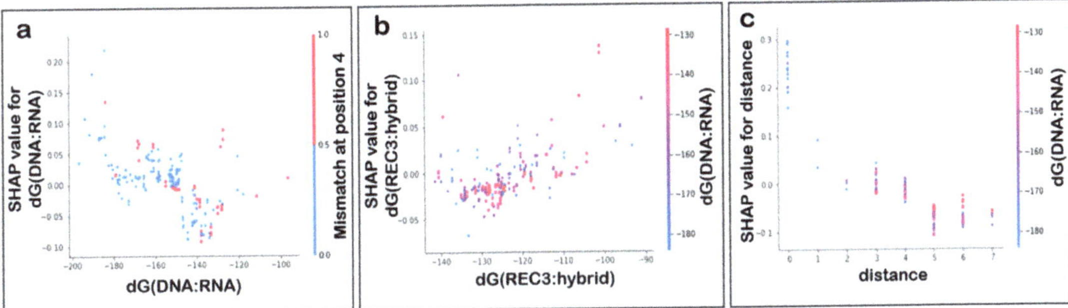

Figure 3. SHAP feature dependence plot. The plots show dependence between (**a**) dG(DNA:RNA) and a mismatch at position 4, (**b**) dG(REC3:hybrid) and dG(DNA:RNA) and (**c**) distance and dG(DNA:RNA). The vertical axis marks the SHAP values for the chosen feature, while the horizontal axis shows spread of the values of the feature. The reference feature was selected by the algorithms automatically and was used to colour the dots that indicate value of the primary feature for an instance. No clear trend can be observed in (**a**,**b**). In (**c**), vertical clusters at individual values indicate a correlation with dG(DNA:RNA) values, and the plot also shows a negative correlation of the values of the distance with the output variable.

3.5. Classifier Model Selection and Performance Assessment

The classifier models were built to study the contribution of the binding energy features to machine learning models that can distinguish between positive (sequences that are off-target sites in experiments) and negative datasets (sequences predicted to be off-targets but were not found in experiments). Various classification models were trained on the dataset, optimised for each type of model (the best performing model's accuracy summarised in Table S2). Since the random forest classifier performed well on the 25–75 test-train split, the model was evaluated after 5-fold cross validation. The classifier yielded good accuracy and was implemented for further analysis. The model metrics for the random forest classifier model are summarised in Table 2.

Table 2. Model performance of the random forest classifier, measured on test dataset.

Model Metrics	Score on Test Data	Overall Score
Accuracy	0.86	0.97
Precision	0.88	0.98
Recall	0.94	0.96
F1 score	0.91	0.97

The accuracy, precision, recall and F1 scores are calculated as mentioned in the Methods section. The accuracy reported is after 5-fold cross validation. The overall score is for combined test and train datasets.

The performance of the random forest classifier was tested using various parameters as shown in Figure 4. The model predicted the correct classes for each label reliably. The precision–recall curve and receiver operating characteristic (ROC) cover over 95% area under the curve, indicating a robust classification model. The next best performing model (support vector machine classifier) did not perform better, even on 5-fold cross validation, and hence was not evaluated further. Since the study aimed not to build an off-target determination model, but rather discern the importance of energy features, more complex models were not tested.

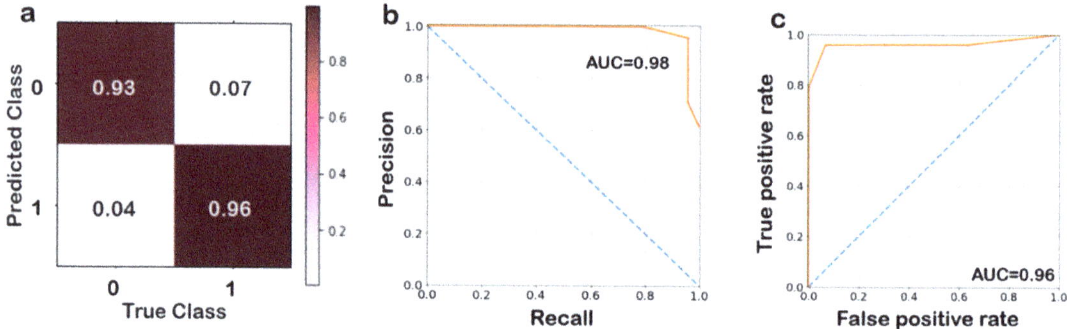

Figure 4. (**a**) Confusion matrix for the random forest classifier, vertical axis is for predicted labels and the horizontal axis states the true labels. The values are ratios of the number of instances predicted to the total instances in the class. (**b**) Precision–recall curve, shown in orange which has an area under the curve of 0.98 for the whole dataset, (**c**) receiver operating characteristic (ROC) also shown in orange for the test dataset, which plots the true positive rate against the false positive rate. The area under the curve (AUC) is 0.96. The dashed blue line across the diagonal shows 50% accuracy.

3.6. *Explaining Feature Importance for the Classifier*

The importance of the features in a well-performing classification model that can learn the difference between the positive and negative datasets will determine if the binding energy features play a significant role in determining the model output. The SHAP value plots for each instance are not shown for lack of space, but three examples are shown in Figure 5. The base value, determined as the average from the training dataset, is influenced by the features listed in order of magnitude of impact. Features in blue lower the output, while features in red increase the output. In all instances, energy features play an important role. However, since feature importance for each datapoint varies, it is important to see each feature's global impact, which is shown in Figure 6.

This SHAP value plot ranks the features in decreasing order of importance, while the spread across the horizontal determines the impact on the model for higher values (in red) and lower values (in blue). As is shown in Figure 6, the energy features are ranked high. Lower values of both binding energies are characteristic of the positive dataset. Hence, lower values of the binding energy tend to result in a positive impact on the model output; here, it is the classification in the positive dataset.

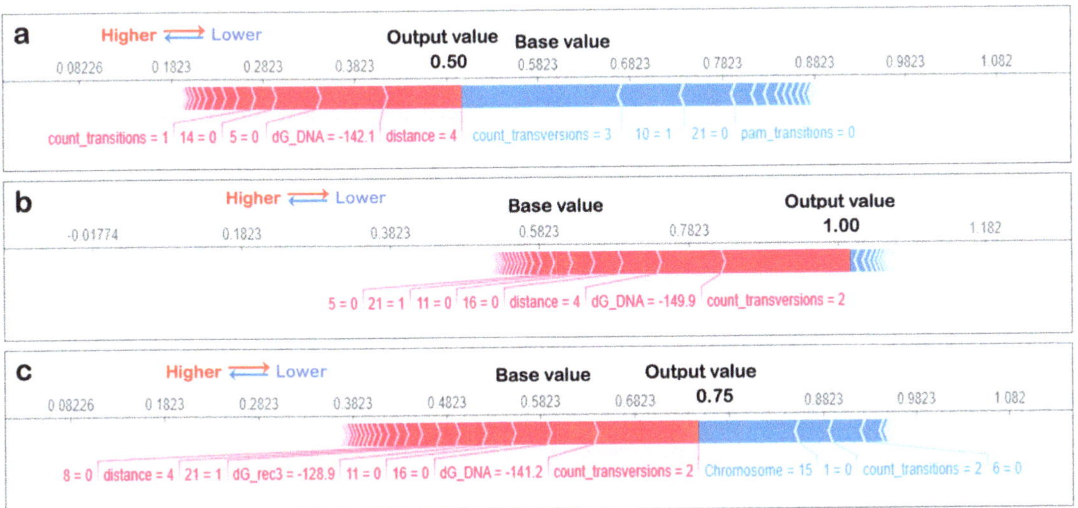

Figure 5. SHAP value plots for singular datapoints. Examples shown are for dataset indices (**a**) 10, (**b**) 17 and (**c**) 21, and are chosen randomly. The base value shown increases by features shown in red and decreases because of features shown in blue. Each feature impacts the value in magnitude indicated by SHAP values labelled alongside for each instance.

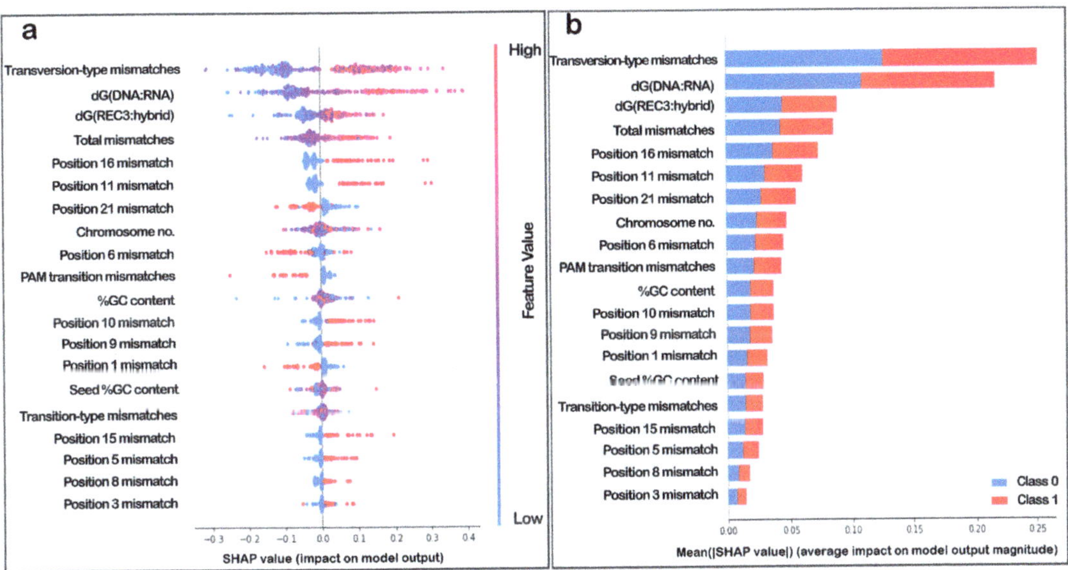

Figure 6. (**a**) SHAP value plot indicating global impact n model output. Each dot is an instance for a datapoint, the colour represents if the value for that instance is low (blue) or high (red). The spread indicates the magnitude of impact on the model output. (**b**) SHAP summary plot shows the impact of the features on each model output, negative class shown in blue and positive class shown in red, as stacked bars, in decreasing order of impact on output.

4. Discussion

The accurate prediction of CRISPR/Cas9 activity is crucial to not only designing experiments for various applications but also understanding the mechanism of Cas9 activity in vivo. Computational methods for predicting activity, off-targets and guide design have advanced significantly in recent times, yet there remains room for improvement regarding precision and accuracy. Prediction models would also benefit from improved and more sensitive Cas9 off-target detection methods to better distinguish between sequences likely to be acted upon by Cas9 (here, the positive dataset). This study reported that the incorporation of novel features allows for creating reliable prediction models. Moreover, the identification of novel features also sheds light on the factors influencing Cas9 activity in vivo.

The two major binding events responsible for Cas9 activity are (1) the binding of the Cas9 protein to the guide RNA, allowing DNA interrogation for complementary sequences, (2) followed by binding to the complementary sequence, which allows nuclease activation and a subsequent DNA double-stranded break [47]. Significantly accelerated by the availability of X-ray and cryo-EM structures, computational methods, such as QM/MM and molecular dynamics (MD), have elucidated the pre-catalytic and catalytic structures of Cas9 [48,49]. Enhanced MD simulations have shed light on the concerted mechanisms of HNH and RuvC domain activities [50–52]. The HNH domain via an Mg^{2+} ion cuts the target strand, while the RuvC domain houses two metal ions coordinated by conserved residues, which mediate a break in the non-target strand [52]. The varying tolerance of the mismatches across the guide-target heteroduplex has also been investigated [18,53,54]. The REC3 domain is known to interact with the guide RNA-target DNA complex, investigate the complementarity between the two, and tolerate mismatches [55,56]. Mismatches were seen to be tolerated towards the centre of the guide–target hybrid [53]. In contrast, mismatches towards the end of the hybrid induced an extended opening of the heteroduplex and leading to a conformational lock with the "L2" loop region [54]. Hence, the interactions of the guide RNA with target DNA and the heteroduplex with the REC3 domain of Cas9 protein have been shown to play a decisive role in nuclease activation, leading to Cas9 activity. The introduction of mismatches alters the interactions, leading to altered Cas9 activity. Understanding the factors that govern the RNA:DNA interactions is critical to elucidating biological function that it is involved in [57–60]. Hence, to quantify the interactions, DNA-RNA hybrid binding energy and Cas9-hybrid binding energy were estimated and analysed. The scores were then included as features alongside sequence features, and machine learning models were built for Cas9 activity prediction. Well-performing models were selected to analyse the importance of the new energy-based features, if any.

The random forest algorithm outperformed the others tested on both classification and regression tasks. The improved performance could be attributed to the limited number of features on each split. When compared to individual decision trees, which have a higher bias, random forests tend to perform better because of the variance reduction due to bagging. The features used, as the results describe, have minimum redundancy. The energy features prove vital in driving model output in both regression and classification tasks. This feature importance was also observed in the second-best performing classification model: a support vector-based machine classifier (a second regressor was not evaluated due to the performance being subpar, not reliable enough to study feature importance). The importance of the number of mismatches in the seed region has already been established in multiple studies [61,62]. Interestingly, a higher number of transversions was shown not to be tolerated in the experimental dataset, indicating a preference in the sequences (Figure 6a). However, a bigger dataset is required to be tested to establish this. The "distance" feature's trend may also be inferred intuitively since lower values of total mismatches are likely to be observed in the positive dataset. The energy features' contribution was novel and ranked high consistently in multiple results, enough to be considered important. The performance of the reported random forest classifier was also compared against existing methods for off-target prediction and was found to perform better (Figure S4).

5. Conclusions

In this study, the binding energy of the Cas9 REC3 domain and the 20-nucleotide DNA-RNA hybrid, and the binding energy of the 20 nucleotides of target DNA to guide RNA were novel features and proposed to be important for Cas9 activity. In the regression model, which predicts Cas9 cleavage frequency, and the classification model, which predicts Cas9 activity, both these features were shown to be important in driving model output. The same importance of the features was observed in the classification model, which can reliably distinguish between experimentally likely and unlikely off-target sequences. The other features used in the model were standard features used in most studies: the number and position of mismatches and type of mismatch, among others. The binding energy features were not redundant and did not show correlation with the other features, and hence they can be implemented in future algorithms for improved off-target prediction and guide-RNA design algorithms.

Supplementary Materials: The following supporting information [23,24,28,63–65] can be downloaded at: https://www.mdpi.com/article/10.3390/biom12081123/s1, Figure S1: Features of the predicted dataset. On the vertical axis is the number of predicted sequences and on the horizontal axis is the number of experimentally validated sequences. (a) The number of CRISPcut predictions vs. CIRCLE-seq off-targets which shows poor correlation as can be determined by the low R^2 value of 0.22. Each dot represents a unique target sequence for which the number of experimentally validated off-targets, plotted on the *X*-axis are compared against the number of predicted off-targets using the CRISPcut tool, as plotted on the *Y*-axis (b) Shows only the number of accessible predictions plotted against number of CIRCLE-seq off-target sites for a guide, high correlation denoted by an R^2 value of 0.84 can be observed here; Figure S2: Correlation plot of the features used for model training. The dark blue diagonal indicates self-correlation. There is a poor correlation between most feature pairs, but a few high correlation islands in dark blue and yellow colour can be seen. Since cell lines are mutually exclusive, the correlation between the cell lines will be negative. The dark blue islands are between PAM mismatches, PAM transitions and PAM mismatch positions, which can be expected; Figure S3: The mean absolute error (MAE) multiplied by 10, and R^2 value plotted for each model tested, various models were tested with increasing n_estimators and random states. The dashed grey line marks the maximum R2 and minimum error instance, which corresponds to n_estimators of 18 and a random state of 6; Figure S4: The best-performing random forest classifier was compared with the existing off-target prediction models for predicting the off-targets of a randomly selected EMX1 locus. The precision was calculated against experimentally validated sequences obtained from CIRCLE-seq. The off-targets were obtained from the CRISPOR [1,2], CRISTA [3], Elevation [4], ge-CRISPR [5] and CRISPcut [6] webservers (accessed on 13 June 2021); Table S1: Results of the two sample Mann–Whitney U test; Table S2: Random forest classification model performance summary; Table S3: Details of negative dataset; Table S4: Complete set of features used in the model learning process.

Author Contributions: D.S.V.: conceptualization, methodology, formal analysis, writing—original draft; Y.V.: investigation, software; D.S.: conceptualization, supervision, writing—review and editing. All authors have read and agreed to the published version of the manuscript.

Funding: This research received no external funding.

Institutional Review Board Statement: Not applicable.

Informed Consent Statement: Not applicable.

Data Availability Statement: All input files and python scripts used for the data generation and analysis are available on GitHub (https://github.com/TeamSundar/crispr-cas9-dG-study).

Conflicts of Interest: The authors declare no conflict of interest.

References

1. Jinek, M.; Chylinski, K.; Fonfara, I.; Hauer, M.; Doudna, J.A.; Charpentier, E. A programmable dual-RNA–guided DNA endonuclease in adaptive bacterial immunity. *Science* **2012**, *337*, 816–821. [CrossRef]
2. Cong, L.; Ran, F.A.; Cox, D.; Lin, S.; Barretto, R.; Habib, N.; Hsu, P.D.; Wu, X.; Jiang, W.; Marraffini, L.A.; et al. Multiplex genome engineering using crispr/cas systems. *Science* **2013**, *339*, 819. [CrossRef] [PubMed]
3. Mali, P.; Yang, L.; Esvelt, K.M.; Aach, J.; Guell, M.; DiCarlo, J.E.; Norville, J.E.; Church, G.M. RNA-guided human genome engineering via cas9. *Science* **2013**, *339*, 823–826. [CrossRef] [PubMed]
4. Porteus, M. Genome editing: A new approach to human therapeutics. *Annu. Rev. Pharmacol. Toxicol.* **2016**, *56*, 163–190. [CrossRef] [PubMed]
5. Gasiunas, G.; Barrangou, R.; Horvath, P.; Siksnys, V. Cas9–crrna ribonucleoprotein complex mediates specific DNA cleavage for adaptive immunity in bacteria. *Proc. Natl. Acad. Sci. USA* **2012**, *109*, 15539. [CrossRef] [PubMed]
6. Garneau, J.E.; Dupuis, M.-È.; Villion, M.; Romero, D.A.; Barrangou, R.; Boyaval, P.; Fremaux, C.; Horvath, P.; Magadán, A.H.; Moineau, S. The crispr/cas bacterial immune system cleaves bacteriophage and plasmid DNA. *Nature* **2010**, *468*, 67–71. [CrossRef]
7. Nishimasu, H.; Ran, F.A.; Hsu, P.D.; Konermann, S.; Shehata, S.I.; Dohmae, N.; Ishitani, R.; Zhang, F.; Nureki, O. Crystal structure of cas9 in complex with guide rna and target DNA. *Cell* **2014**, *156*, 935–949. [CrossRef]
8. Hsu, P.D.; Scott, D.A.; Weinstein, J.A.; Ran, F.A.; Konermann, S.; Agarwala, V.; Li, Y.; Fine, E.J.; Wu, X.; Shalem, O. DNA targeting specificity of rna-guided cas9 nucleases. *Nat. Biotechnol.* **2013**, *31*, 827. [CrossRef] [PubMed]
9. Fu, Y.; Foden, J.A.; Khayter, C.; Maeder, M.L.; Reyon, D.; Joung, J.K.; Sander, J.D. High-frequency off-target mutagenesis induced by crispr-cas nucleases in human cells. *Nat. Biotechnol.* **2013**, *31*, 822. [CrossRef] [PubMed]
10. Tsai, S.Q.; Zheng, Z.; Nguyen, N.T.; Liebers, M.; Topkar, V.V.; Thapar, V.; Wyvekens, N.; Khayter, C.; Iafrate, A.J.; Le, L.P. Guide-seq enables genome-wide profiling of off-target cleavage by crispr-cas nucleases. *Nat. Biotechnol.* **2015**, *33*, 187. [CrossRef]
11. Tsai, S.Q.; Nguyen, N.T.; Malagon-Lopez, J.; Topkar, V.V.; Aryee, M.J.; Joung, J.K. Circle-seq: A highly sensitive in vitro screen for genome-wide crispr–cas9 nuclease off-targets. *Nat. Methods* **2017**, *14*, 607. [CrossRef] [PubMed]
12. Wang, X.; Wang, Y.; Wu, X.; Wang, J.; Wang, Y.; Qiu, Z.; Chang, T.; Huang, H.; Lin, R.-J.; Yee, J.-K.J.N.b. Unbiased detection of off-target cleavage by crispr-cas9 and talens using integrase-defective lentiviral vectors. *Nat. Biotechnol.* **2015**, *33*, 175–178. [CrossRef]
13. Wienert, B.; Wyman, S.K.; Richardson, C.D.; Yeh, C.D.; Akcakaya, P.; Porritt, M.J.; Morlock, M.; Vu, J.T.; Kazane, K.R.; Watry, H.L.J.S. Unbiased detection of crispr off-targets in vivo using discover-seq. *Science* **2019**, *364*, 286–289. [CrossRef] [PubMed]
14. Kim, D.; Kim, J.-S.J.G.r. Dig-seq: A genome-wide crispr off-target profiling method using chromatin DNA. *Genome Res.* **2018**, *28*, 1894–1900. [CrossRef] [PubMed]
15. May, A.P.; Cameron, P.; Settle, A.H.; Fuller, C.K.; Thompson, M.S.; Cigan, A.M.; Young, J.K. SITE-Seq: A Genome-Wide Method to Measure Cas9 Cleavage. 2017. Available online: https://protocolexchange.researchsquare.com/article/nprot-5889/v1 (accessed on 12 July 2022).
16. Doench, J.G.; Hartenian, E.; Graham, D.B.; Tothova, Z.; Hegde, M.; Smith, I.; Sullender, M.; Ebert, B.L.; Xavier, R.J.; Root, D.E. Rational design of highly active sgrnas for crispr-cas9–mediated gene inactivation. *Nat. Biotechnol.* **2014**, *32*, 1262. [CrossRef] [PubMed]
17. Doench, J.G.; Fusi, N.; Sullender, M.; Hegde, M.; Vaimberg, E.W.; Donovan, K.F.; Smith, I.; Tothova, Z.; Wilen, C.; Orchard, R. Optimized sgrna design to maximize activity and minimize off-target effects of crispr-cas9. *Nat. Biotechnol.* **2016**, *34*, 184. [CrossRef] [PubMed]
18. Klein, M.; Eslami-Mossallam, B.; Arroyo, D.G.; Depken, M.J.C.r. Hybridization kinetics explains crispr-cas off-targeting rules. *Cell Rep.* **2018**, *22*, 1413–1423. [CrossRef] [PubMed]
19. Xu, X.; Duan, D.; Chen, S.-J. Crispr-cas9 cleavage efficiency correlates strongly with target-sgrna folding stability: From physical mechanism to off-target assessment. *Sci. Rep.* **2017**, *7*, 143. [CrossRef]
20. Cui, Y.; Xu, J.; Cheng, M.; Liao, X.; Peng, S. Review of crispr/cas9 sgrna design tools. *Interdiscip. Sci. Comput. Life Sci.* **2018**, *10*, 455–465. [CrossRef] [PubMed]
21. Yennmalli, R.; Kalra, S.; Srivastava, P.A.; Garlapati, V.K. Computational tools and resources for crispr/cas 9 genome editing method. *MOJ Proteom. Bioinform.* **2017**, *5*, 00164.
22. Lin, J.; Wong, K.-C. Off-target predictions in crispr-cas9 gene editing using deep learning. *Bioinformatics* **2018**, *34*, i656–i663. [CrossRef]
23. Listgarten, J.; Weinstein, M.; Kleinstiver, B.P.; Sousa, A.A.; Joung, J.K.; Crawford, J.; Gao, K.; Hoang, L.; Elibol, M.; Doench, J.G. Prediction of off-target activities for the end-to-end design of crispr guide rnas. *Nat. Biomed. Eng.* **2018**, *2*, 38–47. [CrossRef]
24. Abadi, S.; Yan, W.X.; Amar, D.; Mayrose, I. A machine learning approach for predicting crispr-cas9 cleavage efficiencies and patterns underlying its mechanism of action. *PLoS Comp. Biol.* **2017**, *13*, e1005807. [CrossRef]
25. Wang, J.; Zhang, X.; Cheng, L.; Luo, Y. An overview and metanalysis of machine and deep learning-based crispr grna design tools. *RNA Biol.* **2020**, *17*, 13–22. [CrossRef]
26. Chuai, G.; Ma, H.; Yan, J.; Chen, M.; Hong, N.; Xue, D.; Zhou, C.; Zhu, C.; Chen, K.; Duan, B. Deepcrispr: Optimized crispr guide rna design by deep learning. *Genome Biol.* **2018**, *19*, 80. [CrossRef]
27. Luo, J.; Chen, W.; Xue, L.; Tang, B. Prediction of activity and specificity of crispr-cpf1 using convolutional deep learning neural networks. *BMC Bioinform.* **2019**, *20*, 332. [CrossRef] [PubMed]

28. Dhanjal, J.K.; Radhakrishnan, N.; Sundar, D. Crispcut: A novel tool for designing optimal sgrnas for crispr/cas9 based experiments in human cells. *Genomics* **2019**, *111*, 560–566. [CrossRef] [PubMed]
29. Liu, Q.; Di He, L.X. Prediction of off-target specificity and cell-specific fitness of crispr-cas system using attention boosted deep learning and network-based gene feature. *PLoS Comp. Biol.* **2019**, *15*, e1007480. [CrossRef] [PubMed]
30. Alkan, F.; Wenzel, A.; Anthon, C.; Havgaard, J.H.; Gorodkin, J. Crispr-cas9 off-targeting assessment with nucleic acid duplex energy parameters. *Genome Biol.* **2018**, *19*, 177. [CrossRef]
31. Jensen, K.T.; Fløe, L.; Petersen, T.S.; Huang, J.; Xu, F.; Bolund, L.; Luo, Y.; Lin, L. Chromatin accessibility and guide sequence secondary structure affect crispr-cas9 gene editing efficiency. *FEBS Lett.* **2017**, *591*, 1892–1901. [CrossRef] [PubMed]
32. Chen, Y.; Zeng, S.; Hu, R.; Wang, X.; Huang, W.; Liu, J.; Wang, L.; Liu, G.; Cao, Y.; Zhang, Y. Using local chromatin structure to improve crispr/cas9 efficiency in zebrafish. *PLoS ONE* **2017**, *12*, e0182528. [CrossRef] [PubMed]
33. Uusi-Mäkelä, M.I.; Barker, H.R.; Bäuerlein, C.A.; Häkkinen, T.; Nykter, M.; Rämet, M. Chromatin accessibility is associated with crispr-cas9 efficiency in the zebrafish (danio rerio). *PLoS ONE* **2018**, *13*, e0196238. [CrossRef] [PubMed]
34. Jacobson, M.P.; Friesner, R.A.; Xiang, Z.; Honig, B. On the role of the crystal environment in determining protein side-chain conformations. *J. Mol. Biol.* **2002**, *320*, 597–608. [CrossRef]
35. Sastry, G.M.; Adzhigirey, M.; Day, T.; Annabhimoju, R.; Sherman, W. Protein and ligand preparation: Parameters, protocols, and influence on virtual screening enrichments. *J. Comput.-Aided Mol. Des.* **2013**, *27*, 221–234. [CrossRef]
36. DeLano, W.L. Pymol molecular viewer: Updates and refinements. In *Abstracts of Papers of the American Chemical Society*; American Chemical Society: Washington, DC, USA, 2009.
37. Jacobson, M.P.; Pincus, D.L.; Rapp, C.S.; Day, T.J.; Honig, B.; Shaw, D.E.; Friesner, R.A. A hierarchical approach to all-atom protein loop prediction. *Proteins Struct. Funct. Bioinform.* **2004**, *55*, 351–367. [CrossRef]
38. Genheden, S.; Ryde, U. The mm/pbsa and mm/gbsa methods to estimate ligand-binding affinities. *Expert Opin. Drug Discov.* **2015**, *10*, 449–461. [CrossRef]
39. Kollman, P.A.; Massova, I.; Reyes, C.; Kuhn, B.; Huo, S.; Chong, L.; Lee, M.; Lee, T.; Duan, Y.; Wang, W.; et al. Calculating structures and free energies of complex molecules: Combining molecular mechanics and continuum models. *Acc. Chem. Res.* **2000**, *33*, 889–897. [CrossRef] [PubMed]
40. Hou, T.; Wang, J.; Li, Y.; Wang, W. Assessing the performance of the mm/pbsa and mm/gbsa methods. 1. The accuracy of binding free energy calculations based on molecular dynamics simulations. *J. Chem. Inf. Modeling* **2011**, *51*, 69–82. [CrossRef] [PubMed]
41. Vallat, R. Pingouin: Statistics in python. *J. Open Source Softw.* **2018**, *3*, 1026. [CrossRef]
42. Pedregosa, F.; Varoquaux, G.; Gramfort, A.; Michel, V.; Thirion, B.; Grisel, O.; Blondel, M.; Prettenhofer, P.; Weiss, R.; Dubourg, V. Scikit-learn: Machine learning in python. *J. Mach. Learn. Res.* **2011**, *12*, 2825–2830.
43. Lundberg, S.M.; Lee, S.-I. A Unified Approach to Interpreting Model Predictions. In Proceedings of the Advances in Neural Information Processing Systems 2017, Long Beach, CA, USA, 19 May 2017; pp. 4765–4774.
44. Lundberg, S.M.; Erion, G.; Chen, H.; DeGrave, A.; Prutkin, J.M.; Nair, B.; Katz, R.; Himmelfarb, J.; Bansal, N.; Lee, S.-I. From local explanations to global understanding with explainable ai for trees. *Nat. Mach. Intell.* **2020**, *2*, 56–67. [CrossRef] [PubMed]
45. Dhanjal, J.K.; Dammalapati, S.; Pal, S.; Sundar, D. Evaluation of off-targets predicted by sgrna design tools. *Genomics* **2020**, *112*, 3609–3614. [CrossRef] [PubMed]
46. Lundberg, S.M.; Nair, B.; Vavilala, M.S.; Horibe, M.; Eisses, M.J.; Adams, T.; Liston, D.E.; Low, D.K.-W.; Newman, S.-F.; Kim, J.; et al. Explainable machine-learning predictions for the prevention of hypoxaemia during surgery. *Nat. Biomed. Eng.* **2018**, *2*, 749–760. [CrossRef] [PubMed]
47. Jiang, F.; Doudna, J.A. Crispr–cas9 structures and mechanisms. *Annu. Rev. Biophys.* **2017**, *46*, 505–529. [CrossRef]
48. Jiang, F.; Taylor, D.W.; Chen, J.S.; Kornfeld, J.E.; Zhou, K.; Thompson, A.J.; Nogales, E.; Doudna, J.A. Structures of a crispr-cas9 r-loop complex primed for DNA cleavage. *Science* **2016**, *351*, 867–871. [CrossRef]
49. Huai, C.; Li, G.; Yao, R.; Zhang, Y.; Cao, M.; Kong, L.; Jia, C.; Yuan, H.; Chen, H.; Lu, D. Structural insights into DNA cleavage activation of crispr-cas9 system. *Nat. Commun.* **2017**, *8*, 1375. [CrossRef]
50. Zhao, L.N.; Mondal, D.; Warshel, A. Exploring alternative catalytic mechanisms of the cas9 hnh domain. *Proteins Struct. Funct. Bioinform.* **2020**, *88*, 260–264. [CrossRef]
51. Casalino, L.; Nierzwicki, Ł.; Jinek, M.; Palermo, G. Catalytic mechanism of non-target DNA cleavage in crispr-cas9 revealed by ab initio molecular dynamics. *ACS Catal.* **2020**, *10*, 13596–13605. [CrossRef]
52. Palermo, G. Structure and dynamics of the crispr–cas9 catalytic complex. *J. Chem. Inf. Modeling* **2019**, *59*, 2394–2406. [CrossRef]
53. Mitchell, B.P.; Hsu, R.V.; Medrano, M.A.; Zewde, N.T.; Narkhede, Y.B.; Palermo, G.J.F.i.m.b. Spontaneous embedding of DNA mismatches within the rna: DNA hybrid of crispr-cas9. *Front. Mol. Biosci.* **2020**, *7*, 39. [CrossRef]
54. Ricci, C.G.; Chen, J.S.; Miao, Y.; Jinek, M.; Doudna, J.A.; McCammon, J.A.; Palermo, G.J.A.c.s. Deciphering off-target effects in crispr-cas9 through accelerated molecular dynamics. *ACS Cent. Sci.* **2019**, *5*, 651–662. [CrossRef] [PubMed]
55. Nierzwicki, Ł.; Arantes, P.R.; Saha, A.; Palermo, G. Establishing the allosteric mechanism in crispr-cas9. *Wiley Interdiscip. Rev. Comput. Mol. Sci.* **2021**, *11*, e1503. [CrossRef] [PubMed]
56. Bravo, J.P.K.; Liu, M.-S.; Hibshman, G.N.; Dangerfield, T.L.; Jung, K.; McCool, R.S.; Johnson, K.A.; Taylor, D.W. Structural basis for mismatch surveillance by crispr–cas9. *Nature* **2022**, *603*, 343–347. [CrossRef] [PubMed]
57. Cheatham, T.E.; Kollman, P.A. Molecular dynamics simulations highlight the structural differences among DNA: DNA, rna: Rna, and DNA: Rna hybrid duplexes. *J. Am. Chem. Soc.* **1997**, *119*, 4805–4825. [CrossRef]

58. Nadel, J.; Athanasiadou, R.; Lemetre, C.; Wijetunga, N.A.; Broin, P.Ó.; Sato, H.; Zhang, Z.; Jeddeloh, J.; Montagna, C.; Golden, A. RNA: DNA hybrids in the human genome have distinctive nucleotide characteristics, chromatin composition, and transcriptional relationships. *Epigenet. Chromatin* **2015**, *8*, 46. [CrossRef]
59. Palermo, G. Dissecting structure and function of DNA rna hybrids. *Chem* **2019**, *5*, 1364–1366. [CrossRef]
60. Terrazas, M.; Genna, V.; Portella, G.; Villegas, N.; Sánchez, D.; Arnan, C.; Pulido-Quetglas, C.; Johnson, R.; Guigó, R.; Brun-Heath, I. The origins and the biological consequences of the pur/pyr DNA· rna asymmetry. *Chem* **2019**, *5*, 1619–1631. [CrossRef]
61. Semenova, E.; Jore, M.M.; Datsenko, K.A.; Semenova, A.; Westra, E.R.; Wanner, B.; Van Der Oost, J.; Brouns, S.J.; Severinov, K. Interference by clustered regularly interspaced short palindromic repeat (crispr) rna is governed by a seed sequence. *Proc. Natl. Acad. Sci. USA* **2011**, *108*, 10098–10103. [CrossRef]
62. Boyle, E.A.; Andreasson, J.O.; Chircus, L.M.; Sternberg, S.H.; Wu, M.J.; Guegler, C.K.; Doudna, J.A.; Greenleaf, W.J. High-throughput biochemical profiling reveals sequence determinants of dcas9 off-target binding and unbinding. *Proc. Natl. Acad. Sci. USA* **2017**, *114*, 5461–5466. [CrossRef]
63. Haeussler, M.; Schönig, K.; Eckert, H.; Eschstruth, A.; Mianné, J.; Renaud, J.-B.; Schneider-Maunoury, S.; Shkumatava, A.; Teboul, L.; Kent, J.; et al. Evaluation of off-target and on-target scoring algorithms and integration into the guide RNA selection tool CRISPOR. *Genome Biol.* **2016**, *17*, 148. [CrossRef]
64. Concordet, J.-P.; Haeussler, M. CRISPOR: Intuitive guide selection for CRISPR/Cas9 genome editing experiments and screens. *Nucleic Acids Res.* **2018**, *46*, W242–W245. [CrossRef] [PubMed]
65. Kaur, K.; Gupta, A.; Rajput, A.; Kumar, M. ge-CRISPR—An integrated pipeline for the prediction and analysis of sgRNAs genome editing efficiency for CRISPR/Cas system. *Sci. Rep.* **2016**, *6*, 30870. [CrossRef] [PubMed]

Article

Entropy and Variability: A Second Opinion by Deep Learning

Daniel T. Rademaker [1], Li C. Xue [1], Peter A. C. 't Hoen [1] and Gert Vriend [1,2,*]

1 Centre for Molecular and Biomolecular Informatics (CMBI), Radboudumc, 260 Nijmegen, The Netherlands
2 Baco Institute for Protein Science (BIPS), Mindoro 5201, Philippines
* Correspondence: vriendgert@gmail.com

Abstract: Background: Analysis of the distribution of amino acid types found at equivalent positions in multiple sequence alignments has found applications in human genetics, protein engineering, drug design, protein structure prediction, and many other fields. These analyses tend to revolve around measures of the distribution of the twenty amino acid types found at evolutionary equivalent positions: the columns in multiple sequence alignments. Commonly used measures are variability, average hydrophobicity, or Shannon entropy. One of these techniques, called entropy–variability analysis, as the name already suggests, reduces the distribution of observed residue types in one column to two numbers: the Shannon entropy and the variability as defined by the number of residue types observed. Results: We applied a deep learning, unsupervised feature extraction method to analyse the multiple sequence alignments of all human proteins. An auto-encoder neural architecture was trained on 27,835 multiple sequence alignments for human proteins to obtain the two features that best describe the seven million variability patterns. These two unsupervised learned features strongly resemble entropy and variability, indicating that these are the projections that retain most information when reducing the dimensionality of the information hidden in columns in multiple sequence alignments.

Keywords: MSA; entropy; variability; deep learning; amino acids; Philip Bourne; FAIR; bioinformatics

Citation: Rademaker, D.T.; Xue, L.C.; 't Hoen, P.A.C.; Vriend, G. Entropy and Variability: A Second Opinion by Deep Learning. *Biomolecules* **2022**, *12*, 1740. https://doi.org/10.3390/biom12121740

Academic Editors: Cameron Mura and Lei Xie

Received: 31 August 2022
Accepted: 19 November 2022
Published: 23 November 2022

1. Introduction

In a recent article [1], Phil Bourne provocatively asked, "Is bioinformatics dead?". As always with such after-dinner questions, the answer is "Yes and No", a conclusion that Phil himself already drew implicitly. Phil credits Florian Markowetz for starting the whole discussion [2]. Phil's article and the three references he cites mention the fourth paradigm—data science—that is to follow the first three: empirical evidence, scientific theory, and computational science [3]. Using Google Scholar, we found many suggestions for the fifth paradigm after data science, e.g., letting computers decide which will be the best experiment to perform next, the mismatch between data-intensive and computer-intensive work, brain computer integration, network pairing for small sample sizes, or a whole lot more. Zubarev and Pitera's [4] definition of the fifth paradigm is arguably among the most inclusive and integrative: "cognitive systems seamlessly integrate information from human experts, experimental data, physics-based models, and data-driven models to speed discovery".

Working with data requires that the data are well-annotated and well-curated, and several articles have been written about the 10 rules for (biological) data storage [5,6]. We agree with Florian Markowetz and Phil Bourne that data science is key to understanding biology, but we also have to deal with the reality that a Google Scholar search for FAIR [7] data results in three million hits, while finding articles that are the result of harvesting multiple FAIR-compliant databases in the bioscience domain is a bit of a challenge. Clearly, a lot of FAIR-related work still needs to be conducted in the worlds of the first three paradigms. The FAIR principles have been applied rigorously to a series of large data collections that are maintained by institutions such as EBI or NCBI, and Phil is one of those whom we should thank for that. Indeed, access to protein, DNA, and RNA sequence data is at the basis of most of today's understanding of biology and biomedicine. Markowetz [2], for example, asked the question of how one can quantify the genetic heterogeneity that was suggested to be related to the outcome of anticancer therapies. He concluded that "Computational biology excels at distilling huge amounts of complex data into something testable..." and we believe this to be a step towards a fifth paradigm: data-based science (biology) with a human domain expert at the helm. We are starting to see an increasing number of fifth paradigm examples that illustrate the power of the trinity of data science, deep learning, and human domain expertise. It is imperative that artificial intelligence—especially deep learning approaches—will be a tool that is equally as important for this fifth paradigm's helmsmen as the data science taught in Phil's school.

The alphafold2 three-dimensional structure prediction algorithm [8], for example, implements human insights and numerous innovations in a deep learning architecture that analyses correlations in multiple sequence alignments (MSAs) to determine which amino acids are in close proximity in three-dimensional space. This follows the philosophy that if it sits together, it evolves together [9,10]. A notable example that demonstrates the power of the data trinity comes from Wang et al., who used deep learning to generate protein scaffolds for user-defined protein functional sites [11]. Mirhoseini et al. demonstrated that the data trinity can even drive AI progress itself by letting a deep learning model create the next generation of Google's AI accelerators, reducing months of human effort to a few hours [12].

Frameworks have been designed that allow non-AI bioscience domain experts—the helmsmen of the trinity—to combine data with deep learning to answer biomedical questions. An example is DeepRank, a general protein–protein interface analysis framework that outperforms the state-of-the-art algorithms in ranking protein docking models and in classifying biological versus crystallographic interfaces [13].

The alphafold2 experiment lent support to one of the classical ideas in biology that all the data needed to determine the three-dimensional structure of a protein is available in its sequence. Two decades earlier, Laerte Oliveira asked the question of whether the functional role of each amino acid could be extracted from an MSA [14,15]. He showed that this was indeed possible, but in those days, databases were small, computers were slow, and the term 'deep learning' still had to be invented. We show here, as an example of the fifth paradigm, that with deep learning, human domain knowledge, and a large set of MSAs, we can reconstruct Laerte's proposed features to determine the functional properties per amino acid given a protein's MSA.

The study of protein sequence–structure–function relations has always been a central theme of bioinformatics, and next-generation sequencing has only strengthened this interest. As there are many more protein sequences available than experimentally determined protein structures, multiple sequence alignments (MSAs) dominate this field, as is, for example, illustrated by information systems such as the GPCRDB [16,17] or 3DM [18].

Two principally different philosophies are in vogue to produce MSAs. Classically, MSAs are produced to best represent what happened to the underlying genes during evolution, and the more sequences that can be included, the more information that can be extracted. MSAs that are used in information systems for protein engineering, drug design, and DNA diagnostics, on the other hand, work best if they are centred on one sequence and if all aligned sequences are of similar length. Correlated mutation analysis (CMA) for the purpose of alphafold2-style structure prediction seems to work best using the broader MSAs [9,19,20], while protein engineering and the prediction of residue function normally require one-sequence-centred MSAs.

The information extracted from MSAs is often visualized low-dimensionally using phylogenetic trees or networks of residues that show a high level of mutation correlation. In DNA diagnostics, for example, the degree of conservation at the amino acid residue position where the disease-causing mutation is observed is the single most important factor underlying all analyses [21]; a fully conserved residue position is very important, while a residue found in a maximally variable MSA column is unlikely to be causative for a patient's disease state. The variability observed in a column in an MSA has been described in many ways, with the Shannon entropy ($\Sigma_{i=1,20}$ $p_i \times \log(p_i)$) with p_i being the fraction of each of the 20 amino acid types i in column p in an MSA) probably being the most popular. Oliveira et al. [14,15] found large functional differences between columns with similar entropy but different numbers of observed residue types and introduced entropy–variability (EV) plots to combine these two features. These plots proved to be a powerful tool to learn about the function of individual residues.

The use of the EV method to answer biological questions is well documented in the scientific literature. Vollan et al. [22] used the EV approach, for example, to determine the multimeric state of porins. Gaspari et al. [23] used the methods of Oliveira et al. [14,15] to analyse and extend the Pacifastin protease inhibitor family. Wang et al. [24] predicted the early risk of ophthalmopathy in Graves' disease patients using EV analyses on a patient's T cell receptor repertoire. Samsonova et al. [25] used the EV method to understand the role of individual residues in the function of olfactory G protein-coupled receptors. Abascal et al. [26] made their model for residue variability among Arthropoda fit the concepts behind the EV method. Bywater [27] used a variant of EV that includes the use of Kolmogorov complexity to extract protein structural features from multiple sequence alignments. These are just a few of the many applications.

EV plots illustrate that residues with either similar entropy or similar variability can still have radically different functions (see Figure 1). Although their method worked nicely for a large series of well-studied proteins, Oliveira et al. could not prove that EV plots were the best way to represent MSAs in two dimensions.

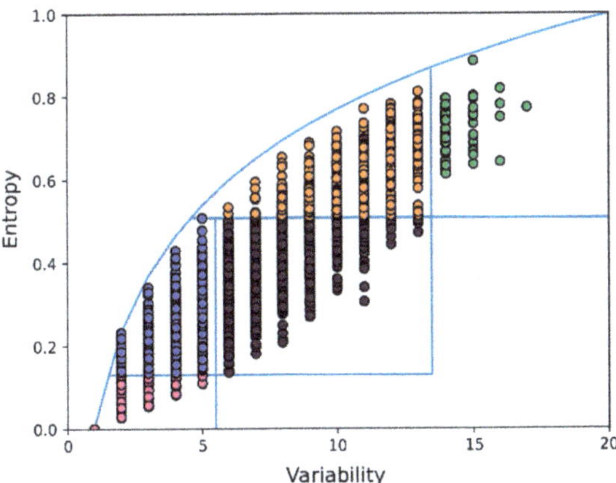

Figure 1. Example EV plot. Colours correspond to functional classes. Each circle represents one column in the MSA, and in this example, thus also one position in the structure. Oliveira et al. divided the EV plot in five areas that—from bottom left to top right—are called Box 11 (pink), 12 (blue), 22 (grey), 23 (orange), and 33 (green). Residue positions in Box 11 were mostly involved in the protein's main function, while Box 12 residue positions were found in the 3D structure around Box 11 residues. Residue positions in Box 23 were generally associated with modulation (such as ligand-binding residues in receptors, calcium-binding residues in calcium-modulated proteins, etc.). Residue positions in Box 22 tended to be in the 3D structure between residue positions from Box 12 at the one side and residue positions from Box 23 at the other. Residue positions in Box 33, finally, tended to have no discernible function.

Modern machine learning methods, such as deep learning, have shown to significantly outperform previous methods in many fields [8,28,29]. The power of deep learning lies in the fact that, given enough data, it can fully automatically and unsupervised learn complex features from raw input alone, thereby bypassing the need to create hand-crafted features using the knowledge of a domain expert.

Deep learning models in the biosciences tend to be heavily parameterized, often using large numbers of data types as input and normally using deep learning in a supervised manner for classification purposes. We asked the question of which features would result from a fully unsupervised reduction of the twenty dimensions of an MSA. Using an autoencoder architecture, we stepwise reduced the dimensionality from 20 to 15 to 10 to 5 to 2, while taking great care that at each step the information loss was kept minimal. The features remaining after reduction to two dimensions resemble entropy and variability remarkably well.

2. Materials and Methods

Multiple sequence alignments were extracted from the human genome HSSP files [30,31]. This dataset was filtered to remove individual columns where the 20 canonical amino acids contributed for less than 75%. The remaining 7,033,530 columns were each converted to a vector **p** of twenty elements p_i that are the fraction of the twenty amino acids i in that column. The elements p_i were sorted from high to low.

We combined elements from several well-known techniques [32–34] into an autoencoder that is optimal for MSA variability signal reduction. This autoencoder consisted of an encoder with layers of size 20-15-10-5-2 and a symmetric decoder [34]. The input to and the output from the autoencoder are the 20-dimensional vector of the relative frequency for each amino acid; the output vector is the best reconstruction possible of the input vector after passage through the 2-dimensional bottleneck. The network does not make use of tied weights. Batch normalization with parameters [33] was used for all hidden layers. The sigmoid function was applied after batch normalization to all units. Due to the small bottleneck of two neurons, the training procedure consisted of a greedy layerwise pretraining finished with fine-tuning [34]. Parameter optimization was performed via stochastic gradient descent using ADAM [35] with a learning rate of 10^{-3} and a batch size of 128 (i.e., 128 column vectors of p). The design of this autoencoder allowed all training steps to use the fast binary cross-entropy loss function. The binary cross-entropy function (not to be confused with the sequence entropy in columns in an MSA) measures the difference between the 20-dimensional input vector and its reconstruction.

The entropy values in neural plots (neuron1 with respect to neuron2) are normalized to the maximum entropy per variability.

The code was written in Python using the PyTorch library [36]. The resulting autoencoder software is available from GitHub: https://github.com/cmbi/EntVar/, accessed on 1 July 2022.

3. Results

Figure 2a is the classical EV plot for the SPG11 protein. Figure 2b is the neural representation for this same protein, i.e., each column from the SPG11 MSA is now represented in the two dimensions according to the autoencoder's data reduction. In Figure 2b, each residue is coloured as in Figure 2a. Figure 2c,d are the same as Figure 2b, but coloured by the variability and the entropy, respectively. It is remarkable how well residues of the same colour cluster in the three neural plots. Oliveira et al. analysed well-studied protein families and mapped experimentally determined residue functions on the EV plots. They then drew boundaries between areas where certain types of function were predominantly observed. These boundaries were somewhat arbitrary, and their optimal location depended, for example, on the number of sequences aligned, the average pairwise sequence identity between the aligned sequences, and the function of the protein family. Oliveira et al. realized that the functional classes are not very sharply divided over the EV plot and that it would be better to see the boundaries as guidelines. The mapping of the EV plot colours on the three neural plots supports this latter idea. The neural plots show a clear gradient when going from low to high entropy or variability. The only exceptions are columns with variability 1 or 2, which are separate groups at the bottom left of the neural plots. Since neural networks operate in continuous space, the discrete character of variability is expected to be blurred out in the plot. Columns with variability 1 or 2 are found at distinct locations, while columns with high variability and high entropy tend to not be separated well. In Figure 2d, entropy values are normalized to the maximal value attainable at each of the twenty variability values. When the entropy values are not normalized, the colour gradient in neural plot 2d does not run from bottom to top, but from bottom left to top right.

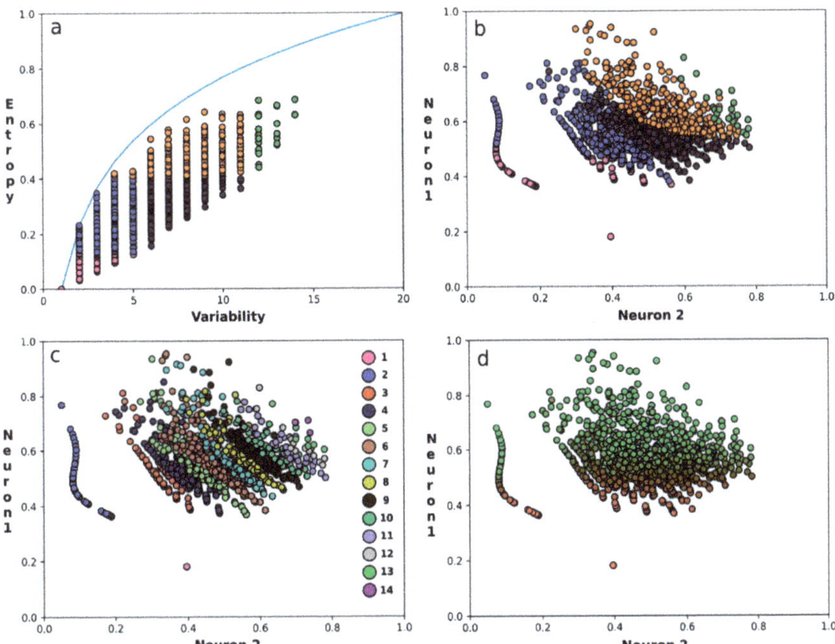

Figure 2. Variability reduction of an MSA. Each circle represents one column from the SPG11 [37] MSA. (**a**) The EV plot according to Oliveira et al. using today's MSA (colours corresponding to functional classes as in Figure 1). (**b**) Neural residue plot in which each residue is coloured as in A. (**c**) The same as B but coloured by variability. The column of circles at the right-hand side indicates the colour used for the variability values from 1 till 14. (**d**) The same as B, but coloured by relative entropy on a gliding scale from red to green. Neuron 1 and neuron 2 are the two elements of the 2-dimensional bottleneck vector of the autoencoder.

4. Discussion and Conclusions

Oliveira et al. had to read nearly a thousand articles to obtain the data needed to functionally classify residue positions in five well-studied protein families. Their EV plots were an attempt to map residue functions on a human readable representation. We used an autoencoder that completely unsupervised, and without the need to spend years of human effort on feature creation, to obtain essentially the same results.

We used an autoencoder with layers 20-15-10-5-2. Alternate layer schemes such as 20-16-8-4-2, 20-19-18 … 4-3-2, 20-64-32-16-8-4-2, etc., all produced highly similar results.

A reduction to three rather than two dimensions resulted in a three-dimensional distribution of MSA columns (and thus residue positions in the protein's structure) that we could not relate to anything biologically meaningful. This is partly caused by the fact that there is no literature available in which variability patterns are reduced to three features by either supervised or unsupervised methods. The three features were not entropy, variability, and a third term. Entropy and variability mapped seemingly randomly on the three-dimensional plot.

The 7,033,530 columns were all sorted with the highest residue frequency first to ensure that the autoencoder analysed variability patterns. When the vectors were not sorted so that the twenty elements p_i always represented the frequencies of Ala, Cys, Asp, Glu, etc., the two dimensions represented the amino acid types, their characteristics, and their mutabilities in ways that are not surprising to bioinformaticians from Phil's generation. These results are shown in Figure 3.

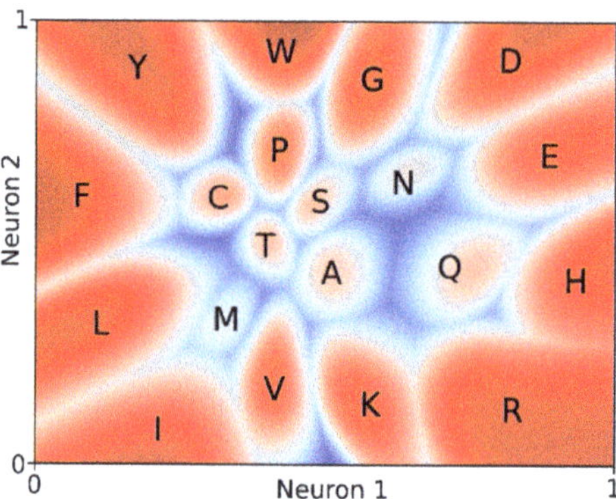

Figure 3. Autoencoder results for the MSA column reduction with fixed order of the twenty amino acid types over the column. These results are beyond the scope of this article but will be discussed extensively on the associated website: https://swift.cmbi.umcn.nl/gv/EV/index.html, accessed on 1 July 2022. Neuron 1 and neuron 2 are the two elements of the 2-dimensional bottleneck vector of the autoencoder.

Figures 2 and 3 illustrate that the autoencoder software can represent variability patterns in MSAs in two dimensions in ways that correspond well to human knowledge about amino acids and protein sequence–structure–function relations. However, the autoencoder remains a black box. It is impossible to determine how it obtained its results. For example, when the dimensionality of the data gets reduced from twenty to three, no discernible patterns emerged, but with two dimensions the classical EV plot emerges. In Figure 3 we observe that residues with similar biophysical characteristics land close to each other. As these biophysical characteristics are not one single continuum, Figure 3 principally must contain exceptions. Indeed, we observe that the largest residue, tryptophan (W), lies adjacent to the smallest one, glycine (G). Other than the knowledge that hydrophobicity is the most important parameter when comparing amino acid types (and thus that hydrophobicity is 'more important' than residue size) we cannot learn from the autoencoder why this is true. Therefore, even though the autoencoder beautifully describes the information in the data, a human expert must still place this information in the wider context of our knowledge, confirming the need for a domain expert in our fifth-paradigm bioinformatics trinity.

Author Contributions: D.T.R. conceived of and designed the study as well as drafted the manuscript. G.V., P.A.C.'t.H., and L.C.X. revised the manuscript. All authors have read and agreed to the published version of the manuscript.

Funding: D.T.R. and L.C.X. acknowledge financial support from the Hypatia Fellowship from Radboudumc (Rv819.52706).

Institutional Review Board Statement: Not applicable.

Informed Consent Statement: Not applicable.

Data Availability Statement: HSSP database downloading instructions can be found at: https://swift.cmbi.umcn.nl/gv/hssp/, accessed on 1 July 2022.

Acknowledgments: We thank Valere Lounnas and Martijn Huynen for critically reading the manuscript. We would also like to thank Gayatri Ramakrishnan for her useful suggestions. Laerte Oliveira is the father of EV analyses.

Conflicts of Interest: The authors declare no conflict of interest.

References

1. Bourne, P.E. Is "bioinformatics" dead? *PLoS Biol.* **2021**, *19*, e3001165. [CrossRef] [PubMed]
2. Markowetz, F. All biology is computational biology. *PLoS Biol.* **2017**, *15*, e2002050. [CrossRef] [PubMed]
3. Wikipedia. The Fourth Paradigm. 19 March 2021. Available online: https://en.wikipedia.org/w/index.php?title=The_Fourth_Paradigm&oldid=1012968154 (accessed on 28 July 2022).
4. Zubarev, D.Y.; Pitera, J.W. Cognitive materials discovery and onset of the 5th discovery paradigm. In *ACS Symposium Series*; Pyzer-Knapp, E.O., Laino, T., Eds.; American Chemical Society: Washington, DC, USA, 2019; Volume 1326, pp. 103–120. [CrossRef]
5. Babbitt, P.C.; Bagos, P.G.; Bairoch, A.; Bateman, A.; Chatonnet, A.; Chen, M.J.; Craik, D.J.; Finn, R.D.; Gloriam, D.; Haft, D.H.; et al. Creating a specialist protein resource network: A meeting report for the protein bioinformatics and community resources retreat. *Database* **2015**, *2015*, bav063. [CrossRef] [PubMed]
6. Parker, M.S.; Burgess, A.E.; Bourne, P.E. Ten simple rules for starting (and sustaining) an academic data science initiative. *PLoS Comput. Biol.* **2021**, *17*, e1008628. [CrossRef]
7. Wikipedia. FAIR Data. 30 June 2022. Available online: https://en.wikipedia.org/w/index.php?title=FAIR_data&oldid=1095813033 (accessed on 28 July 2022).
8. Jumper, J.; Evans, R.; Pritzel, A.; Green, T.; Figurnov, M.; Ronneberger, O.; Tunyasuvunakool, K.; Bates, R.; Žídek, A.; Potapenko, A.; et al. Highly accurate protein structure prediction with AlphaFold. *Nature* **2021**, *596*, 583–589. [CrossRef]
9. Marks, D.S.; Hopf, T.A.; Sander, C. Protein structure prediction from sequence variation. *Nat. Biotechnol.* **2012**, *30*, 1072–1080. [CrossRef]
10. Jones, D.T.; Buchan, D.W.A.; Cozzetto, D.; Pontil, M. PSICOV: Precise structural contact prediction using sparse inverse covariance estimation on large multiple sequence alignments. *Bioinformatics* **2012**, *28*, 184–190. [CrossRef]
11. Wang, J.; Lisanza, S.; Juergens, D.; Tischer, D.; Watson, J.L.; Castro, K.M.; Ragotte, R.; Saragovi, A.; Milles, L.F.; Baek, M.; et al. Scaffolding protein functional sites using deep learning. *Science* **2022**, *377*, 387–394. [CrossRef]
12. Mirhoseini, A.; Goldie, A.; Yazgan, M.; Jiang, J.W.; Songhori, E.; Wang, S.; Lee, Y.-J.; Johnson, E.; Pathak, O.; Nazi, A.; et al. A graph placement methodology for fast chip design. *Nature* **2021**, *594*, 207–212. [CrossRef]
13. Renaud, N.; Geng, C.; Georgievska, S.; Ambrosetti, F.; Ridder, L.; Marzella, D.F.; Réau, M.F.; Bonvin, A.M.J.J.; Xue, L.C. DeepRank: A deep learning framework for data mining 3D protein-protein interfaces. *Nat. Commun.* **2021**, *12*, 7068. [CrossRef]
14. Oliveira, L.; Paiva, A.C.M.; Vriend, G. Correlated Mutation Analyses on Very Large Sequence Families. *ChemBioChem* **2002**, *3*, 1010–1017. [CrossRef] [PubMed]
15. Oliveira, L.; Paiva, P.B.; Paiva, A.C.M.; Vriend, G. Identification of functionally conserved residues with the use of entropy-variability plots. *Proteins* **2003**, *52*, 544–552. [CrossRef] [PubMed]
16. Pándy-Szekeres, G.; Munk, C.; Tsonkov, T.M.; Mordalski, S.; Harpsøe, K.; Hauser, A.S.; Bojarski, A.J.; Gloriam, D.E. GPCRdb in 2018: Adding GPCR structure models and ligands. *Nucleic Acids Res.* **2018**, *46*, D440–D446. [CrossRef] [PubMed]
17. Munk, C.; Isberg, V.; Mordalski, S.; Harpsøe, K.; Rataj, K.; Hauser, A.S.; Kolb, P.; Bojarski, A.J.; Vriend, G.; E Gloriam, D. GPCRdb: The G protein-coupled receptor database—An introduction. *Br. J. Pharmacol.* **2016**, *173*, 2195–2207. [CrossRef]
18. Kuipers, R.K.; Joosten, H.-J.; van Berkel, W.; Leferink, N.; Rooijen, E.; Ittmann, E.; van Zimmeren, F.; Jochens, H.; Bornscheuer, U.; Vriend, G.; et al. 3DM: Systematic analysis of heterogeneous superfamily data to discover protein functionalities. *Proteins Struct. Funct. Bioinform.* **2010**, *78*, 2101–2113. [CrossRef] [PubMed]
19. Senior, A.W.; Evans, R.; Jumper, J.; Kirkpatrick, J.; Sifre, L.; Green, T.; Qin, C.; Žídek, A.; Nelson, A.W.R.; Bridgland, A.; et al. Protein structure prediction using multiple deep neural networks in the 13th Critical Assessment of Protein Structure Prediction (CASP13). *Proteins Struct. Funct. Bioinform.* **2019**, *87*, 1141–1148. [CrossRef] [PubMed]
20. Rao, R.M.; Liu, J.; Verkuil, R.; Meier, J.; Canny, J.; Abbeel, P.; Sercu, T.; Rives, A. MSA transformer. In Proceedings of the 38th International Conference on Machine Learning, Virtual, 18–24 July 2021; pp. 8844–8856. Available online: https://proceedings.mlr.press/v139/rao21a.html (accessed on 30 August 2022).
21. Mooney, S.D.; Klein, T.E. The functional importance of disease-associated mutation. *BMC Bioinform.* **2002**, *3*, 24. [CrossRef]
22. Vollan, H.S.; Tannæs, T.; Vriend, G.; Bukholm, G. In Silico Structure and Sequence Analysis of Bacterial Porins and Specif-ic Diffusion Channels for Hydrophilic Molecules: Conservation, Multimericity and Multifunctionality. *Int. J. Mol. Sci.* **2016**, *17*, 599. [CrossRef]
23. Gáspári, Z.; Ortutay, C.; Perczel, A. A simple fold with variations: The pacifastin inhibitor family. *Bioinformatics* **2004**, *20*, 448–451. [CrossRef]
24. Wang, Y.; Liu, Y.; Yang, X.; Guo, H.; Lin, J.; Yang, J.; He, M.; Wang, J.; Liu, X.; Shi, T.; et al. Predicting the early risk of ophthalmopathy in Graves' disease patients using TCR repertoire. *Clin. Transl. Med.* **2020**, *10*, e218. [CrossRef]
25. Samsonova, E.V.; Krause, P.; Bäck, T.; Ijzerman, A.P. Characteristic amino acid combinations in olfactory G protein-coupled receptors. *Proteins Struct. Funct. Bioinform.* **2007**, *67*, 154–166. [CrossRef] [PubMed]
26. Abascal, F.; Posada, D.; Zardoya, R. MtArt: A New Model of Amino Acid Replacement for Arthropoda. *Mol. Biol. Evol.* **2006**, *24*, 1–5. [CrossRef] [PubMed]
27. Bywater, R.P. Prediction of Protein Structural Features from Sequence Data Based on Shannon Entropy and Kolmogorov Complexity. *PLoS ONE* **2015**, *10*, e0119306. [CrossRef] [PubMed]
28. Min, S.; Lee, B.; Yoon, S. Deep learning in bioinformatics. *Brief. Bioinform.* **2016**, *18*, bbw068. [CrossRef] [PubMed]

29. Stepniewska-Dziubinska, M.M.; Zielenkiewicz, P.; Siedlecki, P. Development and evaluation of a deep learning model for protein–ligand binding affinity prediction. *Bioinformatics* **2018**, *34*, 3666–3674. [CrossRef] [PubMed]
30. Dodge, C.; Schneider, R.; Sander, C. The HSSP database of protein structure—Sequence alignments and family profiles. *Nucleic Acids Res.* **1998**, *26*, 313–315. [CrossRef]
31. Sander, C.; Schneider, R. Database of homology-derived protein structures and the structural meaning of sequence alignment. *Proteins Struct. Funct. Bioinform.* **1991**, *9*, 56–68. [CrossRef]
32. Goodfellow, I.; Bengio, Y.; Courville, A. *Deep Learning*; MIT Press: Cambridge, MA, USA, 2016.
33. Ioffe, S.; Szegedy, C. Batch Normalization: Accelerating Deep Network Training by Reducing Internal Covariate Shift. *arXiv* **2015**, arXiv:1502.03167.
34. Hinton, G.E.; Salakhutdinov, R.R. Reducing the Dimensionality of Data with Neural Networks. *Science* **2006**, *313*, 504–507. [CrossRef]
35. Kingma, D.P.; Ba, J. Adam: A Method for Stochastic Optimization. *arXiv* **2017**, arXiv:1412.6980.
36. Paszke, A.; Gross, S.; Chintala, S.; Chanan, G.; Yang, E.; DeVito, Z.; Lin, Z.; Desmaison, A.; Antiga, L.; Lerer, A. Automatic Differentiation in PyTorch. October 2017. Available online: https://openreview.net/forum?id=BJJsrmfCZ (accessed on 28 July 2022).
37. Crimella, C.; Arnoldi, A.; Crippa, F.; Mostacciuolo, M.L.; Boaretto, F.; Sironi, M.; D'Angelo, M.G.; Manzoni, S.; Piccinini, L.; Turconi, A.C.; et al. Point mutations and a large intragenic deletion in SPG11 in complicated spastic paraplegia without thin corpus callosum. *J. Med. Genet.* **2009**, *46*, 345–351. [CrossRef] [PubMed]

Article

DL-TODA: A Deep Learning Tool for Omics Data Analysis

Cecile M. Cres [1], Andrew Tritt [2,3], Kristofer E. Bouchard [2,4,5] and Ying Zhang [1,*]

[1] Department of Cell and Molecular Biology, College of the Environment and Life Sciences, University of Rhode Island, Kingston, RI 02881, USA
[2] Lawrence Berkeley National Laboratory, Scientific Data Division, Berkeley, CA 94720, USA
[3] Lawrence Berkeley National Laboratory, Applied Mathematics & Computational Research Division, Berkeley, CA 94720, USA
[4] Lawrence Berkeley National Laboratory, Biological Systems & Engineering Division, Berkeley, CA 94720, USA
[5] Redwood Center for Theoretical Neuroscience, Helen Wills Neuroscience Institute, University of California, Berkeley, CA 94720, USA
* Correspondence: yingzhang@uri.edu

Abstract: Metagenomics is a technique for genome-wide profiling of microbiomes; this technique generates billions of DNA sequences called reads. Given the multiplication of metagenomic projects, computational tools are necessary to enable the efficient and accurate classification of metagenomic reads without needing to construct a reference database. The program DL-TODA presented here aims to classify metagenomic reads using a deep learning model trained on over 3000 bacterial species. A convolutional neural network architecture originally designed for computer vision was applied for the modeling of species-specific features. Using synthetic testing data simulated with 2454 genomes from 639 species, DL-TODA was shown to classify nearly 75% of the reads with high confidence. The classification accuracy of DL-TODA was over 0.98 at taxonomic ranks above the genus level, making it comparable with Kraken2 and Centrifuge, two state-of-the-art taxonomic classification tools. DL-TODA also achieved an accuracy of 0.97 at the species level, which is higher than 0.93 by Kraken2 and 0.85 by Centrifuge on the same test set. Application of DL-TODA to the human oral and cropland soil metagenomes further demonstrated its use in analyzing microbiomes from diverse environments. Compared to Centrifuge and Kraken2, DL-TODA predicted distinct relative abundance rankings and is less biased toward a single taxon.

Keywords: deep learning; DNA sequencing; read classification; metagenomics

Citation: Cres, C.M.; Tritt, A.; Bouchard, K.E.; Zhang, Y. DL-TODA: A Deep Learning Tool for Omics Data Analysis. *Biomolecules* **2023**, *13*, 585. https://doi.org/10.3390/biom13040585

Academic Editors: Cameron Mura and Lei Xie

Received: 8 December 2022
Revised: 7 March 2023
Accepted: 22 March 2023
Published: 24 March 2023

1. Introduction

A microbiome defines a community of microorganisms and their activities in a given environment. This term encompasses the microbial species themselves, but also the collection of molecules they produce such as metagenomes [1,2]. Microbiome studies can be useful to different fields such as medicine or environmental protection. For example, the human gut microbiome is being extensively analyzed to uncover how its composition is linked to various disorders [3], while the ocean microbiome provides information on the potential impact of climate change on marine biodiversity [4].

The metagenomic study of microbiomes has gained a lot of interest due to the progress made in DNA sequencing technology. While the history of DNA sequencing started a decade after many proteins were already sequenced and RNA sequencing was being apprehended [5], it quickly evolved in the late 1970s when Sanger and Gilbert independently developed methods that allowed sequencing of 50 and 100 nucleotides, respectively [6,7]. The automation of Sanger's technique combined with the desire to sequence large fragments of DNA brought various improvements that led to the development of efficient machines able to perform DNA sequencing in a particularly parallel fashion. Current high-throughput sequencing methods can produce billions of DNA fragments simultaneously during a single run. In addition, high-throughput sequencing technology offers speed and a decrease in cost per base, but also offers high sequencing depth that comes with better sensitivity. This provides the means to study uncultivable microorganisms and to detect low abundance microorganisms of a microbial community.

In a typical metagenomic study, the genetic material in all organisms contained in a given sample is fragmented, and the DNA fragments sequenced are identified as reads. Following sequencing, diverse bioinformatic tools are used to remove low-quality sequences and to assemble overlapping reads into contiguous DNA segments, also called contigs. Contigs are then arranged through scaffolding into longer segments to eventually reconstruct genomes present in the sample. This complex process of de novo sequence assembly is further challenged when dealing with short-read sequences and the high sequencing depth that is required to differentiate similar or repetitive sequences. The recent development of third-generation sequencing platforms enabled the determination of long-read sequences. With a length of 10–25 kb [8] from the Pacific Biosciences (PacBio) and 10–100 kb from Oxford Nanopore Technologies (ONT) platforms, de novo assembly will be greatly facilitated and improved as it can already be seen [9,10].

A complementary approach to analyze metagenomic data and provide information on the composition of microbial communities is the taxonomic classification of reads. This method involves assigning a taxonomic group to every read with the goal of classifying as many sequences as possible and identifying species present in the sample. One strategy for taxonomic classification consists of comparing k-mer signatures in metagenomic reads to a database of categorized k-mers. One of the state-of-the-art tools for metagenomic classification is Kraken [11], which relies on a database of k-mers with each k-mer associated with the lowest common ancestor of all genomes containing that specific k-mer. Kraken has been criticized for employing a memory-intensive algorithm [12–14], prompting its designers to release Kraken2, which features a more memory-efficient data structure [15]. An alternative method to efficiently store and query the database of k-mers is a modified implementation of the FM-index, employed by Centrifuge [16]. Both Kraken2 and Centrifuge have been praised in the literature for providing high accuracy and rapid runtimes [13,17].

The rapid development of deep learning techniques has inspired new applications in the analysis of metagenomic data. Deep learning models rely on artificial neural networks designed based on the structure and function of neurons in the human brain [18]. Complex deep learning models containing many layers have the ability to extract relevant features and find abstract patterns in data, allowing them to achieve high accuracy. For example, the desire to push forward the capacities of deep neural networks has led to the development of new techniques and architectures to classify images, which can reach an accuracy of 99.84% [19] on the MNIST handwritten digit classification dataset.

The first study to consider deep learning algorithms in the classification of DNA sequences built a convolutional neural network (CNN) to classify 16S small subunit ribosomal RNA (rRNA) genes which are commonly used for the identification of bacteria [20]. This study employed the bag of words technique to represent reads simulated from 16S rRNA reference sequences in terms of k-mer occurrences, thus obtaining sparse matrices as input vectors for their neural network. A k-mer size of 7 was used to restrict the storage and computational complexities that occur with sparse input vectors. Despite these limitations, they reported an accuracy of 91% across 100 bacterial genera on an artificial validation dataset. Another CNN method was proposed to classify short reads of 16S rRNA genes across 2768 genera, and achieved better sensitivity compared to Kraken2 at the genus level on 100 bp and 200 bp synthetic reads generated using 16S rRNA genes as templates [21]. While the tools mentioned above would support a high classification rate with amplicon sequencing data that targets specific genetic regions such as the 16S rRNA genes, other software have been designed to analyze the entire genetic materials sequenced from a sample. One such method called GeNet shows the improvement of training a CNN model with long DNA sequences by recording better classification of long metagenomic reads from a mock community consisting of ten microbial species, with comparable performances with Kraken and Centrifuge at the species and genus levels [22]. A more recent tool called DeepMicrobes targets 2505 bacterial species from the human gut and implements a bidirectional long short-term memory (LSTM) in addition to a self-attention mechanism [23]. DeepMicrobes outperforms other traditional taxonomic classification tools at the genus level on mock communities, suggesting the potential of LSTM in metagenomic read classification. However, LSTM is significantly slower than CNN. Finally, a recent model called BERTax, based on the state-of-the-art model BERT for natural language processing, classifies DNA sequences at the superkingdom, phylum, and genus taxonomic levels and shows generalization on unknown data compared to other approaches mentioned previously [24]. For a more in depth analysis of the deep learning techniques applied to taxonomic classification, we recommend a review published recently by [25].

Here, we present DL-TODA, a deep learning model based on CNN that classifies short metagenomic reads from over 3000 bacterial species. Compared to the aforementioned tools, DL-TODA is trained with a modified version of the deep neural network AlexNet, a successful CNN in computer vision. A training dataset containing 250 bp reads simulated from all complete bacterial genomes available in the NCBI Reference Sequence database was used for training the DL-TODA model. This enabled the identification of bacteria originating from a wide range of free-living and host-associated habitats. DL-TODA classifies each read at the species level and supports the inference of higher-order taxa based on NCBI or GTDB taxonomy. A probability score is generated for each prediction, hence permitting the quality control of prediction results based on probability thresholds.

2. Materials and Methods

An overview of all steps involved in the training, validation, and testing of the DL-TODA model is presented in Figure S1. Below, we provide detailed descriptions of the corresponding steps.

2.1. Bacterial Genome Selection

A total of 9859 complete bacterial genomes representing 3313 different species isolated from diverse free-living and host-associated environments were selected from the genome taxonomy database (GTDB) release 95 and the NCBI RefSeq database, downloaded on 7 March 2020. The genomes selected are not derived from metagenome or environmental samples and have a size equal to or above 500 kb. For each species, 70% of the genomes were randomly assigned for model training and the remaining 30% for model testing. In the cases of species with a single genome, the genome in question was automatically appointed to the training set. Additionally, all representative genomes from GTDB were automatically assigned for training. In total, we have 7405 and 2454 genomes assigned for training and testing, respectively. Table 1 provides a summary of the number of taxa represented at species, genus, family, order, class and phylum levels in training, validation and testing sets, and for both GTDB and NCBI taxonomy. A smaller number of taxa are represented in the NCBI classification due to different assignments with the GTDB classification. For example, amongst the 537 genomes classified as *Escherichia coli* by NCBI, 363, 93, 80 and 1 genomes are assigned by GTDB to *Escherichia flexneri, Escherichia coli, Escherichia dysenteriae,* and *Escherichia coli_C*, respectively. Additionally, 244 genomes lack specific assignments in at least one of the given taxonomic ranks in the NCBI taxonomy database. For example, genome GCA_000317655.1 is not assigned a class in the NCBI taxonomy but is assigned to the class of Cyanobacteria in the GTDB taxonomy.

Table 1. Taxonomic distribution of training/validation and testing datasets based on the GTDB or NCBI taxonomy databases.

	Training and Validation Sets		Testing Sets	
Database	**GTDB**	**NCBI**	**GTDB**	**NCBI**
Species	3313	3053	709	639
Genus	1414	1136	331	289
Family	465	387	138	146
Order	224	171	79	77
Class	100	74	38	33
Phylum	45	43	24	19

2.2. Reads Simulation

Paired-end reads of 250 bp were simulated using ART Illumina read simulator (version 2.5.8) [26]. A coverage of 7 and 3 was used for read simulations using training and testing genomes, respectively. A mean fragment length of 300 bp and a standard deviation of fragment length of 10 bp were chosen according to ART Illumina usage information. The built-in error profile of MiSeq v1 (MSv1) was used for simulation. The command for running ART Illumina is art_illumina -ss 'MSv1' -i <input fasta file> -d <reads prefix id> -na -s <standard deviation of fragment length> -m <mean fragment length> -l <read length> -f <fold coverage> -p -o <output file>.

2.3. Training, Validation and Testing Sets

Paired-end reads obtained from training genomes were randomly shuffled and split into 70% for training and 30% for validation. The forward and reverse reads from testing genomes were treated separately and classified independently. Identical reads between the training and testing data were identified by clustering the training and testing reads using Mmseqs2 easy-linclust (version 13.45111) with a minimum sequence identity of 1.0 and a fraction of aligned residues of 1.0. To avoid biases in testing, testing reads that are identical to the training reads were removed from the testing set. Table 2 summarizes the final number of reads included in the training, validation and testing sets of this study. The number of training reads allocated to each species in the NCBI taxonomy had a median of 80,067 and ranges between 10,359 and 56,838,380 (Figure 1A). The number of testing reads allocated to each species in the NCBI taxonomy had a median of 56,839 and ranges between 6455 and 14,223,296 (Figure 1B). The genome coverage represented in the training data was calculated for each species based on Equation (1), where the "number of training reads" are the number of reads assigned to a given species label in the training data, and the "average training genomes size" accounts for the average length of training genomes of the given species.

$$genome\ coverage = 250\ *number\ of\ training\ reads\ /\ average\ training\ genome\ size \quad (1)$$

Table 2. The total number of simulated reads in training, validation and testing datasets.

Dataset	Number of Reads
Training	563,434,720
Validation	241,467,730
Testing	109,851,839

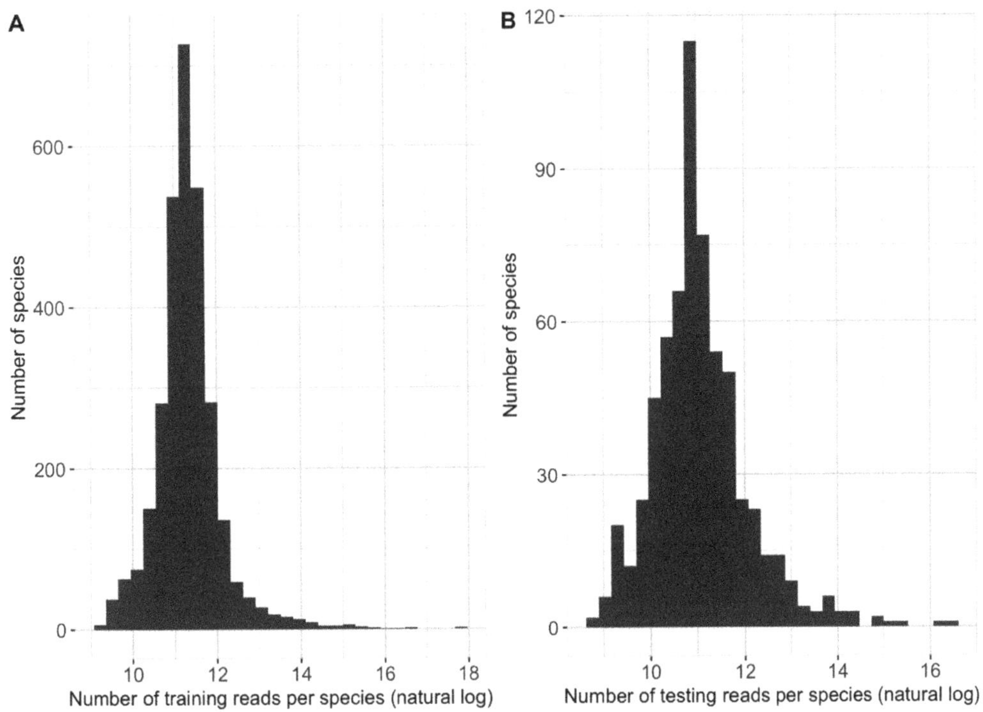

Figure 1. Distribution of number of training (**A**) and testing (**B**) reads per species based on the NCBI taxonomy in the natural log scale.

2.4. Deep Learning Neural Network

2.4.1. Reads Representation

DL-TODA represents each read as a vector of k-mers, using a sliding-window of size 12 across the 250 bp read sequence. Reads shorter than 250 bp were padded with 0s before representation of the k-mers. A vector of 239 integers was then used to represent each read based on a k-mer size of 12 and an indexed vocabulary of 12-mers (described in the section below). The read vectors were then stored in TensorflowRecord (TFRecord) files alongside labels corresponding to the species assignment (i.e., ground truth), and presented to the embedding layer.

2.4.2. K-mer Embedding

The DL-TODA model embeds each k-mer by choosing only the canonical form in a pair consisting of the k-mer and its reverse complement. The canonical k-mer corresponds to the k-mer that appears first, according to the alphabetical order. This strategy allows us to reduce the vocabulary learned by the neural network and therefore lower the complexity of the model. The number of all possible canonical 12-mers is 8,390,656, defined as $\frac{(4^k+4^{(k/2)})}{2}$ $(k = 12)$. The vocabulary of DL-TODA included all the canonical 12-mers and two additional digits, one accounting for unknown 12-mer with characters different from the four universal bases (i.e., A, T, G, C), and another for padded 0s to the right of sequences shorter than 250 bp. Following the vocabulary definition, each 12-mer was assigned an index between 0 and 8,390,657 in order to retrieve a vector of 60 real values from a list. These vectors were initiated in the Tensorflow embedding layer, with each real value drawn from the He Normal distribution [27], and were updated during training.

2.4.3. DL-TODA Neural Network

The deep neural network architecture of DL-TODA is a modified version of AlexNet [28] (Figure 2) with a trainable embedding layer generating an (8,390,658 × 60) embedding matrix. The input layer of this neural network is a (239 × 60) matrix consisting of 239 rows of 12-mers embedded as 60 real value vectors (described above). The input data are then processed by five convolutional layers, two max pooling layers and three fully connected layers. The rectified linear unit (ReLU) activation function is applied throughout the neural network, except in the last layer, in which the softmax function transforms the output from the fully connected layer to a probability distribution over the 3313 species.

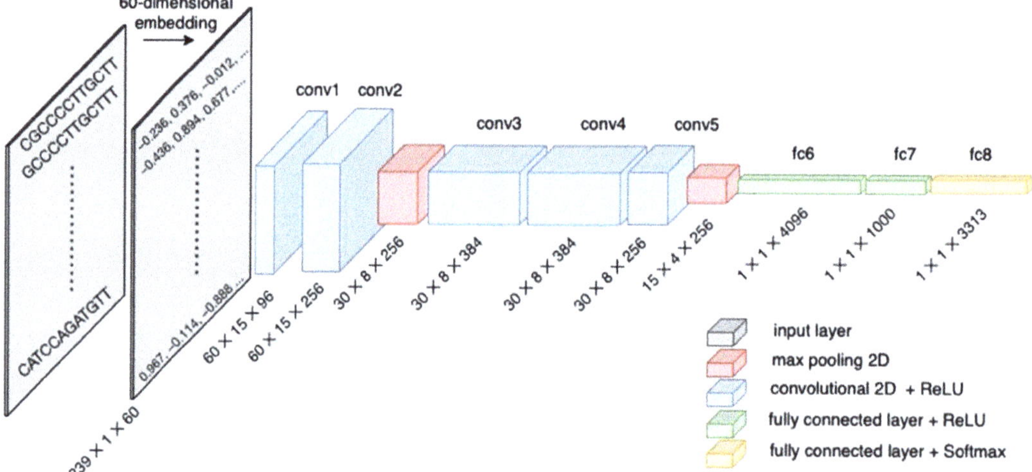

Figure 2. Convolutional neural network architecture used to build the taxonomic read classifier DL-TODA. Each read is represented as an input layer (239 × 1 × 60) by embedding 12-mers into vectors of 60 real values. The input layer is then processed by five convolutional layers, two max pooling layers and three fully connected layers.

2.4.4. Loss Function and Probability Scores

The following cross entropy loss function (Equation (2)) was used to compute the difference between the species desired output (0 or 1) and the estimated probability of correct prediction for a given species for one example.

$$Cross\ Entropy\ Loss = -\sum_{i=1}^{3313} actual\ value\ of\ Species_i * log(predicted\ probability\ of\ Species_i) \quad (2)$$

The estimated probability of every species is obtained by applying the softmax function [29] to an output vector of 3313 real numbers.

2.5. Training and Testing

Data loading to the neural network was performed using the Nvidia Data Loading Library (DALI). Shuffling was carried out exclusively for the training and validation sets. Distributed training was executed by dispatching batches of 512 reads to four different GPUs (global batch size of 2048). Each GPU computed gradient updates independently; these were then averaged together and finally applied to the model. The accuracy and loss computed with the training and validation sets were monitored and saved throughout the training to create learning curves (Figure 3). Additionally, the model was saved at the end of every epoch. Testing and applications to the oral and soil metagenomes were carried out similarly with a batch size per GPU of 512 reads distributed among four GPUs and using the trained model saved at epoch 14.

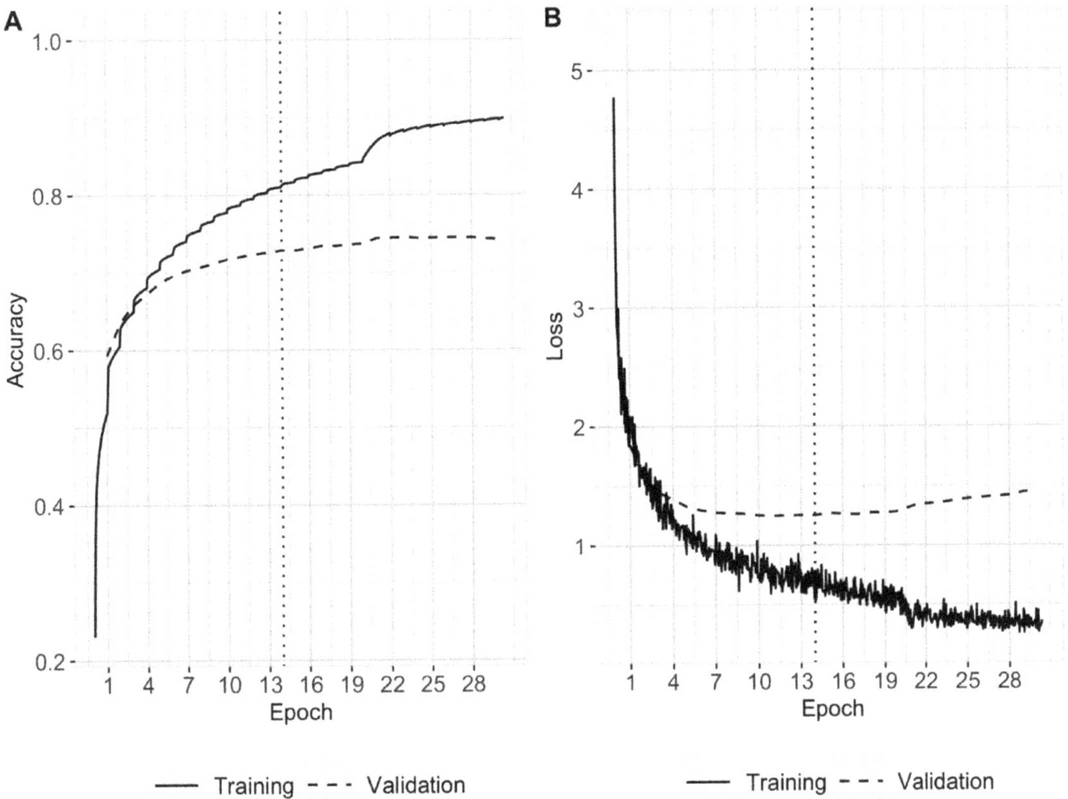

Figure 3. Learning curves representing the predictive performance of DL-TODA during training in terms of accuracy (**A**) and loss (**B**). The training loss and accuracy (solid line), validation loss and accuracy (dashed line), and epoch 14 at which the model was tested (dotted line) are reported.

2.6. Evaluation of Model Performance

The performance of DL-TODA was assessed with the overall classification accuracy, defined in Equation (3), at different taxonomic ranks including species, genus, family, order, class and phylum.

$$Accuracy = \#\, reads\ correctly\ classified\ /\ \#\, reads\ classified \quad (3)$$

At the species level, the number of correctly classified reads was directly obtained from the neural network. At higher taxonomic ranks, the number of correctly classified reads was calculated with the sum of all reads that were correctly assigned to the species within each taxon.

The percentage of classified vs. unclassified reads was also examined with the application of different thresholds on the predicted probability of species. The selection of threshold settings was guided by the overall distribution of probability scores among the correct or incorrect classification in the testing dataset (Figure 4). The eqgamma function of the R package EnvStats (version 2.7.0) was used for identifying confidence intervals based on a gamma distribution for the elimination of incorrect predictions. The precision

(Equation (4)), recall (Equation (5)) and F1-score (Equation (6)) were obtained for each species. The macro and micro average of each metric (Equations (7)–(12)) were computed to provide a comparison of the performance between DL-TODA, Kraken2 and Centrifuge. The number of true positives (TP), false positives (FP) and false negatives (FN) per species required to compute precision, recall and F1-score were obtained based on the generation of a confusion matrix.

$$Precision = TP/(TP + FP) \tag{4}$$

$$Recall = TP/(TP + FN) \tag{5}$$

$$F1 - score = 2 * \ Precision \ * \ Recall/(Precision + Recall) \tag{6}$$

$$Macro\ average\ precision \ = sum\ of\ Precision\ for\ each\ species/number\ of\ species \tag{7}$$

$$Micro\ average\ precision = sum\ of\ TP/(sum\ of\ \ TP + \ sum\ of\ FP) \tag{8}$$

$$Macro\ average\ recall = sum\ of\ Recall\ for\ each\ species/number\ of\ species \tag{9}$$

$$Micro\ average\ recall = sum\ of\ TP/(sum\ of\ \ TP + sum\ of\ FN) \tag{10}$$

$$Macro\ average\ F1 - score = sum\ of\ F1 - score\ for\ each\ species/number\ of\ species \tag{11}$$

$$Micro\ average\ F1 - score = sum\ of\ TP/(sum\ of\ \ TP + 1/2\ * (sum\ of\ FN + sum\ of\ FP)) \tag{12}$$

2.7. Comparison with Kraken2 and Centrifuge

We evaluated the performance of DL-TODA in comparison with Kraken2 version 2.0.8 and Centrifuge version 1.0.3. For both programs, an index was built with the training genomes as references to classify the simulated reads in the testing set using the default settings. Given that both Kraken2 and Centrifuge classify reads to the NCBI taxonomy database, we used the NCBI taxonomy for analyzing the results from DL-TODA. Centrifuge provides multiple possible predictions per pair of reads or unpaired reads. Here, the top hit was systematically used as the predicted taxon.

2.8. Classification of Metagenomic Data

The functionality of DL-TODA was determined by classifying metagenomes obtained from sampling two distinct environments, human oral cavity and cropland soil. The human oral cavity datasets were identified following [30]. The cropland soil datasets (NCBI accessions: ERR5004682, ERR5003895, ERR5003204, ERR5001925 and ERR4995171) were identified from the National Microbiome Data Collaborative (NMCD) data portal [31], using "soil" as the keyword for ecosystem type and "cropland ecosystem" as the keywords for broad-scale environmental context. The metagenomic reads were retrieved using the SRA Toolkit from NCBI, converted to TFRecords and classified by DL-TODA with a probability score threshold above 0.5 (i.e., reads with probability scores below or equal to 0.5 were counted as unclassified). The relative abundance of each taxon was measured by dividing the number of reads classified to that taxon by the total number of reads in the metagenome (Equation (13)). The DL-TODA classification was compared with Kraken2 and Centrifuge classifications of the same metagenomes, using the training genomes as references.

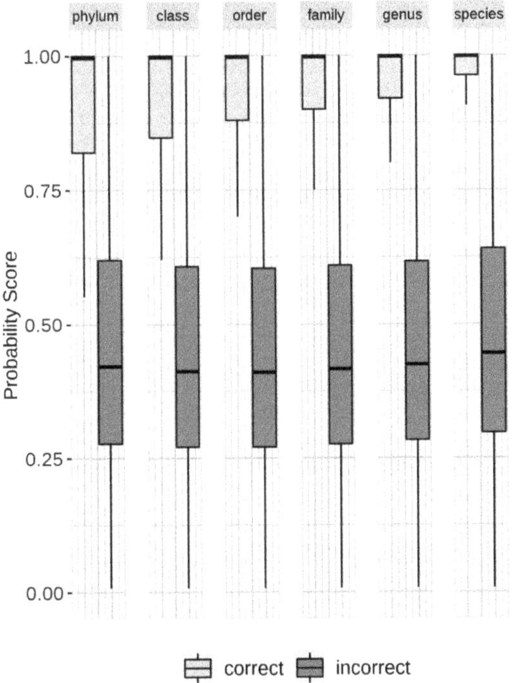

Figure 4. Distribution of probability scores in DL-TODA for correct and incorrect predictions obtained on the entire testing set. The visualization is made in the form of a box plot. The median values are indicated with a thick horizontal line in the rectangle boxes. Lower and upper edges of the rectangle boxes indicate the first and third quartiles, respectively. The thin vertical lines indicate the upper and lower whisker limits, defined as $Q3 + 1.5 \times IQR$ and $Q1 - 1.5 \times IQR$, respectively, where $Q1$ is the first quartile, $Q3$ is the third quartile, and IQR is the inter quartile range from $Q1$ to $Q3$. Outlier data points beyond the upper and lower whisker limits are not shown in the box plot.

$$Relative\ Abundance = number\ of\ reads\ classified\ to\ a\ taxon\ /\ total\ number\ of\ reads \quad (13)$$

2.9. Computational Requirements

The DL-TODA model was trained and tested on a compute node with 768 GB of High Performance DDR4 2666 MHz ECC system memory, 48 Intel Xeon Cascade Lake Scalable Cloud Ready Processor Cores/2.2 GHz processors and four Nvidia A100/40 GB HBM2 Memory GPUs. Kraken2 and Centrifuge were run on a compute node with 24 Intel(R) Xeon(R) CPU E5-4607 0/2.20 GHz processors and 512 GB of memory. The deep learning model was implemented with TensorFlow as a Python3 script, Horovod was used to distribute training across multiple GPUs and the Nvidia DALI was used to load the TFRecord files.

3. Results

3.1. Model Training and Testing

Training of DL-TODA was conducted on a GPU node with four GPUs and was terminated when the model had reached 31 epochs, as no improvements in the validation accuracy were observed (Figure 3A). The model saved at epoch 14 was subsequently selected to perform testing on the testing set, as the model started memorizing the training data after that point, as shown by the progressive increase in the validation loss (dashed line on Figure 3B). Furthermore, additional testing carried out at other checkpoints did not show significant accuracy improvement.

DL-TODA is designed to provide a vector of probability scores in the prediction of every read, with each score corresponding to the probability that the read should be assigned to a given taxon. A taxon with a score of 0.5 has an equal probability of being the true or false assignment of the read analyzed, while a score between 0.5 and 1.0 gives a higher confidence that the read can be truly assigned to the taxon. The DL-TODA prediction of each testing read was designated as either correct or incorrect based on whether the highest probability score was assigned to the ground truth taxon. Of the 109,851,839 reads tested, over 82%, 88%, 90%, 92%, 94%, and 96% were correctly assigned to the corresponding ground truth taxa at the taxonomic ranks of species, genus, family, order, class, and phylum, respectively. The distributions of probability scores among correct and incorrect classifications were plotted in Figure 4. The probability scores of incorrect predictions had median values under 0.5 across all taxonomic ranks, aligning with the expectation that a probability of 0.5 or lower represents predictions with low confidence. In contrast, the probability scores of correct predictions had median values above 0.99 for all taxonomic ranks, and the 25th percentile ranging from 0.82 at the phylum level to 0.96 at the species level. Given the high number of correct taxonomy assignments even with the simple application of top-ranking probability scores, along with the observed clear separation of probability score distributions among correct predictions compared to incorrect predictions, we hypothesize that a decision threshold can be applied on the top-ranking probability scores to further enhance the prediction accuracy of DL-TODA.

3.2. *Optimization of Probability Threshold*

To guide the selection of an optimal threshold, we visualized the species-level precision of DL-TODA predictions in the testing data, given a series of cutoff values. The probability scores below 0.5, 0.57, 0.66, 0.8 and 0.93 correspond to 60%, 70%, 80%, 90% and 95% of incorrect predictions, respectively, based on fitting a gamma distribution over the probability scores of the incorrect assignments. The elimination of low confidence assignments (by assigning predictions only to reads with probability score higher than a designated threshold) greatly enhances the overall precision of DL-TODA predictions for the 639 species tested (Figure 5A). With a threshold of 0.93, the median precision across all species was 0.98, which is 9% higher than the median precision of 0.89 obtained with a threshold of 0.5. The higher thresholds, however, could potentially limit the number of classified reads. Of the thresholds tested, the percentage of classified reads ranged from 87% under 0.5 to 66% under 0.93 (Figure 5B). To balance the gains of precision on species-level predictions and the losses on the number of classified reads, we decided to choose a threshold of 0.8, which gives a median precision of 0.95 across the individual species while still classifying 73% of all the testing reads with high confidence.

Despite the overall high performance, DL-TODA obtained relatively low precision scores in the prediction of a small number of species (Figure 5A). A close examination of these poorly predicted species revealed that each species was represented by only one or a few genomes in the training data, suggesting a general lack of training depth in the deep learning model. Figure S2 elucidates the correlations between training genome coverage and model performance. With genome coverage higher than 55 ($\sim e^4$), DL-TODA consistently reported high precision (e.g., greater than 0.75) in the prediction of corresponding species. Under lower training coverage, however, the minimum precision scores were positively correlated with the training coverage. It was also noted that many species, despite having a training genome coverage of less than 7 ($\sim e^2$), achieved high precision of above 0.9, suggesting that a high coverage is not required for all species in the DL-TODA training.

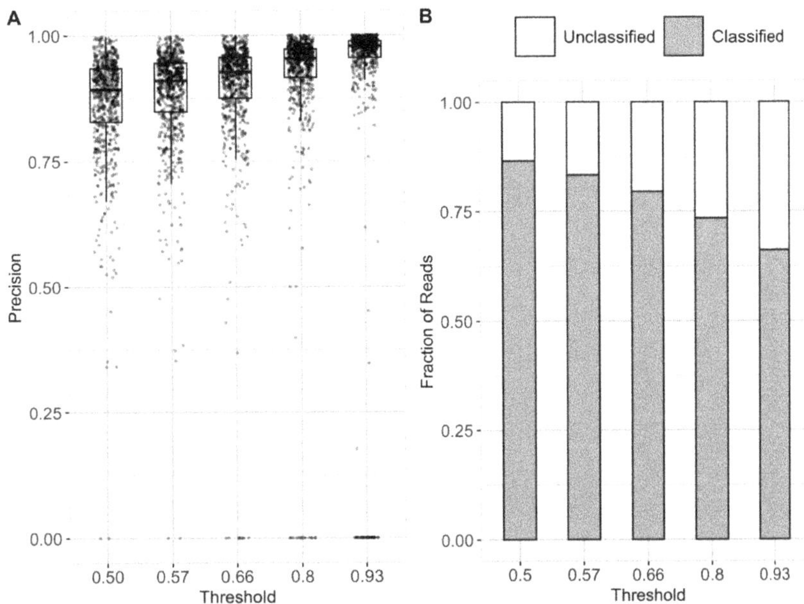

Figure 5. DL-TODA precision for the 639 species in the testing set (**A**) and fraction of unclassified and classified reads at the species level (**B**) at different decision thresholds (0.5, 0.57, 0.66, 0.8 and 0.93).

3.3. Comparison with Kraken2 and Centrifuge

Kraken2 and Centrifuge were applied to the same testing set to assess the performance of DL-TODA amongst taxonomic classification tools. Both Kraken2 and Centrifuge require the construction of reference databases. In order to make a fair comparison, all genomes seen by DL-TODA during training were used to build the indexed reference database for both tools. The average accuracy obtained on ten subsets of the testing data is shown in Figure 6. The ten subsets were obtained by randomly shuffling the testing reads and splitting the testing dataset into nine subsets with 11,000,000 reads and 1 subset with 10,851,839 reads. Comparable performances were observed among all three tools at taxonomic ranks above the genus level, with the overall accuracy averaging above 0.98. At the species level, DL-TODA reached a higher average accuracy of 0.97, compared to 0.93 and 0.85, respectively, achieved with Kraken2 and Centrifuge (Figure 6). The micro average and macro average of precision, recall and F1-score obtained for the 639 species on the entire testing set are shown in Table 3. DL-TODA has higher micro average precision, recall and F1-score, which suggests that DL-TODA makes better overall predictions than Kraken2 and Centrifuge, regardless of the species compared. On the other hand, the macro average metrics for DL-TODA are lower than the corresponding micro average metrics, indicating that DL-TODA performs better for some species compared to others, especially with regard to the performance of recall. For example, with a probability threshold of 0.8, 14 species obtained a recall of 0 due to the removal of predictions with low probability scores, although the majority of other species were predicted with high precisions (greater than 0.95) and recalls (greater than 0.85) by DL-TODA. As a contrast, Kraken2 and Centrifuge appear to manifest similar performances for all species, as their macro average metrics are largely consistent with the corresponding micro average metrics, with the exception that Centrifuge shows variability across species in terms of the recall.

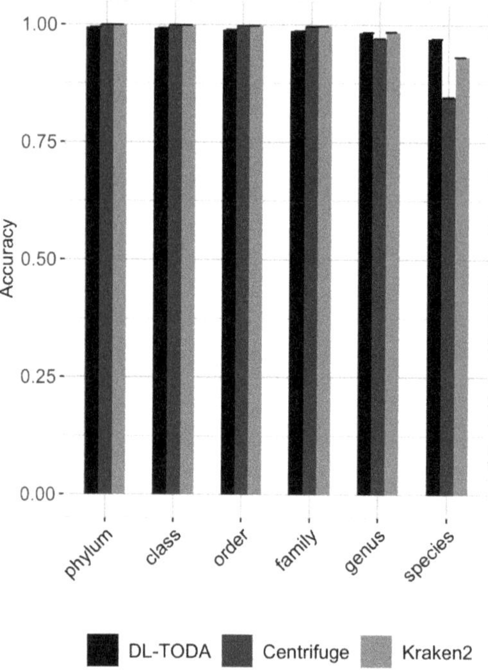

Figure 6. Accuracy across taxa at different taxonomic ranks obtained by running DL-TODA, Centrifuge and Kraken2 on ten subsets of the testing set. The error bar is plotted at the top of each bar. Results for DL-TODA are reported in the presence of a decision threshold of 0.8.

Table 3. Micro average and macro average of precision, recall and F1-score obtained for the 639 species in the testing set for DL-TODA, Kraken2 and Centrifuge. The DL-TODA metrics were calculated with testing reads classified with a probability score higher than 0.8.

	DL-TODA	**Kraken2**	**Centrifuge**
	Micro average		
Precision	0.98	0.97	0.97
Recall	0.97	0.93	0.85
F1-score	0.98	0.95	0.90
	Macro average		
Precision	0.91	0.96	0.97
Recall	0.76	0.92	0.80
F1-score	0.80	0.93	0.82

3.4. *Taxonomic Profiling of Metagenomic Data*

The performance of DL-TODA on metagenomic data was assessed based on a probability threshold of greater than 0.5, using two sets of metagenomes. The first dataset was taken from the human oral microbiome [30] and the second dataset was taken from the soil microbiome [32], with a total count of 3,417,111,096 and 52,290,557 reads, respectively, for the two environments. The relative abundance of reads classified by DL-TODA, Kraken2, and Centrifuge are summarized at the species and genus levels (Table 4). In the oral microbiome, a similar percentage of metagenomic reads (20–30%) was classified by all three tools. While a similar number of taxa was identified by the three tools, DL-TODA identified the highest number of species (452 species) with a relative abundance above 0.01% over the entire set of metagenomes. This is in contrast to Centrifuge, which classified the highest percentage of reads (33%, largely driven by the assignment of classifications to read pairs) but identified a lower number of species (114 species) with a relative abundance above 0.01%. Kraken2 assigned a highest percentage of reads to unknown species compared to the other tools, suggesting a relatively low resolution at the species level. In the soil microbiome, the percentage of metagenomic reads classified by the three tools differed greatly, ranging from 20% of total metagenomic reads identified by Centrifuge to merely 4–5% identified by Kraken2. The latter also had the highest percentage of reads assigned unknown at both species and genus levels; this is similar to what was observed in the analysis of oral microbiome data. DL-TODA classified around 15% of the reads in the soil metagenome and identified 283 species with a relative abundance above 0.01%, which is slightly lower than the Centrifuge predictions but higher than the Kraken2 predictions.

Table 4. Summary of species and genus level classifications made by DL-TODA, Centrifuge and Kraken2 on the human oral and soil metagenomes. The number (#) of taxa observed with relative abundances (r.a.%) ≥ 0.01% or <0.01% is reported in the table. Relative abundances represent the percentage of classified reads over the total number of reads in the metagenomes. Unknown taxa represent groups at a given taxonomic level that are not named in the NCBI taxonomy.

		Tool	# Taxa ≥ 0.01%	Sum of r.a.% ≥ 0.01%	# Taxa with r.a. < 0.01%	Sum of r.a.% < 0.01%	Sum r.a. % of Classified Reads	Unknown Taxa r.a%
oral	Species	DL-TODA	452	19.60	2571	3.76	23.35	0.0024
		Kraken2	85	20.97	2942	1.32	22.29	2.89
		Centrifuge	114	29.19	3066	4.32	33.50	0.036
	Genus	DL-TODA	281	21.27	853	1.57	22.84	0.527
		Kraken2	47	23.56	1075	0.85	24.41	0.78
		Centrifuge	111	31.22	1025	2.07	33.29	0.25
soil	Species	DL-TODA	283	11.80	2648	3.04	14.84	0.012
		Kraken2	62	1.13	2941	3.78	4.92	2.60
		Centrifuge	697	15.81	2451	4.81	20.63	0.096
	Genus	DL-TODA	206	13.11	918	1.57	14.68	0.18
		Kraken2	119	3.84	1002	1.66	5.50	2.02
		Centrifuge	345	18.65	786	1.81	20.46	0.26

Further examination of the classification results was based on the visualization of taxonomic compositions at the class rank (Figure 7). A general consistency was observed in the predicted classes by all three tools in both the oral and soil metagenomes, while the ranking of each class's relative abundance may vary among the different tools. The most abundant classes identified by DL-TODA in the human oral microbiome (Figure 7A) included Gammaproteobacteria (4.8%), Bacilli (3.9%), Actinomycetia (2.1%) and Clostridia (2.2%). In comparison, Clostridia was only found in a small percentage of reads (0.4% and 0.14%, respectively) by Centrifuge and Kraken2. The taxa most seen by both Centrifuge and Kraken2 are Actinomycetia (12.0% and 9.1%), Bacilli (4.9% and 4.3%), Betaproteobacteria (4.1% and 3.4%), Gammaproteobacteria (3.3% and 2.4%) and Bacteroidia (2.3% and 2%). The results obtained with the soil metagenome show similar trends. Kraken2 and Centrifuge manifest similar outcomes with Kraken2 classifying a much lower number of reads (Figure 7B). Actinomycetia, Alphaproteobacteria, Betaproteobacteria and Gammaproteobacteria are amongst the top-ranking classes observed by Kraken2 and Centrifuge, with relative abundances ranging from 0.6% to 6.6%. These bacterial taxa are also predicted by DL-TODA with different relative abundances varying between 1.7% and 2.9%. Additionally, DL-TODA identified Coriobacteriia and Clostridia with relative abundances of 1.4% and 0.9%, respectively, while the relative abundance for Coriobacteriia was 0.08% with Centrifuge and 0.02% with Kraken2, and the relative abundance for Clostridia was 0.15% with Centrifuge and 0.05% with Kraken2.

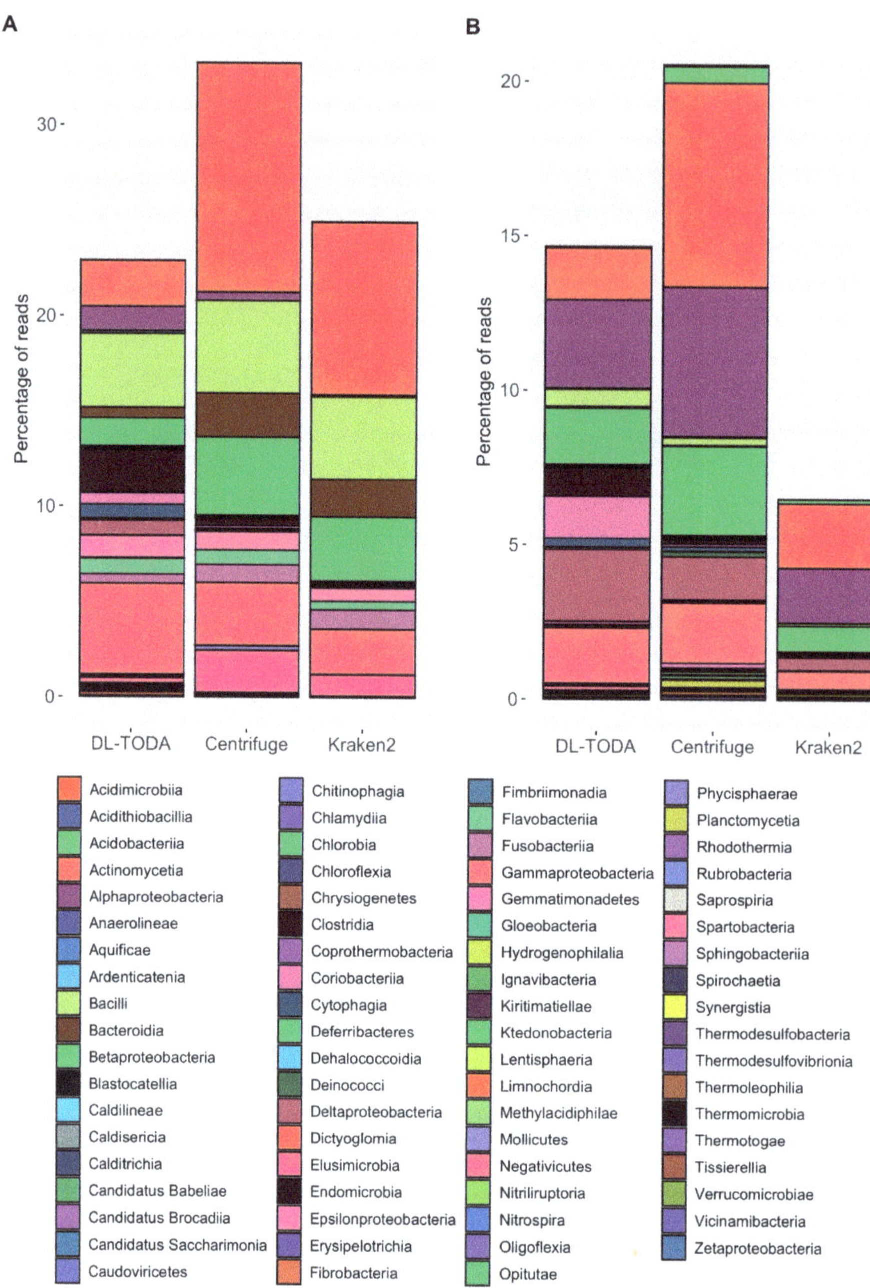

Figure 7. Taxonomic distribution of metagenomic reads at the class rank based on predictions made by DL-TODA, Centrifuge and Kraken2 in the human oral (**A**) and soil (**B**) metagenomes. The Y-axis indicates the percentage of reads over the entire metagenome. The two panels are color coded with the same color pallet so that the same color indicates identical taxa across the different stacked bars.

4. Discussion

Taxonomic classification of billions of short sequencing reads is an important step in the analysis of metagenomic data, shedding light into the function and diversity of microbiomes. Such analysis can be performed by several existing programs but still has room for improvement. K-mer based approaches, such as Kraken2 and Centrifuge, are the most common strategies to classify metagenomic data. While both Kraken2 and Centrifuge rely on the construction of reference databases, the use of a deep learning model in DL-TODA permits the extraction of features during model construction, hence circumventing the requirement of a reference database.

An accuracy similar to higher classification was achieved by DL-TODA compared to Kraken2 and Centrifuge on an independent test set of over a hundred million simulated metagenomic reads (Figure 6). A look at the precision, recall and F1-score (Table 3) further demonstrated the better performance of DL-TODA, as it carried a higher micro average on all three metrics compared to Kraken2 and Centrifuge. However, lower macro than micro averages were observed in DL-TODA, indicating potential differences in how well it recognizes different species. In contrast, Kraken2 and Centrifuge appeared to perform more equally across species, as their macro average metrics are comparable to the corresponding micro average metrics.

One possible reason why DL-TODA may have performed poorly on some species may be the lack of sufficient training data. This is supported by the positive correlations between depth of training genome coverage and minimum precisions observed (Figure S2). For example, when the coverage is greater than 55 ($\sim e^4$), the precision values are consistently higher than 0.75, suggesting that a higher and potentially more diverse set of training data may lead to an enhanced performance of DL-TODA. However, we note that some species, despite having a low number of training reads, reached high precisions in DL-TODA predictions. This may indicate that the DL-TODA model is efficient at extracting traits from these species for label classification. While reaching high performances on a majority of the species tested (Figure 5), DL-TODA seems to assign low probability scores to reads from a few species, resulting in low precisions approaching zero for the prediction of these species, especially when a probability threshold is used. Given the variability in the classification of different species, the probability threshold may be individually adjusted for each species to optimize the performance of DL-TODA. A careful selection of the probability threshold may require more benchmarking efforts to maximize the prediction accuracy while minimizing the fraction of unclassified data; this may be a topic of future research using diverse test cases. Future studies that seek to reveal the correlations between different genomic features (e.g., GC content, tetranucleotide frequency, distribution of mobile genetic elements, etc.) and the outcomes of read classifications can also help guide the further advancement of DL-TODA models and enhance their precision across all species.

The application of DL-TODA to the human oral and cropland soil metagenomes supports a general consensus on the prediction of top-ranking taxa, but distinct predictions on the relative abundance of different taxonomic groups compared to Kraken2 and Centrifuge (Figure 7). In the human oral metagenomes, DL-TODA identified a higher proportion of Clostridia, which is known to be abundant and diverse in the human oral microbiome [33] compared to Centrifuge and Kraken2. Likewise, in the cropland soil metagenomes, a higher proportion of Clostridia and Coriobacteriia was identified by DL-TODA compared to Centrifuge or Kraken2. The abundance of Clostridia and Coriobacteriia, as predicted by DL-TODA, aligns well with prior studies of diverse agricultural related soil types [34–36]. Due to the lack of ground truth data, it is difficult to fully assess the accuracy of different tools on the metagenomes. However, the Centrifuge and Kraken2 predictions seem to be highly skewed towards assigning large proportions to a small number of taxa. For example, the class Actinomycetia was assigned the highest proportions by both Centrifuge and Kraken2 in both the oral and soil metagenomes, suggesting the potential biases of Centrifuge and Kraken2 towards classifying certain taxa. In contrast, the prediction of DL-TODA is less biased towards a single taxon, and it predicted different rankings of the dominant taxa between the human oral cavity and cropland soil, two highly distinct environments. The total number of reads classified remains low across all three tools, and the percentage of classified reads varies among the two environments tested (Table 4). Large differences were observed with Kraken2, which classified over 20% of reads in the oral metagenome but only around 5% of reads in the soil metagenome. Centrifuge seems to have classified the highest proportion of reads among all three tools in both the oral and soil metagenomes. Considering that Centrifuge assigns the same taxa to paired reads, similar strategies may be employed by DL-TODA to leverage the read pairs for enhancing the number of classified reads. It is noted that the DL-TODA predictions were based on a probability threshold higher than 0.5 which was uniformly applied to all taxa. Based on discussions in the above paragraph, further optimization of the probability threshold, together with the introduction of more training data, especially for some underrepresented species, will likely further enhance the number of classified reads in the metagenomes.

Overall, DL-TODA is a new deep learning-based model for the taxonomic classification of metagenomic reads. The model showed a high accuracy in classifying synthetic reads and demonstrated the potential of recognizing a wide range of taxonomic groups from diverse environments. Besides DL-TODA, several other deep learning models have recently been created for the classification of metagenomic data, showing varied accuracy and generalizability, usually at the genus or higher taxonomic levels [23,24]. DL-TODA is distinct from these deep learning-based read classification tools. It uses a convolutional neural network designed based on the architecture of AlexNet and classifies metagenomic reads at the species level. DL-TODA has the ability to classify over 3000 bacterial species, covering all the phyla represented in the current GTDB and NCBI databases. An additional advantage of DL-TODA is the possibility to resume training with new data without needing to reanalyze the previous training sets. This allows the model to be efficiently updated with newly discovered genomes. DL-TODA also supports the calibration of classification results based on a probability score associated with each taxonomic assignment. The implementation of DL-TODA is designed to support high efficiency in processing high volumes of metagenomic data. By making use of Horovod, DL-TODA distributes the training and testing tasks across multiple GPUs in parallel, faster than with the data distribution strategy provided by TensorFlow. This feature, in addition to loading data directly to the GPU memory using the Nvidia DALI library, creates an efficient pipeline for dealing with large datasets. Future developments will include investigating solutions to reduce the size and number of parameters in DL-TODA to further accelerate the training and testing processes. Given the rapid growth of deep learning applications in metagenomic data analysis, future benchmarking studies would provide useful guidelines for the application of different deep learning tools and will likely nurture the engagement of a broader scientific community.

Supplementary Materials: The following supporting information can be downloaded at: https://www.mdpi.com/article/10.3390/biom13040585/s1, Figure S1: A detailed illustration of the DL-TODA pipeline; Figure S2: Precision of DL-TODA predictions over 639 species in the testing set plotted against the depth of training set coverage for each corresponding species.

Author Contributions: Conceptualization, Y.Z. and K.E.B.; methodology, C.M.C. and A.T.; software, C.M.C.; validation, C.M.C. and Y.Z.; formal analysis, C.M.C.; investigation, C.M.C. and Y.Z.; resources, Y.Z. and K.E.B.; data curation, C.M.C. and Y.Z.; writing—original draft preparation, C.M.C. and Y.Z.; writing—review and editing, C.M.C., A.T., K.E.B. and Y.Z.; visualization, C.M.C. and Y.Z.; supervision, Y.Z.; project administration, Y.Z. and K.E.B.; funding acquisition, Y.Z. and K.E.B. All authors have read and agreed to the published version of the manuscript.

Funding: This project was supported by the National Science Foundation under grant DBI-1553211. Y.Z. and C.M.C. acknowledge partial support from the Exascale Computing Project (17-SC-20-SC).

Institutional Review Board Statement: Not applicable.

Informed Consent Statement: Not applicable.

Data Availability Statement: Source code of DL-TODA and the data presented in this study have been deposited in https://github.com/zhanglab/dl-toda (accessed on 28 February 2023) and 10.6084/m9.figshare.22184821 (accessed on 24 March 2023).

Acknowledgments: The authors acknowledge use of the resources of the URI Center for Computational Research and the Massachusetts Green HPC Center (MGHPCC) for this work.

Conflicts of Interest: The authors declare no conflict of interest.

References

1. Berg, G.; Rybakova, D.; Fischer, D.; Cernava, T.; Vergès, M.-C.C.; Charles, T.; Chen, X.; Cocolin, L.; Eversole, K.; Corral, G.H.; et al. Correction to: Microbiome Definition Re-Visited: Old Concepts and New Challenges. *Microbiome* **2020**, *8*, 119. [CrossRef]
2. Whipps, J.M.; Lewis, K.; Cooke, R.C. Mycoparasitism and Plant Disease Control. In *Fungi in Biological Control Systems*; Burge, M.N., Ed.; Manchester University Press: Manchester, UK, 1988; pp. 161–187. ISBN 9780719019791.
3. Fan, Y.; Pedersen, O. Gut Microbiota in Human Metabolic Health and Disease. *Nat. Rev. Microbiol.* **2021**, *19*, 55–71. [CrossRef] [PubMed]
4. Sunagawa, S.; Coelho, L.P.; Chaffron, S.; Kultima, J.R.; Labadie, K.; Salazar, G.; Djahanschiri, B.; Zeller, G.; Mende, D.R.; Alberti, A.; et al. Ocean Plankton. Structure and Function of the Global Ocean Microbiome. *Science* **2015**, *348*, 1261359. [CrossRef] [PubMed]
5. Shendure, J.; Balasubramanian, S.; Church, G.M.; Gilbert, W.; Rogers, J.; Schloss, J.A.; Waterston, R.H. DNA Sequencing at 40: Past, Present and Future. *Nature* **2017**, *550*, 345–353. [CrossRef] [PubMed]
6. Sanger, F.; Coulson, A.R. A Rapid Method for Determining Sequences in DNA by Primed Synthesis with DNA Polymerase. *Mol. Biol.* **1975**, *94*, 441–448. [CrossRef]
7. Maxam, A.M.; Gilbert, W. A New Method for Sequencing DNA. *Proc. Natl. Acad. Sci. USA* **1977**, *74*, 560–564. [CrossRef]
8. Hon, T.; Mars, K.; Young, G.; Tsai, Y.-C.; Karalius, J.W.; Landolin, J.M.; Maurer, N.; Kudrna, D.; Hardigan, M.A.; Steiner, C.C.; et al. Highly Accurate Long-Read HiFi Sequencing Data for Five Complex Genomes. *Sci. Data* **2020**, *7*, 399. [CrossRef]
9. Xie, H.; Yang, C.; Sun, Y.; Igarashi, Y.; Jin, T.; Luo, F. PacBio Long Reads Improve Metagenomic Assemblies, Gene Catalogs, and Genome Binning. *Front. Genet.* **2020**, *11*, 516269. [CrossRef]
10. Jain, M.; Koren, S.; Miga, K.H.; Quick, J.; Rand, A.C.; Sasani, T.A.; Tyson, J.R.; Beggs, A.D.; Dilthey, A.T.; Fiddes, I.T.; et al. Nanopore Sequencing and Assembly of a Human Genome with Ultra-Long Reads. *Nat. Biotechnol.* **2018**, *36*, 338–345. [CrossRef]
11. Wood, D.E.; Salzberg, S.L. Kraken: Ultrafast Metagenomic Sequence Classification Using Exact Alignments. *Genome Biol.* **2014**, *15*, R46. [CrossRef]
12. Knutson, T.P.; Velayudhan, B.T.; Marthaler, D.G. A Porcine Enterovirus G Associated with Enteric Disease Contains a Novel Papain-like Cysteine Protease. *J. Gen. Virol.* **2017**, *98*, 1305–1310. [CrossRef] [PubMed]
13. Ye, S.H.; Siddle, K.J.; Park, D.J.; Sabeti, P.C. Benchmarking Metagenomics Tools for Taxonomic Classification. *Cell* **2019**, *178*, 779–794. [CrossRef] [PubMed]
14. Meiser, A.; Otte, J.; Schmitt, I.; Grande, F.D. Sequencing Genomes from Mixed DNA Samples—Evaluating the Metagenome Skimming Approach in Lichenized Fungi. *Sci. Rep.* **2017**, *7*, 14881. [CrossRef]
15. Wood, D.E.; Lu, J.; Langmead, B. Improved Metagenomic Analysis with Kraken 2. *Genome Biol.* **2019**, *20*, 257. [CrossRef]
16. Kim, D.; Song, L.; Breitwieser, F.P.; Salzberg, S.L. Centrifuge: Rapid and Sensitive Classification of Metagenomic Sequences. *Genome Res.* **2016**, *26*, 1721–1729. [CrossRef] [PubMed]
17. Meyer, F.; Fritz, A.; Deng, Z.-L.; Koslicki, D.; Lesker, T.R.; Gurevich, A.; Robertson, G.; Alser, M.; Antipov, D.; Beghini, F.; et al. Critical Assessment of Metagenome Interpretation: The Second Round of Challenges. *Nat. Methods* **2022**, *19*, 429–440. [CrossRef]
18. McCulloch, W.S.; Pitts, W. A Logical Calculus of the Ideas Immanent in Nervous Activity. *Bull. Math. Biol.* **1943**, *5*, 115–133. [CrossRef]
19. Mazzia, V.; Salvetti, F.; Chiaberge, M. Efficient-CapsNet: Capsule Network with Self-Attention Routing. *Sci. Rep.* **2021**, *11*, 14634. [CrossRef] [PubMed]
20. Fiannaca, A.; La Paglia, L.; La Rosa, M.; Lo Bosco, G.; Renda, G.; Rizzo, R.; Gaglio, S.; Urso, A. Deep Learning Models for Bacteria Taxonomic Classification of Metagenomic Data. *BMC Bioinform.* **2018**, *19*, 198. [CrossRef]
21. Busia, A.; Dahl, G.E.; Fannjiang, C.; Alexander, D.H.; Dorfman, E.; Poplin, R.; McLean, C.Y.; Chang, P.-C.; DePristo, M. A Deep Learning Approach to Pattern Recognition for Short DNA Sequences. *bioRxiv.* **2019**, 353474. [CrossRef]
22. Rojas-Carulla, M.; Tolstikhin, I.; Luque, G.; Youngblut, N.; Ley, R.; Schölkopf, B. GeNet: Deep Representations for Metagenomics. *bioRxiv.* **2019**, 537795. [CrossRef]
23. Liang, Q.; Bible, P.W.; Liu, Y.; Zou, B.; Wei, L. DeepMicrobes: Taxonomic Classification for Metagenomics with Deep Learning. *NAR Genom. Bioinform.* **2020**, *2*, lqaa009. [CrossRef]
24. Mock, F.; Kretschmer, F.; Kriese, A.; Böcker, S.; Marz, M. Taxonomic Classification of DNA Sequences beyond Sequence Similarity Using Deep Neural Networks. *Proc. Natl. Acad. Sci. USA* **2022**, *119*, e2122636119. [CrossRef] [PubMed]
25. Mathieu, A.; Leclercq, M.; Sanabria, M.; Perin, O.; Droit, A. Machine Learning and Deep Learning Applications in Metagenomic Taxonomy and Functional Annotation. *Front. Microbiol.* **2022**, *13*, 811495. [CrossRef] [PubMed]
26. Huang, W.; Li, L.; Myers, J.R.; Marth, G.T. ART: A next-Generation Sequencing Read Simulator. *Bioinformatics* **2012**, *28*, 593–594. [CrossRef] [PubMed]

27. He, K.; Zhang, X.; Ren, S.; Sun, J. Delving Deep into Rectifiers: Surpassing Human-Level Performance on ImageNet Classification. In Proceedings of the 2015 IEEE International Conference on Computer Vision (ICCV), Santiago, Chile, 7–13 December 2015.
28. Krizhevsky, A.; Sutskever, I.; Hinton, G.E. ImageNet Classification with Deep Convolutional Neural Networks. In Proceedings of the 25th International Conference on Neural Information Processing Systems, Lake Tahoe, NV, USA, 3 December 2012; Curran Associates Inc.: Red Hook, NY, USA, 2017; Volume 1, pp. 1097–1105.
29. Bridle, J.S. Training Stochastic Model Recognition Algorithms as Networks Can Lead to Maximum Mutual Information Estimation of Parameters. In Proceedings of the 2nd International Conference on Neural Information Processing Systems, Denver, CO, USA, 27–30 November 1989; Touretzky, D., Ed.; MIT Press: Cambridge, MA, USA, 1989; Volume 2, pp. 211–217.
30. Shaiber, A.; Willis, A.D.; Delmont, T.O.; Roux, S.; Chen, L.-X.; Schmid, A.C.; Yousef, M.; Watson, A.R.; Lolans, K.; Esen, Ö.C.; et al. Functional and Genetic Markers of Niche Partitioning among Enigmatic Members of the Human Oral Microbiome. *Genome Biol.* **2020**, *21*, 292. [CrossRef]
31. Wood-Charlson, E.M.; Anubhav; Auberry, D.; Blanco, H.; Borkum, M.I.; Corilo, Y.E.; Davenport, K.W.; Deshpande, S.; Devarakonda, R.; Drake, M.; et al. The National Microbiome Data Collaborative: Enabling Microbiome Science. *Nat. Rev. Microbiol.* **2020**, *18*, 313–314. [CrossRef] [PubMed]
32. Shaffer, J.P.; Nothias, L.-F.; Thompson, L.R.; Sanders, J.G.; Salido, R.A.; Couvillion, S.P.; Brejnrod, A.D.; Lejzerowicz, F.; Haiminen, N.; Huang, S.; et al. Multi-Omics Profiling of Earth's Biomes Reveals Patterns of Diversity and Co-Occurrence in Microbial and Metabolite Composition across Environments. *bioRxiv.* **2021**, 446988. [CrossRef]
33. Dewhirst, F.E.; Chen, T.; Izard, J.; Paster, B.J.; Tanner, A.C.R.; Yu, W.-H.; Lakshmanan, A.; Wade, W.G. The Human Oral Microbiome. *J. Bacteriol.* **2010**, *192*, 5002–5017. [CrossRef] [PubMed]
34. Neupane, S.; Davis, T.; Nayduch, D.; McGregor, B.L. Habitat Type and Host Grazing Regimen Influence the Soil Microbial Diversity and Communities within Potential Biting Midge Larval Habitats. *Env. Microbiome* **2023**, *18*, 5. [CrossRef]
35. Pathan, S.I.; Roccotelli, A.; Petrovičovà, B.; Romeo, M.; Badagliacca, G.; Monti, M.; Gelsomino, A. Temporal Dynamics of Total and Active Prokaryotic Communities in Two Mediterranean Orchard Soils Treated with Solid Anaerobic Digestate or Managed under No-Tillage. *Biol. Fertil. Soils* **2021**, *57*, 837–861. [CrossRef]
36. Custodio, M.; Espinoza, C.; Peñaloza, R.; Peralta-Ortiz, T.; Sánchez-Suárez, H.; Ordinola-Zapata, A.; Vieyra-Peña, E. Microbial Diversity in Intensively Farmed Lake Sediment Contaminated by Heavy Metals and Identification of Microbial Taxa Bioindicators of Environmental Quality. *Sci. Rep.* **2022**, *12*, 80. [CrossRef] [PubMed]

MDPI
St. Alban-Anlage 66
4052 Basel
Switzerland
www.mdpi.com

Biomolecules Editorial Office
E-mail: biomolecules@mdpi.com
www.mdpi.com/journal/biomolecules

www.ingramcontent.com/pod-product-compliance
Lightning Source LLC
LaVergne TN
LVHW071305170726
843515LV00006BA/1493

* 9 7 8 3 0 3 6 5 8 6 1 1 3 *